Plant Biotechnology
&
Its Applications in Tissue Culture

Plant Biotechnology & Its Applications in Tissue Culture

Editors

Ashwani Kumar
Professor
Department of Botany, University of Rajasthan
Jaipur (Rajasthan)

Shikha Roy
Asst. Professor
Department of Botany, University of Rajasthan
Jaipur (Rajasthan)

Consulting Editor

Sudhir K. Sopory
Group Leader
Plant Molecular Biology
International Centre for Genetic Engineering and Biotechnology
New Delhi

I.K. International Pvt. Ltd.

NEW DELHI • BANGALORE

Published by

I.K. International Pvt. Ltd.
S-25, Green Park Extension
Uphaar Cinema Market
New Delhi–110 016 (India)
E-mail:info@ikinternational.com
Website: www.ikbooks.com

ISBN 81-88237-50-7

Reprint 2013

Published by Krishan Makhijani for I.K. International Pvt. Ltd., S-25, Green Park Extension, Uphaar Cinema Market, New Delhi–110 016 and Printed by Rekha Printers Pvt. Ltd., Okhla Industrial Area, Phase II, New Delhi–110 020.

Foreword

This book is a compilation of papers written by scientists engaged in the field of plant cell and tissue cultures. The articles presented in it cover practically every aspect of plant biotechnology, with respect to practical applications. Between others, considered are the propagation of plants for breeding purposes, gene technology, and selection of strains tolerant to environmental stress and the use of plant cell cultures producing compounds of the secondary metabolism of medicinal interest. It has to be appreciated that mainly plants of significance for India are considered in the book. These contributions indicate the broad acceptance of plant biotechnology in India, a country with a long tradition in the field, where continuous and successful efforts are on to be at the forefront in the research of this exciting branch of Botany. It is hoped that the support of official and private sponsors will continue to help maintain its highly appreciated present standard.

This book is recommended for all those interested not only in this field, but also to the entire scientific community at large. I extend my hearty congratulations to the editors for producing this book and wish them all the very best for its reception by the audience.

Hungen-Bellersheim, August 2005

Prof. Dr. K.-H. Neumann
Justus – Liebig - Universität
Gießen, Germany

Preface

During the last decade, biological sciences have experienced a period of unprecedented progress and the excitement of this new era is more than evident in the field of plant tissue culture and its applications in plant biotechnology. The current emphasis is on plant tissue culture and molecular biology, in particular, the potential applications of genetic engineering and plant biotechnology. Recombination DNA techniques provide new tools for understanding gene regulation of various life processes and their applications in the field of agriculture, forestry, pharmacology, horticulture and basic sciences. Needless to say, this fast, near dazzling pace of discovery has permeated the making of this book. However, the need to understand the basic components of living organisms to interact at cellular level will not be eliminated by detailed characterisation of genome. On the contrary, enriched understanding of gene regulation in plants has generated a wealth of new questions concerning plant functions.

Joining us in this endeavour are some outstanding scientists, each an internationally recognized authority in his or her area of expertise. Professor Sudhir Sopory of ICGEB, New Delhi and Secretary, Plant Tissue Culture Association, has played a catalytic role in bringing about this meeting of minds. The contributors have shown not only the capacity to convey the excitement and direction of contemporary research but also their ability to identify and pursue specific problems of plant tissue culture and its applications in plant biotechnology.

Even though numerous publications worldwide report embryogenesis in plant tissue culture systems, yet its regulation at molecular level has been a matter of in-depth investigation. This book attempts to provide answers to some of these questions based on several decades of research being carried out under the leadership of Professor Dr. Neumann, at Justus Liebig Universität, Giessen, Germany in the opening article. A step further has been taken in understanding the genetic factors regulating embryogenesis and stress resistance at molecular level and we take pride in being associated with such a group led by Professor Dr. Neumann and Professor Dr. Sven Schubert, at JLU, Giessen, Germany.

The organization of the book is modular in the sense that the given topics are self-contained and self-explanatory, while at the same time providing cross-references. The fundamental concepts of plant tissue culture have been provided in the introductory chapter. Cellular differentiation and genetic transformation studies in Indian legumes, crucifers, mulberry, cereals and crop plants have been dealt with in detail. Marker genes play an important role and their characterization is an important aspect covered in this book. Genomics is important with respect to important plants like bamboo and its characterization using RAPDs provides an insight into genetic diversity.

Micropropagation of different plants of horticultural and forestry importance is an important aspect of plant tissue culture application. Several important plants like grapes, pomegranate, chir pine, chickpea, Garcinia have been covered extensively. Polyamines have an important role to play in differentiation and hence, have also been dealt with in detail.

Applications of plant tissue culture in the biotechnology of production of secondary metabolites including biotechnology of medicinal plants, steroid factories and steroidal sapogenins, Plumbagin, Bacosides, metabolites in plants of semi-arid regions cultured *in vitro* are in focus. Stress resistance is encountered in plant systems. The use of plant tissue culture to understand the basic process and develop salt-resistant plants has immense potential. Several new techniques are also dealt with exclusively in this book.

Much of the credit for the present work goes to my teacher Professor H.C. Arya, who introduced me to this field and Professor Uma Kant for his constant encouragement from time to time. We are also indebted to all our colleagues in the Department of Botany and Biotechnology, University of Rajasthan, Jaipur, for sharing their scientific ideas and fruitful discussions. Professor S.L. Kothari, Honorary Director Life Sciences and Head of the Department of Botany, Professor P.C. Trivedi deserve special mention. Thanks are also due to Dr. Neetu Vijay, senior research scholar in our laboratory for her critical reading of the manuscripts and valuable suggestions.

We are also grateful to Ms Sonia Mamgain, from the publishing team of I.K. International for maintaining high professional standards during the preparation of the book.

Editors

Introduction

Biotechnology is an area of production and research in which biological systems and biological principles are employed to solve technological problems.

In recent years, plant cells are being increasingly utilized as model systems to unravel the cellular and molecular mysteries of nature and understand the basic mechanisms of developmental biology, cellular and molecular interactions at the cell and tissue level, gene expression and regulation leading to genetic modifications and areas of genomics.

The following areas have been highlighted in the book:

1. Embryogenesis, regeneration and micropropagation.
2. Genomics and genetic transformation systems.
3. Secondary metabolites, metabolic engineering, and molecular biology.
4. Resistance to biotic and abiotic stress.
5. Tree biotechnology.

In recent times, many laboratories have been using somatic embryogenesis as an essential part of raising transgenic plants, without being completely aware of its cytological, physiological or biochemical background. Despite its practical applications being obvious, it is fundamental to understand differentiation as one of the pertinent features of biological systems.

Two important aspects dealt in detail in this book include (Neumann,):

- Molecular organization of the program of somatic embryogenesis and its realization.
- The stimulus contained in competent cells in order to induce the program of embryogenesis.

An understanding of the basic mechanisms of differentiation and molecular cell biology is necessary to undertake genetic modification of plants. Multiplication and androgenesis of the tree species *Albizzia lebbeck, Cassia siamea* and *C. fistula,* and chemical factors promoting enhanced DNA uptake and expression of the introduced gene in protoplasts of *Brassica, Arabidopsis*, wheat and rice have been elucidated. Modern-day hexaploid wheat (*Triticum aestivum*) is generally used for bread making but it lacks certain agronomically important features such as resistance to drought and disease. In contrast, tetraploid emmer wheat (*Triticum dicoccum*) shows high adaptability to dry, parched lands and exhibits resistance to drought and salinity, but its poor visco-elastic properties and hardness of the grain reduces its consumer acceptability. From the nutritional perspective, *T. durum* is the only wheat with a rich source of β-carotene, and *T. dicoccum* has lower gluten content. With the advent of transgene technology and practical amenability of cereals it should be possible to introduce advantageous traits in different wheats to augment their nutritive value and consumer acceptance. However, genetic enrichment and improvement of wheat though highly desirable, prerequisites a well-established regeneration protocol coupled with molecular amenability. Unfortunately, wheat amongst grasses is considered to be least amenable to tissue culture techniques and leaf protoplasts exhibit notorious recalcitrance. Tissue culture procedures in wheat have generally been raised from embryogenic explants such as immature embryos, scutella, their calli and are regularly employed as the target tissue for various transformation strategies due to their high regeneration efficiency. However, tissues of non-embryogenic origin such as leaf or basal segments are rarely employed for transformation studies owing to their low regeneration response. Our work reveals the unusually high morphogenic potential of the basal leaf segment of various wheat cultivars. Of particular interest is its extreme sensitivity to short-term treatments of 2, 4-D for rapid induction of somatic embryogenesis (Khurana, 15-23).

Agrobacterium-mediated transformation of chickpea (*Cicer arietinum* L.) study revealed that multiple shoots can be successfully induced from embryo axes explants of chickpea and *Agrobacterium*-mediated transformation protocol can be used in the future to develop insect-resistant transgenic lines of chickpea (Tripathi *et al.*, 25-32). Somatic embryogenesis in chickpea is also reported (Anju *et al.*, 59-67).

Sorghum (*Sorghum bicolour* L.) is an important dry-land crop and is used as food, fodder and feed. Thus, it is important to develop and deploy novel technologies such as genetic transformation to generate sorghums with highly enhanced levels of specific end products (for example, quality protein) and resistance to pests that are difficult to breed by conventional means. Alternative approaches like transformation of shoot apices, followed by induction of multiple shoots or direct rooting, were found useful. Details of study on such new explant sources, genetic transformation tools that are applicable for improvement of sorghum along with the current worldwide scenario are presented by Sai Kishore *et al.* (30-40).

Ever increasing number of transgenic plants do possess marker genes that are often antibiotic resistant or herbicide tolerant. There is a growing debate on the safety of the marker genes (antibiotic and other marker genes) in genetically modified food organisms as they might present an unpredictable hazard to the human health and ecosystem. Based on these critical considerations there has been an increasing interest on the removal of selectable marker genes. As there is a need to introduce more gene combinations, either one-step introduction of gene fusions, artificial operons or consecutive transformation steps are required. During the consecutive transformation, a selectable marker gene should be used that is different from the one introduced together with the first gene of interest. This necessitates a wide range of alternative selection systems. Most of the selection systems currently being used are based on the principle of negative selection: the selective agent effectively kills untransformed cells. It is understood that during such selection regime, most transformed cells may not regenerate into plants with desired efficiency as the dying non-transformed cells release growth inhibitors and toxic substances that can compromise the uptake of essential minerals and vitamins from the culture medium. During the past few years, advances have been made in the development of alternate marker. This has either led to the removal of marker genes or the use of alternative marker genes. Several strategies have been developed for the removal of these genes including multi-auto-transformation (MAT) system, co-transformation, site-specific recombination, targeted gene replacement, intragenomic relocation of transgenes through transposable elements. During the past few years, advances have been made in the development of alternative marker systems that are based on the concept of positive selection and in several important crop plants these have been successfully employed in transgenesis (Suprasanna and Bapat, 41-49).

In India, grapes are grown on an area of 34,000 ha with a production of about 1,00,000 tonnes, 85% of which are mainly table grapes. Seedless table and raisin grapes are preferred by consumers all over the world. However, Thompson Seedless is highly susceptible to downy mildew (*Plasmopara viticola* Beri. and deToni), one of the most devastating fungal diseases affecting the grapevine. Taxonomically, grapes are divided into two sub-genera, *Euvitis* Planch. (2n=38) and *Muscadinia* Planch. (2n=40) (Einset and Pratt, 1975). Most commercial grapes belong to subgenus *Euvitis* with *Vitis vinifera* being the most widely grown species. However, success in terms of useful cultivars is yet to be realized, mainly due to the differences in chromosome number between the two genera and embryo abortion. Thus, *in ovulo* embryo rescue provides the solution not only for obtaining good interspecific crosses, but also allows the use of seedless vines as female parents. The present study was taken up mainly to introgress downy mildew resistance in Thompson Seedless using breeding and *in ovulo* embryo rescue methods. It is the first instance in India where a seedless variety has been used as female parent (Bharathy *et al.*, 51-58).

Efficient plant regeneration via organogenesis or somatic embryogenesis is very important for plant multiplication as well as the development of transgenic crops. In fact, good regeneration procedures have been developed for a large number of plant species. Regeneration was achieved in the long-term cultures of a few plant systems, yet regeneration from long-term cell and tissue cultures is still a problem in many plants. There are several reasons behind the loss of regeneration in such long-term cultures and the role of polyamines has been studied by Rajam (69-76).

Pines constitute one of the most divergent and economically important group of species, which provide valuable natural resources and contribute significantly to the local and industrial economy of the country. A detailed study of conditions necessary for the induction of somatic embryogenesis and multiplication carried out in chir pine are reported (Arya *et al.*, 77-89).

Random amplified polymorphic DNA (RAPD) analysis has been employed with 32 rice genotypes including 29 agronomically superior maintainer lines with *eui* gene for good panicle exsertion. The 12 primers generated 102 amplification products, of which 69 were polymorphic. The genetic divergence among the maintainer lines specifically bred with improved plant type, duration, yield potential and grain quality at the G.B. Pant University of Agriculture & Technology, Pantnagar and some varieties adapted by employing RAPD (Random Amplified Polymorphic DNA) markers has been studied. This will aid in finding out new vistas for greater exploitation of heterosis in hybrid rice improvement programme (Rai *et al.*, 91-100)

India is the second largest bamboo resource in the world, next only to China, both in terms of diversity of species and the extent of distribution. *Dendrocalamus strictus* is a bamboo species predominantly found in the Western Ghats and is in great commercial demand. Eleven candidate clumps of *Dendrocalamus strictus,* which were identified based on morphological parameters, have been characterized using random amplified polymorphic DNA- polymerase chain reaction (RAPD-PCR) technique. Protocols for isolation of DNA from field grown clones and PCR reactions to generate RAPD profiling are reported (Yaradoni *et al.*, 101-106).

Pomegranate (*Punica granatum* L.) is an important horticultural crop of the tropical and sub tropical regions grown primarily for dessert fruits. Conventionally, pomegranate varieties are propagated by stem cutting, including both hard and softwood cuttings. However, this method is not commercially viable, as it requires about one year to raise plants. Procedure for micropropagation of promising pomegranate cv. Jyoti, using semi-hardwood nodal segments from the mature tree, has been standardized (Singh and Khawale, 107-113).

Plumbago rosea Linn (synonym *P. indica*) belongs to the family Plumbaginaceae. Plumbagin (5-hydroxy, 2-methyl, 1,4-naphthoquinone) is a yellow coloured naphthoquinone and has anti-cancer, anti-bacterial and anti-fungal properties. Increase in the metabolic yield was found to be 20-25 times higher than control cultures when the cells were subjected to different treatments such as elicitation, permeabilization, immobilization and *in situ* extraction of the product. Suspension and hairy root cultures of *Plumbago rosea* represent an excellent method to produce plumbagin of commercial quality (Komaraiah *et al.,* 123-139)

Arid zone plants are good candidates for the production of various types of secondary metabolites, which make them resistant to drought, salinity and pathogens. Hence, the screening of plants of arid areas *in vivo* and *in vitro* for useful metabolites have been covered by Nag et al. and Tyagi and Nag in this book (141-149).

Bacopa monnieri (L) Wettst (Scrophulariaceae) is an important medicinal herb commonly known as Brahmi or 'Jalnimba' classified as 'Medhya Rasayan' in the Indian traditional system. The medicinal properties of *B. monnieri* have been attributed to the presence of different tetracyclic triterpenoid saponins amongst which, bacosides are proving to be biologically active. The bacosides help repair damaged neurons by augmenting kinase, the protein involved in the synthesis of new neurons to replace old ones. Isolation and identification of bacosides from multiple shoot cultures and tissue cultured plantlets of this important medicinal plant has been reported (Ahuja *et al.*, 155-162)

Almost 50% of the commercial drugs available in the market today are of plant origin. These plants are under threat today because to their ruthless exploitation. The problem is further aggravated due to increased anthroprogenic activities coupled with rapidly eroding ecosystem. Therefore, strategies are required to produce propagules on a large scale and conserve the fast depleting drug-yielding plants. *In vitro* propagation of medicinal plants, already achieved by using various explants, has been reviewed in detail (Narula *et al.*, 163-176).

Oilseed crops in the agricultural economy of India occupy an important place. In terms of acreage, tonnage and value, oilseed is a major agricultural crop, just next to the foodgrains. Among the cultivated edible oilseed crops *Arachis hypogaea* L. is an important one. A rapid method of plant regeneration forms mature zygotic embryo of *Arachis hypogaea* through direct differentiation as well as through de-differentiation via callus formation. Salt tolerant callus of *Arachis hypogaea* has also been developed (Shekhawat and Kumar, 177-191).

Flow-cytometry (FCM), as the name implies, is the process of analysis of the optical properties of cells in a moving fluid stream. The optical and light scattering properties of cells are measured as single cell passes through the flow-cytometer. The fluorescence and light scattering properties depend on the size and granularity of the cells. The main advantage of flow-cytometry is the measurement of at least 3-5 parameters at a time and its swiftness in measuring the number of cells/ nuclei (Amita Pal, 193-204).

The established pharmaceutical products derived from phytosteroids include corticosteroids, contraceptives, sex hormones and anabolic agents. Solasodine is one of the important steroidal precursors, which have been detected and explored for pharmaceutical purposes. There is a need to enhance the production of steroidal alkaloids, which is being done using the tools of biotechnology. Various steroidal precursors, their use and conversion to the steroidal molecule, with special reference to solasodine from various *Solanum* species and the biotechnological advances made in their scale up, extraction, pharming, its future prospects and application have been reviewed by Asha Jacob and Nutan Malpathak (205-241).

In many parts of the world legumes are the most important crops where these are grown and consumed along with cereals. Grain legumes have been hampered by their recalcitrant nature in tissue culture and lack of compatible gene transfer methods. In this context efficient regeneration and transformation protocols are required for the application of biotechnological tools for legume improvement. In the past few years grain legumes have been transformed using various marker and reporter genes, agronomically useful genes imparting resistance to herbicides, insects, viruses and genes responsible for nutritional quality (high fatty acids, lipids and oils, carbohydrates and proteins). Progress achieved in legume transformation is reviewed by Jaiswal et al. (243-283).

Regeneration of *Citrus jambhiri* and *C. reticulata* through nucellar tissue adherent to cotyledons holds promise as large number of plantlets can be produced, without much danger of carrying any viral disease(s). The protocol developed in the two genotypes could be employed for undertaking attempts on their *in vitro* conservation (Khawale and Singh, 285-292)

It is desirable to micropropagate trees using tissues from mature trees since the desired phenotype is expressed at the physiologically mature state. A further problem during micropropagation of trees arises because *in vitro* formed shoots often show leaf shed, shoot tip necrosis or a lack of elongation on subsequent subculture, leading to the loss of micropropagated plants. *In vitro* responses of mature tree tissues of the tropical deciduous tree *Gmelina arborea* Roxb, an important timber-yielding tree, are elucidated (Naik *et al.*, 293-299).

This book is unique in several respects, the most important one being the attempt to deal with all the topics in-depth and to find out the molecular aspects of different processes in cultured plant cells, so as to exploit them for the benefit of plant biotechnology.

Editors

Contents

CHAPTER

1

Some Studies on Somatic Embryogenesis: A Tool in Plant Biotechnology

K.-H. Neumann

INTRODUCTION

Biotechnology is an area of production and research in which biological systems and biological principles are employed to solve technological problems. This includes beer brewing using yeast as well as the production of pharmaceuticals by immobilized enzymes. Some systems in biotechnology were also developed using cultured cells of higher plants to produce economically valuable chemical compounds, mostly of pharmaceutical interest, in fermenter systems upto ten thousand litres. Despite great efforts during the last decades only a few applications of such systems are known in commercial companies. Here economic considerations require that the product for the market should not only have the same price as the one produced conventionally from raw material obtained from agricultural products. In order to initiate a change in the production process the product from the fermentor has to be produced at least at half the production cost as the conventionally produced compound. Another area of biotechnology is micropropagation of plants. The aim of this technique is speedy production of a number of genetically identical plants from a highly valuable mother plant or monosexual male and female plants. These plants can either be directly sold in the market for planting or used for breeding purposes, genetechnology or to carry out basic science studies.

To this end three broad techniques are available in many variations for micropropagation:

- **meristem cultures**, in which apical or axillary meristems are induced to produce shoots, which are later rooted in a suitable medium, usually containing an auxin.
- **organogenesis** in callus material, in which the differentiation of shoots and roots is induced and the two parts grow together to form plants, or possibly on which shoots are grafted following rhizogenesis.

*Address for correspondence: Institut für Pflanzenernährung, Justus-Liebig-Universität, 35392 Gießen, Germany.

- **somatic embryogenesis**, during which a program of differentiation is induced in somatic cell material closely resembling the development of an embryo out of the zygote, i.e. the fertilized egg cell, though some differences exist (Ammirato, 1985).

The first two methods require manual isolation of the original explant and the plants produced and are consequently labour-intensive with high production cost. However, for somatic embryogenesis the possibility of automatization exists. Moreover, artificial seeds for some purposes can be envisaged through this method too. But again the costs involved would limit its application. Still, for some species methods are already available to produce embryos or raise plantlets in fermentor culture. For example out of 3g innocculum of induced carrot cells ca 100 000 plantlets can be produced within two weeks (Imani, 1999).

Another application of plant cell cultures along with gene technological methods is the production of medical agents like antigens to be used for vaccination. For example, in our laboratory we introduced the gene for a coat protein of Hepatitis B into the carrot genome. Through somatic embryogenesis, plants containing the gene and expressing this protein in the tap root of mature plants ready for harvest were raised. Clinical tests feeding these roots are now being conducted.

Somatic Embryogenesis–General Considerations

This article will concentrate on somatic embryogenesis as originally described by Reinert (1964) and reported by the Steward group at Cornell University in the same year (Steward *et al.* 1958), in which the production of somatic embryos out of cells derived from the root explants of carrot were described and which were eventually raised to intact plants. Explants were obtained from these plants to be used again as a source of a second generation of plants.

A peculiarity of eukaryotes is the propagation by zygotic embryogenesis, following the fusion of the male and female gametes, upon fertilization. For higher plants, detours to zygotic embryogenesis for some species, generally summarized as apomixis, became known.

These detours are confined to the generative apex. Still, this natural competence of cells to somatic embryogenesis seems to exist more generally, though usually camouflaged in the intact plant. However here, somatic embryogenesis as practiced in many tissue culture laboratories has to be induced by special conditions, as can be provided *in vitro,* by a suitable environment consisting of a nutrient medium containing a stimulus, usually an auxin, or a cytokinin or both, and some requirements for appropriate temperature and illumination should be fulfilled. In some protocols also an ABA supplement is beneficial. In most plant species this competence is lost during ontogenesis and somatic embryos cannot be produced from explants of other parts of the plant. Thus, these species are heuristically defined as recalcitrant. Still, using cells of embryonic origin often proved to be successful. This example is the case for some economically important cereals. Here mature seeds are germinated in a medium containing high concentrations of 2.4-D (e.g. 10-15 ppm 2,4-D) followed by excessive callus formation, from which a great number of embryos can be derived after a transfer to an auxin free nutrient medium (Imani, 1999).

Some hierarchial order within the plant seems to exist for many species, with highest success using embryonic cell material, followed by that of the hypocotyl, the petiole, the young leaves and finally the root. However, there are some exceptions to this loss of embryogenic competence during ontogenesis and one of these is *Daucus carota*, the common carrot. It is possible to culture intact six to eight weeks old plants aseptically and partly submerged in an appropriate medium containing an auxin, and within about 4 weeks somatic embryos appear from all parts of the shoot. If IAA is used as the auxin adventitious roots emerge about two weeks earlier (Schäfer *et al.*, 1988). In this system the competence to somatic embryogenesis is apparently preserved beyond the embryo stage and it can be activated rather easily by means of a suitable environment.

Such favourable conditions to initiate somatic embryogenesis can also be invoked in special genetic circumstances in intact plants on an inorganic Agar medium. This was observed in about six to eight weeks old plantlets derived by somatic embryogenesis from the fusion products of protoplasts obtained from transgenic plants of the wild carrot and transgenic plants of a domestic variety of *Daucus carota*. Hygromycine resistance was introduced into the wild carrot strain and 5-methyl-tryptophan resistance into the strain of the domestic variety (Rotin). On the petioles and roots of these cybrids a small callus developed on which somatic embryos later appeared. These could be used to obtain plantlets on which this process was observed again. This was repeated for three "generations" (Chinachit, 1991; De Klerk *et al.*, 1997). This kind of development was not observed on sexual crosses of the "parents" of the cybrids or of protoplast fusion products of the genetically unaltered parent genomes. How the introduction of the foreign DNA should have changed the developmental control system of the cybrids, possibly related to the hormonal system, is unclear and no further explanation about these observations can be given at present.

Results of histological examination of embryogenic systems require a distinction between a **direct** and an **indirect** induction. In **direct** somatic embryogenesis the embryos develop out of distinct cells without an interfering callus phase, whereas in the **indirect** process a callus phase characterized by high cell division activity is required apparently during which some process of dedifferentiation occurs. In the latter, genetic obstruction in the "offspring" usually called somaclonal variation, is often reported. In some species like the carrot both of these developmental processes can be observed. This review will be mainly concerned with **direct** embryogenesis and some results describing this process from our own research program would be mentioned.

For further studies to understand this developmental process it is not suitable to use cultured entire plants *in vitro* and the use of explants from various organs would be more practical. In our laboratory we mainly use explants of petioles of about 6 weeks old plants. Here two culture systems are in use. If the rather stable 2.4-D is used as the auxin the cultured explants have to be transferred to an auxin-free medium two weeks after the culture is begun. And the various stages of embryo development can be observed about two weeks later. Though rhizogenic centres histologically comparable to the IAA treatment (as described below), occasionally some adventitious roots too develop. In the other system, after about 10 days of using IAA as an auxin adventitious roots appear and again four weeks later the isolation of the explants embryo development occurs. Since the IAA of the nutrient medium is destroyed after two or three days in the light (Bender and Neumann, 1978 a), transfer to an auxin-free medium is not required. In the former system the number of embryos developed is usually higher than in the second and for many biochemical investigations the inclusion of root tissue is often not desirable. Thus, in many studies the first system is preferred, while in others the second (Neumann, 1995). However, a culture of 48 h in the 2.4-D-medium is sufficient to initiate embryo development at t12 to t14 after the initiation of the culture (Grieb, 1991/92).

The various developmental processes are confined to special areas in the petiole and only the subepidermal cells (sometimes in near proximity to a glandular canal) are truly totipotent and competent to produce somatic embryos in a direct way and without a prerequisite for a callus phase (Li and Neumann, 1985, Neumann and Grieb, 1992, Neumann, 1995). Interestingly, the cells forming the glandular canal contain high concentrations of auxins, as is shown by using transgenic plants containing the auxin sensitive MAS-promoter coupled with the GUS-gene (Grieb *et al.* 1997).

Rhizogenic centres develop near vascular bundles prior to those embryogenic centres. In both the cases some originally vacuolated cells start to produce new cytoplasm before the initiation of cell division. This contrasts with the initiation of cell division in cultured explants of the carrot tap root. Here the first responses are associated with the formation of phragmosomes localized in strings of cytoplasm traversing the vacuoles (Neumann, 1995). In these strings of cytoplasm nuclear division also takes place. The cytological events from an extensive histological investigation are summarized but all stages could not be observed in all the sections obtained. Looking at the longitudinal cuttings of cultured petiole, regenerative centres are irregularly distributed along the axis. Apparently the regenerative cells in these petiole explants contain at least a second competence to that brought about in the original petiole hidden during its development on the intact plant, i.e. rhizogenesis in the cells near the vascular bundles or embryogenesis in those subepidermal cells.

Even though nowadays laboratories mostly use somatic embryogenesis as an essential part of raising transgenic plants, neither its cytological nor its physiological or biochemical background is really understood. Despite its practical applications being obvious, it is an object of basic studies to understand differentiation as one of the significant features of biological systems. What do we know now and on what lines should research concentrate?

First, some essential requirements for somatic embryogenesis are given below and these generally also apply to zygotic embryogenesis, if one comes right at it, actually at all differentiation pathways.

Requirements to induce somatic embryogenesis

- competent cells
- suitable environment
- stimulus

Based on these requirements, the following questions were formulated for further investigations

- What constitutes embryogenic competence at the cytological and the molecular level?
- How are embryogenic competent cells produced during ontogenesis of plants?
- Molecular organization of the program of somatic embryogenesis and its realization.
- What does the stimulus to induce the program of embryogenesis in competent cells consist of?

Of these four questions, this article considers only the third and the fourth.

Genetic Aspects–DNA Organization

Genetic factors play a central role to induce somatic embryos, i.e. by providing competence to the species for the process. Here, great variation can be found even within a genera such as *Daucus*. Eight of twelve *Daucus* species cultured in identical conditions produced somatic embryos (*D. halophilus, D. capillifolius, D. commutatus, D. azoricus, D. gadacei, D. maritimus, D. maximus, D. carota*), whereas four species (*D. montevidensis, D. pussillus, D. muricatus, D. glochidiatus*) were not competent to do so. Since two of the recalcitrant species and three of the competent ones are native to the Mediterranean, the geographic location of origin seems to be of secondary importance (Le Tran Thi, unpublished results of our laboratory).

We used several molecular approaches to characterize the DNA of competent and recalcitrant species and one is the RAPD-Technique, which gives Random Amplified Polymorphic DNA sequences employing PCR, i.e. the polymerase chain reaction. Here selected primers with special sequences of about 10 nucleotides are applied as starters for replication and the number of nucleotides between the primer sequences yield DNA stretches of various lengths, which can be separated on a gel by electrophoresis. The location of the primer sequences on the DNA is genetically fixed and characteristic of the species.

The RAPDs of these *Daucus* species were compared using some 30 primers and for one of these two areas were identical in the embryogenic species, missing in the recalcitrant, i.e. the areas with RAPDs at about 1100 bp and about 650 of primer 3. We do not yet know what the function of these "marker DNA" sequences for an embryogenic potential could be and whether they have anything to do with somatic embryogenesis at all. The two conspicuous bands were isolated and sequenced. Without going into the details of this study (Imani *et al.*, 2001), both bands were quite similar in the embryogenic species with an identity between 70 and 95% in the nucleotide sequence. Neither indicated an open reading frame and it is safe to conclude that no sequences of genes occur in these stretches of DNA. A search in data banks did not help for further characterization, as these DNA-stretches were, apparently, not described previously. Further investigations are required to see whether these bands can be regarded as markers for the ability to produce somatic embryos, at least for the genera *Daucus*.

Going from species to varieties, here too differences exist with respect to competence to produce somatic embryos, as reported some years ago. While petioles from wild carrots, a French variety (Vosgeses) and the old German variety Rote Riesen are moderately competent under the conditions employed, the variety Lobbericher is not embryogenic, and the more recent German variety Rotin is highly competent. The DNA of these varieties was compared by density gradient centrifugation (Dührssen and Neumann, 1980). The GC content of DNA sequences obtained by mechanical shearing of total DNA varies and the density increases with the GC content. GC rich sequences appear as heavy satellites of the main band DNA. The highest number of GC rich satellites (Cs_2SO_4/Ag^+ density gradient centrifugation) was found for wild carrots and in the domestic varieties one or the other satellites are always missing. This indicates that during domestication some DNA sequences are lost or altered in concentration (Dührssen *et al.*, 1984).

The wild carrot is embryogenic and so is the variety Rote Riesen, while Lobbericher and an Italian variety are not. The common denominator of wild carrots and Rote Riesen is the satellite with a density of 1.422 g/cm³, missing in the nonembryogenic. Here again we do not yet know whether the DNA of this satellite has anything to do with somatic embryogenesis, or if it could only be regarded as a "marker" for the potential.

These GC rich satellites are either due to highly or middle repeated sequences generally not coding for proteins and are nowadays often defined as so called "junk DNA". This probably also holds true for the RAPDs discussed before and the function of this "junk DNA" is unknown. But one has to keep in mind that this junk DNA can represent more than 90% of the DNA of an organism. The genes proper could be similar or even identical in these varieties and even species with different embryogenic potential. What distinguishes them could be the organization of the genetic system, i.e. the ways and sequences in which they are activated to produce proteins and enzymes. Here possibly junk DNA plays a crucial role.

Some Remarks on the Physiological Situation of Cultured Cells *in vitro*

Before turning to the endogenous hormonal system as related to the induction of somatic embryogenesis in cultured petioles, we will first consider the physiological situation cultured plant cells are generally in. They are isolated from their natural environment, which provides them with water, nutrients and growth regulators generally supplied by the surrounding cells or even remote tissues in the intact plant, as a result of the many interactions of the various cell and tissue types produced during the development of the intact plant in its environment. This, however, is *in vitro* replaced by the nutrient medium. This aspect is of some disadvantage if results and conclusions derived from tissue culture experiments are used to explain and understand problems of the physiology of intact plants. The advantage of cultured cells *in vitro*, however, can be seen in the opportunity to investigate the significance of one or the other ingredient in plants or other physiological aspects like the pH or osmotic phenomena for the performance of plant cells. Secondly, by obtaining primary explants a wound is set, which provokes physiological reactions by the explanted tissue. Actually it was a natural consequence of tissue culture experiments for Haberlandt, the father of plant tissue culture, to get interested in wound healing later. It is known that wounding promotes the synthesis of ethylene and that, in turn, the synthesis of ethylene can be induced by auxins (e.g. Bender and Neumann, 1978b). The amount of ethylene produced after a few days of culture is closely related to the number of cells produced after a culture period of three weeks. However, the basis of the phenomenon is not yet understood. Also, hardly any data is available on the influences of decaying cells on the environment and viable cells in the same culture vessel. In the intact plants, the number of dying cells, for example in a leaf, should be much less in percentage then in cell suspensions and their influence should be more restricted than in a closed system, as represented by cultured cells *in vitro* in its original sense of the word, namely "glass". Still, reactions observed *in vitro*, like somatic embryogenesis, can also be observed on intact plants *in vitro,* in special circumstances without wounding (s. above, Schäfer *et al.* 1988, Chinachit, 1991, De Klerk *et al.*, 1997). This indicates that wounding is not an essential prerequisite for the growth of primary explants, as often discussed.

The fate of cultured cells is basically determined by three physiological phenomena common also to all cells in general. These are cell growth, cell division and cell differentiation. They are interconnected and it is difficult to separate one from the other in experiments and in the interpretation of results derived from tissue culture experiment, let alone those obtained using intact plants as experimental systems. Still an attempt would be made by referring to some data obtained by one of the most ancient and prominent experimental systems in the field of cell cultures, cultures of carrot explants and cell suspensions derived thereoff. This system was originally introduced by the Steward group at Cornell University in Ithaca, N.Y. in the late forties of the last century.

The Phytohormone System

Competent cells have to be induced to somatic embryogenesis by a trigger, which in the carrot system is an auxin in a suitable nutrient medium such as a modified B5 medium or the NL-Medium developed in our laboratory (Neumann, 1968, 1995). If these induced cells continue to grow in an auxin-free medium, either brought about by photooxidative destruction of IAA or by transfer into an auxin-free medium as in the case of 2.4-D embryo, development will happen. However, if these induced cells are continuously subcultured in the 2.4-D medium, this commitment will be preserved for many years and the realization of the embryogenic program will be prevented till these cells are transferred into an auxin-free medium.

Turning first to the induction process, the hormonal system plays a key role in the induction of somatic embryogenesis. It is known from earlier investigations that cultured cells develop an endogenous hormonal system different from that of the original explants (Bender and Neumann, 1978a, Stiebeling and Neumann, 1987) and thus, as a first approach the concentration of some phytohormones was determined at various stages during the induction of somatic embryogenesis in cultured petiole explants (Grieb *et al.*, 1997). Of those phytohormones determined, ABA dominates the original petiole explants, followed by a much lower concentration of IAA and cytokinins. While the ABA concentration continuously decreases during culture, the IAA concentration reaches its maximum six days after the start of the experiment, along with the start of the development of adventitious roots. Afterwards the concentration of this native auxin continuously decreases until the 14th day, when again a small peak appears. At the tenth or eleventh day, though at a lower level, a maximum of the cytokinin fraction also occurs, dominated by 2iP and its riboside coinciding with the onset of cytoplasmic growth in small, originally vacuolated subepidermal cells to become embryogenically induced. To summarize, the cultured petiole explant produces its own hormonal system with continuous changes in the ratio of concentration of these phytohormones investigated with each other. The importance of such changes for plant development was recognized many years ago by Skoog and Miller (1957) and can possibly play a decisive role too in cultured petiole explants to become embryogenic. To a certain extent, this is confirmed by comparing the data obtained from a highly embryogenic variety with that obtained by a recalcitrant variety, with a completely different course of the concentration of IAA and cytokinins as published earlier (Li and Neumann, 1985). Also in cultured cells the concentration of phytohormones follows a circadian rhythm, which has to be considered an interpretation of results on the phytohormone system *in vitro*. This was at least demonstrated for callus cultures of root origin.

Using petiole explants from transgenic plants containing the auxin responsive MAS promoter linked to the GUS reporter gene, the distribution of auxin within the cultured petiole could be followed during the induction phase of somatic embryogenesis. In the original petiole explant, the auxin is more or less evenly distributed throughout the petiole after five to six days in culture, along with the formation of root primordia near vascular bundles, and IAA gets accumulated in this area of the petiole. After nine days of culture, immediately before root development can be observed, the auxin concentration in this area is drastically reduced and IAA now accumulates in the emerging embryogenic areas. Apparently, in the cultured petioles not only changes in the total concentrations of IAA but also distinguished changes in its distribution related to histogenic events can be observed (Grieb *et al.*, 1997, Imani, 1999).

Although the *Daucus carota* system does not need to be supplied to the medium to induce somatic emryogenesis, cultures of other plant species often require an ABA supplement for the process. In *Daucus carota* cultures an ABA supplement actually seems to be slightly inhibitory. If ABA plays a role in the process then it should do so during the first week of culture when its concentration in the petiole explants is high. Of all the *Daucus* species investigated (s. above) *Daucus carota* exhibits the highest embryogenic potential. To investigate the role of ABA in somatic embryogenesis, first its concentration in the original petiole explants of the various *Daucus* species, embryogenic and nonembryogenic, was determined. The highest concentration of free ABA was found in the highly embryogenic domestic carrot variety Rotin. The other embryogenic species with a much lower potential for the process had considerably lower concentrations of this phytohormone. The number of embryos produced was significantly increased and the time required for the initiation of embryogenesis was clearly reduced by the application of ABA to the medium used for culture (Le and Pleschka, 2005). In yet unknown ways ABA seems to be just as involved in embryogenesis as auxins. Possibly the high potential of *Daucus carota* for somatic embryogenesis could be related to the high concentration of free ABA in the original explants of this species at explantation. Studies on the distribution of ABA in analogy to those on IAA using the SalT promoter, linked again to the GUS reporter gene or the GFP-gene are being taken up. Such transgenic plants will also be good tools to study the concentration of hormones during the development of intact plants.

The Protein System

At some of these stages of the petiole system used for hormonal investigations the protein synthesis patterns were determined (Grieb, 1991/92, Neumann and Grieb, 1992; Grieb *et al.*, 1997). For localization of protein synthesis activities within the cultured petioles ^{14}C leucine was applied and its distribution was followed by histoautoradiography and electrophoresis after protein extraction. Most of the labelling was concentrated in those parts of the petiole, which indicated an accumulation of auxin, i.e. the protein synthesis at one stage or the other should occur in cells engaged in differentiation. For these elaborate studies the cultured petiole explants in the induction medium, i.e. the B5 medium supplemented with 0,5 ppm 2.4-D, were labelled for three hours each with ^{14}C leucine, starting at 0 h, 5 h, then the next at day seven and the last labelling at day 14 of culture. The soluble protein was extracted as described earlier (Gartenbach-Scharrer et al., 1990), and the extracts were separated by 2D-gelelectrophoresis followed by either staining with coomassy brilliant blue to visualize proteins on the gels or subjected to fluorography to detect the distribution of ^{14}C in various proteins. In these investigations about 280 proteins were detected on the gels either by one detection method or both. According to the staining and labelling pattern and various labelling periods, the proteins were arranged in nine groups and set into relation to cytological events during the induction of somatic embryogenesis. A continous change in the composition of the protein moiety occurs with the initiation or termination of the synthesis of proteins of one or the other group in a sequential and hierarchial pattern during the induction of somatic embryogenesis. Some proteins were detected throughout the experimental period and should represent the so called household proteins. Of special interest are the 17 proteins, which could be only detected at day 14 and which should be somehow related to the occurrence of the cytoplasm-rich subepidermal cells destined to be the origin of somatic embryos and/or to the initiation of embryo development (Grieb, 1991/92).

As a first approach to understand the physiological significance of these proteins for somatic embryogenesis based on the isoelectric point and the molecular weight, a search for analogs published in databases like Swiss-Prot or Trembl was undertaken. Of those 17 proteins that synthesized on day 14 after the start of the cultures, only 3 analogs could be found. Interestingly, all three are related to carbohydrate metabolism, namely alpha-amylase, phosphofructokinase and alcoholdehydrogenase (Mashayekhi-Nezamabadi, unpublished results of our laboratory). Another protein described by the Komamine group, as characteristic for an embryogenic status, was already synthesized on the seventh day of culture prior to the cytoplasmic growth of subepidermal cells destined to become embryogenic, usually observed from the 10th or 12th day of the culture. This indicates that the initiation of the embryogenic program starts some time before its histological evidence. For characterization of the function of alpha-amylase synthesis, starch content and starch distribution by histochemistry on the 12th and the 14th day was compared. While upto the 12th day high starch accumulation in parenchyma cells of the petiole was observed, the starch concentration on the 14th day was reduced, presumably the result of the action of the newly synthesized alpha-amylase molecules (unpublished results of our laboratory). This was in tendency confirmed by the enzymatic determination of starch. The glucose resulting from the starch breakdown should be phosphorylated by Hexokinase while preparing for further metabolic processing.

As described for other plant species also cultured cells of carrot can perform photosynthesis (Neumann, 1972, 1995; Bender *et al.*, 1981, 1985). For further studies on the function of free glucose a photoautorophic cell culture strain, also embryogenic, was employed. The cells of this strain are able to grow slowly at ambient CO_2 in the light without differentiating somatic embryos. However, this is the case after a supply of sucrose, glucose and the less pronounced fructose or mannose at low concentrations (Pleschka, 1995, Grieb *et al.*, 1994). To distinguish the nutritive and regulatory effects of these sugars on somatic embryogenesis, cell suspensions of this autotrophic strain were cultured at an elevated CO_2-concentration of 2.3%. Here growth is comparable to what it would be after a supplement of 2% sucrose, and yet somatic embryogenesis could not be observed here too (Pleschka, 1995, Grieb *et al.*, 1994). This is the case after an additional supply of 0.1% of sucrose to the nutrient medium. These results indicate that the requirement of sugar should not be its only function as a nutrient and it should also act as a regulator in the development of embryos

(Pleschka, 1995, Grieb *et al.*, 1994). Glucose can be substituted by glucose-6-phosphate but not by glucose-1-phosphate in petiole cultures, which is the result of starch breakdown by starch-phosphorylase and bypasses Hexokinase for further breakdown. Apparently, for embryo development Hexoskinase activity with free hexose or glucose-6-phosphate as substrates described for animal systems are required for embryo development. To substantiate these ideas the occurrence and metabolic activity of hexokinase has to be investigated further. Especially important shall be the investigations on the histological and cytological distribution of the enzyme. Such investigations are presently in progress.

Cell Cycle Studies

To understand the regulation of embryo development in-depth, the relation of some aspects of the cell cycle to differentiation would have to be considered. But first some related data on rhizogenesis in cultured explants of the secondary phloem of the carrot tap root would be discussed as a basis. As already shown, a kinetin supplement to a nutrient medium containing IAA significantly increases the growth of cultured cells, mainly by cell division (Linser and Neumann, 1968) and concurrently suppresses regeneration, for example adventitious root formation. While those explants growing in a hormone-free nutrient medium show growth by cell expansion and also some cell divisions (data not provided), those supplemented with IAA and m-Inositol without kinetin show almost a tripling of cell number per explant during a three weeks cultural period indicating a clear, though rather small log phase of cell number (Neumann, 1995).

The samples supplemented additionally too with kinetin are characterized by an extensive log phase of high cell division activity and a decrease in cell size. Most remarkably, only the treatment containing IAA and m-Inositol in the medium is able to differentiate cells leading to the formation of adventitious roots. Both the other treatments remain morphologically undifferentiated cell material. Still, the production of a callus also could be defined as a kind of differentiation and after about 4-5 weeks of culture in a medium originally also supplemented with kinetin during a stationary phase of cell division, some roots appear. A similar course of the cell number per explant can also be observed if kinetin and IAA are replaced by the comparitively stable auxin 2.4-D, which strongly promotes growth by cell division. Here to the formation of adventitious roots is prevented. Apparently, some correlation between the possibility of root formation, i.e. differentiation and cell division activity seems to exist.

The number of cells produced per cultured explant during a given unit of time is either due to the number of cells engaged in active cell division or the length of time between successive divisions, i.e. the duration of the cell cycle, or both. This could be demonstrated for a kinetin supplement as resulting in the reduction of the duration of the cell cycle for *Datura* as well as for carrot cell cultures by about 60-90 minutes. This was mainly due to a reduction of the G1 phase of the cell cycle (Blaschke *et al.* 1978, Froese and Neumann, 1997).

Although information is still limited, the following hypothesis was formulated many years ago to explain the correlation between cell division activity and differentiation (Neumann, 1968, s. also Neumann, 1995). After the completion of the M-phase during the G1-phase, genetic information is utilized in a hierarchial sequence till this is interrupted at the onset of the S-phase and which part of the genetic information potential can be realized during the cell cycle depends on the duration of the G1-phase. Apparently, in our system information related to differentiation is later localized in this sequential activation of the genetic information. The initiation of the S-phase and consequently the potential for differentiation here is determined by kinetin. In plus-kinetin treatments, the length of the G1-phase is just not sufficient to activate the information required to bring about root or embryo development. Theses ideas are illustrated by the following scheme.

Kinetin, cell cycle phases and differentiation: (********* genetic information in G_1)

§------M------§*-*-*-*-*-*-*-*-*-*§------S------§------G_2------§------M----

G_1

Without kinetin: Cell cluster, PEMs, rhizogenesis, somatic embryogenesis

§------M------§*-*-*-*-*-§------S------§------G_2------§------M----

G_1

With kinetin: cell cluster, PEMs

The results reported for cell suspension cultures below confirm this old hypothesis to some extent. In the regulation of the hierarchial activation of genetic information during the G1-phase, the cycline system, which has come to be known in more detail recently, seems to play a central role. However, it will not be dealt with here.

A prolongation of the cell cycle duration alone is not sufficient to promote differentiation. Just as an omission of kinetin, as an example at iron deficiency the number of cells produced by the cultured explants is also reduced, however, cell growth is also retarded and no root formation could be observed (Neumann, 1972).

As mentioned before too, in the hormone-free treatment no differentiation was observed. Only about two rounds of cell division occurred in some areas of the explant here and then growth took place only as cell expansion. Cell division activity is also considerably reduced in the treatment supplemented with IAA and m-Inositol as compared to the plus kinetin treatment. However, here a short log phase of cell division activity with a reduced duration of the cell cycle occurs. Apparently, during this log phase some dedifferentiation takes place in dividing cells, later to become rhizogenic. This dedifferentiation may result in some reprogramming of the genetic machinery as well as the cells in toto, possibly both. A continuous protein turnover takes place in cells and if the hypothesis of the relation between the duration of the cell cycle and realization of genetic information has some meaning, then some proteins with coding positions later in the G_1-phase, however present at the beginning of this short log phase with high cell division activity, could be removed from the cells by breakdown without being replaced by newly synthesized molecules due to the reduction of the G1-phase. If the protein moiety of a cell represents its differentiational status then this should be responsible for changes in differentiation at the molecular level of cultured cells engaged in active cell division.

Reprogramming of the genetic machinery could be due to a change in DNA methylation (LoSchiavo *et al.* 1989). During a log phase of cell division in our system, an increase in methylation could be observed and the amplification of some DNA sequences was found, usually associated with de novo differentiation, only at the stationary phase of cell division. Here DNA methylation gets reduced again. As shown, this increase in methylation during the log phase of cell division is not directly related to a kinetin application but occurs due to the reduction of cell cycle duration (Dührssen and Neumann, 1980, De Klerk *et al.*, 1997, Arnholdt-Schmitt *et al.*, 1998). Variation in the amplification of DNA sequences also require mechanisms of DNA breakdown. Years ago, the occurrence of such "metabolic" DNA was reported particularly for differentiating cells employing the pulse/chase technique (Schäfer *et al.*, 1978).

As mentioned above, the higher cell division activity in the embryogenic carrot cell suspension due to the kinetin supplement to the nutrient solution reduces the duration of the cell cycle by about one hour (Froese and Neumann, 1997), again due to the reduced duration of the G1-phase, as was shown for *Datura* cultures several years ago (Blaschke *et al.*, 1978). While in the carrot system in the kinetin-free medium supplemented with IAA and m-inositol (2 mg/l), rhizogenesis and the development of embryos upto plantlets was observed, in the medium supplemented with kinetin (0.1 mg/l) in addition to IAA and m-inositol, no roots were produced and embryo development did not proceed beyond the stage of PEMs (pre-embryogenic masses). It was assumed that here also a correlation between the duration of the G1-phase and the regeneration capacity exists. The determination of protein synthesis pattern during the prolonged G1-phase by labelling of proteins using synchronized cultures (FDU/Thymidine System with ^{14}C leucine as described above for the petiole system) indicated the synthesis of 132 additional proteins in the minus kinetin cultures, prevented from being synthesized in the kinetin treatment due to the immediate transition from the G1-phase to the S-Phase of this treatment. These cells in the kinetin-supplemented medium immediately pass upon the addition of thymidine into the S-phase with strongly reduced protein synthesis activity whereas those of the kinetin-free treatment remain in G1 for up to another 60 to 90 or even 120 minutes (varies from one experiment to the other). During this prolonged G1-phase the synthesis of these additional proteins occurred in a hierarchial sequence, with initiation and termination of the synthesis of individual specimen.

It yet remains to be seen if and to what extent one or the other of these additional proteins are required for embryo development of these induced cultures. At present a similar search in data banks as described above for the petiole system above is being conducted.

Acknowledgements

Thanks are due to Dr. J. Imani for the technical finish of this article and to J. Korb-Müller for many hours spent on the computer (all of our group in Gießen).

REFERENCES

Ammirato P.V. (1985) Patterns of development in culture. In: *Tissue Culture in Forestry and Agriculture.* Henke R.R., Hughes K.W., Constantin M.I. and Hollaender A. (eds.) New York: PLENUM. 9-29.

Arnholdt-Schmitt B., Schäfer C. & Neumann K.-H. (1998) Untersuchungen zum Einfluß physiologisch bedingter Genomvariationen auf RAPD-Analysen bei *Daucus carota. Vortr. Pflanzenzüchtg.* 43: 107-113.

Bender L. & Neumann K.-H. (1978a) Investigations on indole-3-acetic acid metabolism of carrot tissue cultures. *Z. Pflanzenphysiol.* 88: 209-217.

Bender L. & Neumann K.-H. (1978b) Ethylene production and growth of carrot root tissue cultures (*Daucus carota* L.) as influenced by ß-indole-3-acetic acid and kinetin. *Z. Pflanzenphysiol.* 88: 227-234.

Bender L., Kumar A. & Neumann K.-H. (1981) Photoautotrophe pflanzliche Gewebekulturen in Laborfermentern. In: *Fermentation.* (Lafferty R.M. ed.) Springer-Verlag, Wien. 193-203.

Bender L., Kumar A. & Neumann K.-H. (1985) On the photosynthetic system and assimilate metabolism of *Daucus* and *Arachis* cell cultures. *Primary and Secondary Metabolism of Plant Cell Cultures* (Neumann K.-H., Barz W. & Reinhard E., eds.). Springer-Verlag, Berlin, Heidelberg. 24-42.

Blaschke J.R., Forche E. & Neumann K.-H. (1978) Investigations on the cell cycle of haploid and diploid tissue cultures of Datura innoxia Mill. and its synchronization. *Planta.* 144: 7-12.

Chinachit W. (1991) Die somatische Hybridisierung von Wildkarotten und Kulturkarotten (*Daucus carota* L.) und die Charakterisierung der Hybriden. Diss., Justus-Liebig-Universität Gießen.

De Klerk G.-J., Arnholdt-Schmitt B., Lieberei R. & Neumann K.-H. (1997) Regeneration of roots, shoots and embryos: physiological, biochemical and molecular aspects. *Biologia Plantarum.* 39(1): 53-66.

Dührssen and Neumann (1980) Charakterisation of satellite-DNA of *Daucus carota* L. *Z. Pflanzenphysiol.* 100(5): 447-454.

Dührssen E., Lanzendörfer M. & Neumann K.-H. (1984) Comparative investigations on DNA organization of some varieties of *Daucus carota* L. *Z. Pflanzenphysiol.* 113: 223-229.

Froese C. & Neumann K.-H. (1997) The influence of kinetin on the protein synthesis pattern at some stages of the cell cycle of regenerative synchronized cell suspension cultures of *Daucus carota* L. *Angew. Bot.* 71: 111-115.

Gartenbach-Scharrer U., Habib S. & Neumann K.-H. (1990) Sequential synthesis of some proteins in cultured carrot explant (*Daucus carota* L.) cells during callus induction. *Plant Cell Tissue and Organ Culture.* 22: 27-3.

Grieb B. (1991/92) Untersuchungen zur Induktion der Kompetenz zur somatischen Embryogenese in Karottenpetiolenexplantaten (*Daucus carota* L.)—Histologie und Proteinsynthesemuster. Maraun Verlag, Frankfurt/M. (Dissertation, Justus-Liebig-Universität Gießen).

Grieb B., Groß U., Pleschka E., Arnholdt-Schmitt B. & Neumann K.-H. (1994) Embryogenesis of photoautotrophic cell cultures of *Daucus carota* L. *Plant Cell Tissue and Organ Culture.* 38: 115-122.

Grieb B., Schäfer F., Imani J., Nezamabadi Mashayekhi K., Arnholdt-Schmitt B. & Neumann K.-H. (1997) Changes in soluble proteins and phytohormone concentration of cultures carrot petiole explants during induction of somatic embryogenesis (*Daucus carota* L.). *Angew. Bot.* 71: 94-103.

Imani J., Nezamabadi Mashayekhi K. & Neumann K.-H. (1999) Gentechnische Verfahren als Werkzeuge in der pflanzlichen Hormonforschung. http://www.tzm.uni-giessen.de/unigi/ifz/beitrag.cfm?key=30.

Imani J. (1999) In situ- Nachweis der Auxinverteilung in kultivierten Petiolenexplantaten von transgenen Pflanzen während der Induktion der somatischen Embryogenese bei *Daucus carota* L. Diss. Justus Liebig Universität, Gießen, Germany.

Imani, J., L. Tran Thi., G. Langen., B. Arnholdt-Schmitt., S. Roy., C. Lein., A. Kumar., and K.-H. Neumann. (2001). Somatic embryogenesis and DNS organisation of genomes from selected *Daucus* species. *Plant Cell Reports,* 20: 537-541.

Li T. & Neumann K.-H. (1985) Embryogenesis and endogenous hormone content of cell cultures of some carrot varieties (*Daucus carota* L.). *Ber. Dt. Bot. Ges.* 98: 227-235.

Linser H. & Neumann K.-H. (1968) Untersuchungen über Beziehungen zwischen Zellteilung und Morphogenese bei Gewebekulturen von *Daucus carota* L. I. Rhizogenese und Ausbildung ganzer Pflanzen. *Physiol. Plant.* 21: 487-499.

LoSchiavo F., Pitto L., Giulilano G., Torti G., Nuti Ronchi V., Marazziti D., Vergara R., Orselli S. & Terzi M. (1989) DNA methylation of embryogenic carrot cell cultures and its variation as caused by mutation, differentiation, hormones and hypomethylating drugs. *Theor. Appl. Genet.* 77: 325-331.

Neumann K.-H. (1968) Über Beziehungen zwischen hormonalgesteuerter Zellteilungsgeschwindigkeit und Differenzierung, ein Beitrag zur Physiologie der pflanzlichen Ertragsbildung. Habilitation, Justus-Liebig-Universität Gießen.

Neumann K.-H. (1972) Untersuchungen über den Einfluß des Kinetins und des Eisens auf den Nukleinsäure- und Proteinstoffwechsel von Karottengewebekulturen. *Z. Pflanzenern. Bodenkunde.* 131: 211-220.

Neumann K.-H. (1995) Pflanzliche Zell- und Gewebekulturen. Ulmer Verlag, Stuttgart, 304 pages.

Neumann K.-H. & Grieb B. (1992) Somatische Embryogenese bei höheren Pflanzen: Grundlagen und praktische Anwendung, Wiss. Zeitschrift der Humboldt-Universität zu Berlin, R. Mathematik/Naturwiss. 41: 63-80.

Pleschka E. (1995) Untersuchungen zum Einfluß von Saccharose und einigen weiteren Kohlenhydraten auf die somatische Embryogenese von mixotrophen und photoautotrophen Gewebekulturen von *Daucus carota* L. Diss., Justus-Liebig-Universität Gießen.

Reinert J. (1958) Üeber die Kontrolle der Embryogenese und die Induktion von Adventivembryonen an Gewebekulturen aus Karotten. *Planta.* 53: 318-333.

Schäfer A., Blaschke J.R. & Neumann K.-H. (1978) On DNA metabolism of carrot tissue cultures. *Planta* 139: 97-101.

Skoog F. & Miller C.O. (1957) Chemical regulation of growth and organ formation in plant tissues cultured in vivo. *Symp. Soc. Exp. Biol.* 11: 118-131.

Schäfer F., Grieb B. & Neumann K.-H. (1988) Morphogenetic and histological events during somatic embryogenesis in intact carrot plantlets (*Daucus carota* L.) in various nutrient media. *Botanica Acta* 101: 362-365.

Steward F.C., Mapes M.O., Kent A.E. & Holsten R.D. (1964) Growth and development of cultured plant cells. *Science.* 143: 20-27.

Stiebeling B. & Neumann K.-H. (1987) Identification and concentration of endogenous cytokinins in carrots (*Daucus carota* L.) as influenced by development and a circadian rhythm. *J. Plant Physiol.* 127: 111-121.

Tran, Thi L. and Pleshka, E. (2005) Somatic embryogenesis of some *Daucus carota* species influenced by ABA. *J. Appl. Botany* and *Food Quality*. 79: 1-4

CHAPTER

2

Cellular Differentiation and Genetic Transformation Studies in Indian Legumes, Crucifers, Mulberry and Cereals

Paramjit Khurana

In recent years, culture of plant cells and tissues has gained considerable importance as a promising tool in improving and multiplying economically important crop plants. Plant cells are also being increasingly utilized as model systems to unravel the cellular and molecular mysteries of nature and understand the basic mechanisms of developmental biology, cellular and molecular interactions at the cell and tissue level, gene expression and regulation leading to genetic modifications and areas of genomics. The present presentation traces my own experiences with various plant cell systems for understanding the basic mechanisms of differentiation, molecular cell biology and presents results of efforts aimed at genetically modifying important crop plants.

LEGUME BIOTECHNOLOGY

The legumes are an important group of crop plants and consist of not only the important grain legumes, but also of various tree species many of which are important N_2 fixers and thus important in soil reclamation programmes as well. Amongst the various grain legumes extensively cultivated in India, *Lathyrus sativus* – having a relatively high protein content – is one capable of growth in semi-arid and arid zones. Organogenesis and plantlet regeneration was established in this important recalcitrant species from shoot tip cultures (Gharyal and Maheshwari, 1980). Despite a strong genotypic dependence to regeneration, utilizing physiologically activated shoot tips it was feasible to achieve higher levels of morphogenetic response even in the recalcitrant genotypes (Gharyal and Maheshwari, 1983). These also displayed somatic embryogenesis in long-term maintained cultures and resulted in normal, mature plantlets. Regeneration in tissue cultures of the sensitive plant,

*Address for correspondence: Department of Plant Molecular Biology, University of Delhi, South Campus, New Delhi-110021, India.

Mimosa pudica (Gharyal & Maheshwari, 1982a) was initiated with the view to investigate the role of cholinesterases in higher plants (Maheshwari *et al.,* 1982).

Albizzia lebbeck, the "East Indian Walnut", is a leguminous hardwood timber tree of India grown extensively as a plantation and shady tree in India. Due to its vast economic importance, methods for its micropropagation have been standardized. It is of special interest since this plant species displays the two major modes of differentiation in tissue cultures, i.e. organogenesis and somatic embryogenesis. The plant has a high morphogenetic potential and can thus be micropropagated from cotyledons, hypocotyls, roots or leaves (Gharyal and Maheshwari, 1982b). The response is nearly 100% and plantlets can be successfully transferred to the field. Besides differentiation from the seedling explants, methods have also been established for differentiation of explants obtained from fully mature trees, about 80 years old, (Gharyal and Maheshwari, 1989). These plantlets have also been successfully transferred to the field making it feasible for micropropagation of elite plants.

Additionally, the hypocotyl explants display production of somatic embryos directly, without any callus formation (Gharyal and Maheshwari, 1981). This system in which the two major modes of differentiation can be modulated at will is thus exploited for investigating the molecular mechanism of these two diverse pathways of morphogenesis in higher plants (Gharyal, 1992). The effects of various hormones and physical factors on differentiation has been examined and interestingly, light has been found to have a rather specific inductive effect (Gharyal, 1992). This response to light has been characterized at the monochromatic levels and the response appears to be Red/Far-red reversible, thus implicating phytochrome as the possible photoreceptor (Baweja *et al.*, 1995). With respect to the growing importance of tree biotechnology (Khurana & Khurana, 1999), this model system has been tailored for genetic transformation studies and some preliminary work in this direction has been initiated by particle bombardment approaches (Khurana & Khurana, 2000).

Besides *Albizzia*, two other important leguminous avenue trees investigated for their *in vitro* regeneration potential were, *Cassia fistula* and *C. siamea*. It was feasible to regenerate these species from explants obtained from mature trees (Gharyal & Maheshwari, 1989). Androgenesis was also obtained in all the three tree species, namely, *Albizzia lebbeck, Cassia siamea* and *C. fistula* (Gharyal *et al.,* 1983a,b).

Soybean cell suspension cultures have been used to unravel the cellular mysteries of the cell, e.g. the role of pectins in mediating cell wall porosity in higher plants (Baron *et al.*, 1988). Use of FRAP (*see* Schindler *et al.*, 1988) facilitated the demonstration that LPS-mediated cellular recognition occurs during legume-*Rhizobium* symbiosis prior to release of the infection thread in the cortical cells (Gharyal *et al.*, 1989). A non-invasive method for introduction of impermeant macromolecules in living cells was also developed to investigate the role of plasmodesmata during plant development (Meiners *et al.*, 1991).

Model Systems

The genus *Brassica* includes many economically important oil yielding, vegetable, and forage crops. *Arabidopsis*, the plant *Drosophilla,* also belongs to this family. Although protoplast culture has been reported in many species (Maheshwari *et al.*, 1986), a simple protocol for high frequency somatic embryogenesis and plantlet regeneration has been established from the hypocotyl protoplasts of *B. napus* and leaves of *Arabidopsis* (Parihar *et al.*, 1995, 1997). Protoplasts divided at the rate of 30-35% and regeneration was possible by both organogenesis and embryogenesis (Gandhi & Khurana, 2001). The somatic embryos developed into normal plantlets and also displayed secondary and repetitive embryogenesis (Parihar *et al,* 1995, Gandhi *et al.*, 1999). Protoplasts have been used for PEG mediated direct gene transfer and transient gene expression studies (Parihar *et al.*, 1997, Gandhi *et al.*, 1999, Gandhi & Khurana, 2001). Various physical and chemical factors were found to enhance DNA uptake and expression of the introduced gene in protoplasts of *Brassica, Arabidopsis*, wheat and rice (Mahalakshmi *et al.,* 1993, Mahalakshmi & Khurana 1995, Parihar *et al.*, 1997, Gandhi *et al.*, 1999, Gandhi & Khurana, 2001). *Arabidopsis* has also been exploited for the isolation of genes of commercial importance. Towards this direction, genetic and molecular analysis was undertaken at the T1 and T2 progeny levels subsequent to transformation with *Agrobacterium*. Mendelian and non-Mendelian inheritance patterns were obtained (Gandhi *et al.*, 2001). One interesting T-DNA tagged *Polyrosetta* mutant was isolated which showed reversion of flowering and had an enhanced biomass. Molecular analysis revealed gene recombinations to have occurred extensively, thus hampering isolation of this important gene from *Arabidopsis.*

SERIBIOTECHNOLOGY

Mulberry is a highly heterozygous and clonally propagated species that is prone to prolonged juvenile periods. Since the quality of silk production is directly proportional to the quality of leaves used as the exclusive feed for these worms, leaf quality is of utmost importance in sericulture. From the commercial viewpoint, a shorter tree/shrub with enhanced foliar production/vegetative growth and thicker and juicer leaves is highly desirable. Most of these traits are quantitative in nature and hence difficult to manipulate by conventional approaches. Since efficient regeneration is a prerequisite for genetic manipulation and transformation studies, regeneration in mulberry has been attempted from the axillary buds, hypocotyl, cotyledon, leaf and stem explants using different combinations of cytokinins and auxins (Khurana *et al.*, 2001). Micropropagation *via* axillary bud proliferation has been achieved in *M. indica* K2, RFS175, C176, C776 and S1 and *M. multicaulis* cv. Goshoerami by an exogenous requirement for cytokinin for bud-break and bud multiplication. Thidiazuron (TDZ) was found to be more effective than BAP (*see* Tewari *et al.*, 1999, Bhatnagar *et al.*, 2002a). TDZ not only reduced the days required for bud-break but also increased the percentage of bud-break and the number of shoots formed per explant.

Differentiation from seedling explants was also investigated and TDZ induced highest percentage of adventitious bud formation from both the hypocotyl and cotyledon explants. Multiple shoots emerged at the base of the cotyledon and from the cut ends of the hypocotyl (Bhatnagar *et al.*, 2001). Of the various sources of leaf explants, the response from the leaves derived from axillary buds cultured on BAP was almost double than those obtained from *in vivo* or field plants and also from *in vitro* seedlings. Differentiation from mature leaf explants was maximum from the explants obtained from field-grown plants on MS + TDZ followed by the *in vitro* raised explants. Upon regular subculture, regeneration from the callus could be observed over a year with as many as 15-20 shoots/culture. Histological examination indicated regeneration by the organogenesis in these long-term cultures (Kapur *et al.*, 2001). Protoplast isolation and culture was undertaken from leaf mesophyll protoplasts from *in vitro* grown shoot cultures and mature leaf tissues. The protoplasts were cultured in agarose beads/Ca-alginate layers and showed first divisions between 7-10 days of culture and microcolonies were formed by a month of culture (Kapur *et al.*, 2003)

For genetic transformation studies, the highly regenerative callus obtained from the leaf base explants and the multiple shoot forming hypocotyls, cotyledon and leaf explants have been utilized for direct gene delivery by particle bombardment (Khurana *et al.,* 2002, Bhatnagar *et al.*, 2002b) and *Agrobacterium*-mediated co-cultivation. The GUS histochemical assay performed after 3 days of *Agrobacterium* co-cultivation culture showed 80% transformation frequency (Bhatnagar *et al.*, 2003). However, when GUS activity was assayed after 45 days of culture it reduced drastically. The regenerants were raised and multiplied for a period of 3-4 months and GUS assay repeated. PCR analysis was employed for rapid screening of the regenerants (Bhatnagar & Khurana, 2003). The transformants were multiplied for detailed molecular analysis and characterization of the introduced gene. The second phase work is directed towards screening and evaluation of existing germplasm for abiotic responses/tolerance and targeting genetic manipulation of mulberry towards these goals.

CEREAL BIOTECHNOLOGY

Amongst various cereal grains wheat is an important ingredient of the human diet. Different varieties of wheat depending upon their dough quality and other visco-elastic characteristics are used for making a range of food products. Ninety-five percent of the cultivated wheat is of the hexaploid type used for the preparation of bread and other baked products and nearly all of the remaining is durum (tetraploid) wheat, which is used essentially for making pasta and macaroni. Modern-day hexaploid wheat (*Triticum aestivum*) is used generally for bread making but lacks certain agronomically important features such as drought or disease resistance. Tetraploid emmer wheat (*Triticum dicoccum*), in contrast, shows high adaptability to dry, parched lands, exhibits drought and salinity resistance, but its poor visco-elastic properties, hardness of the grain reduces its consumer acceptability. From the nutritional perspective, *T. durum* is the only wheat with a rich source of β-carotene content, and *T. dicoccum* has a lower gluten content. With the advent of transgene technology and practical amenability of cereals it should be possible to introduce advantageous traits in different wheats to augment their nutritive value and consumer acceptance. However, genetic enrichment and improvement of wheat though highly desirable, prerequisites a well-established regeneration protocol coupled with molecular amenability. Unfortunately, wheat amongst grasses is considered to be least amenable to tissue culture techniques and leaf protoplasts exhibit notorious recalcitrance (Mahalakshmi *et al.*, 1993).

Tissue culture procedures in wheat have generally been raised from embryogenic explants such as immature embryos, scutella, their calli and are regularly employed as the target tissue for various transformation strategies due to their high regeneration efficiency. However, tissues of non-embryogenic origin such as leaf or basal segments have been rarely employed for transformation studies owing to their low regeneration response. Our work reveals the unusually high morphogenic potential of the basal leaf segment of various wheat cultivars. Of particular interest is its extreme sensitivity to short-term treatments of 2,4-D for rapid induction of somatic embryogenesis (Mahalakshmi *et al.*, 2003). Wheat leaf base segments treated with 10 μM 2,4-D for 24 h developed somatic embryos without an intervening callus when cultured on basal Murashige and Skoog's medium for less than 2 weeks. This response is auxin specific and genotype independent. The rapid induction of somatic embryogenesis in wheat leaf bases under easily manipulable culture conditions provides a unique opportunity to monitor the molecular events associated with auxin-mediated induction of somatic embryogenesis and also to study the associated developmental events (Chugh & Khurana, 2002). Towards this end, we have made attempts to identify the components that constitute auxin signal transduction pathway leading to changes in gene expression and eventually more over to morphogenic responses. Employing various activators and inhibitors, our preliminary investigation has revealed that the Ca^{++}-calmodulin signalling pathway may be operative in auxin-induced somatic embryogenesis.

Genetic transformation of crop plants by *Agrobacterium*-mediated co-cultivation is an efficient and cost-effective method for gene delivery. Monocotyledonous plants including important cereals were earlier thought to be recalcitrant to *Agrobacterium*-mediated gene transfer (*see* Mahalakshmi & Khurana, 1997), but the scenario has changed in the last few years. Consistent efforts by researchers on cereal crop plants have resulted in development of methodologies for efficient gene delivery via *Agrobacterium* into rice (Gandhi & Khurana, 1999) maize, barley and wheat (Patnaik & Khurana, 2001). One of the key points in these protocols was that tissues consisting of actively dividing cells such as immature embryos and calli induced from immature embryos were co-cultivated with *Agrobacterium* in the presence of potent inducers of virulence genes.

In our laboratory, the regenerative explants have been employed for introduction of herbicide resistance in the selected Indian cultivars based on their regeneration potential. Resistance to L-phosphinothricin, glyphosate and bialaphos has been successfully used to select transgenic calli and plants in wheat. Also use of *bar* gene besides providing an efficient selection strategy renders transformed plants with useful agronomic traits. A comparison has also been attempted by particle bombardment and *Agrobacterium*-mediated transformation protocols and presence of transgene confirmed by PCR and Southern hybridization in the primary transformants.

Most of the experiments on *Agrobacterium*-mediated transformation of wheat employed incubation of suitable explants with the bacterium. Work from this lab by Mahalakshmi and Khurana (1995) reported the transient expression of *gus* gene in meristematic leaf bases, calli, mature seeds and mesocotyl punctured seedlings following co-cultivation with different strains and vectors of *Agrobacterium tumefaciens* (Mahalakshmi & Khurana, 1995). An efficient and reproducible method for *Agrobacterium*-mediated co-cultivation of mature embryos of hexaploid bread wheat (*Triticum aestivum*), tetraploid pasta wheat (*Triticum durum*), and emmer wheat (*Triticum dicoccum*) is also reported (Khurana *et al.*, 2002, Chugh & Khurana, 2003a, Patnaik *et al.*, 2003). Presence of acetosyringone at 200 mM concentration in the bacterial growth medium, inoculation medium, and the co-cultivation medium was essential for achieving a 1.5-2.0-fold increase in transient expression of the introduced *gus* gene. Successful generation of transgenic plants, at a transformation frequency of upto 10% has been achieved following co-cultivation for 2-3 days of mature embryos and mature embryo-derived calli with *Agrobacterium* strain LBA4404 (pBI101::Act1) and LBA4404 (p35SGUSINT), respectively. Paromomycin and phosphinothricin were used as the selection agent as they did not adversely affect plantlet regeneration, and also helped in overcoming the detection of escapes. The activity of *npt*II transgene is demonstrated by dot blot assay and spraying the plants with paromomycin. The presence of transgene has been confirmed by Southern hybridization and PCR amplification of the introduced gene in T_1 generation. Optimization of this method facilitated the introduction of *bar* gene as a selectable marker. Similarly, use of excised mature embryos and mature embryo-derived calli as primary explants for *Agrobacterium*-mediated transformation of the hexaploid, bread wheat (*T. aestivum*) and the tetraploid, macaroni wheat (*T. durum*) resulted in the introduction of potato proteinase inhibitor (*pin2*) gene at a frequency of 8 and 15%, respectively, the successful use of which was previously reported for engineering of *japonica* rice for insect resistance.

Direct gene transformation strategies such as microprojectile bombardment have also been successfully employed for cereals in general and wheat in particular. The biolistic approach has utilized the use of immature embryo calli, isolated scutella and cultured immature embryos as the recipient explants. Particle bombardment is thus emerging as the method of choice for introduction of agronomically important genes for quality improvement (Patnaik & Khurana, 2001).

Till date, most of the studies on wheat transformation have concentrated on few model cultivars and the procedures employed require a constant supply of immature embryos, which is difficult to achieve throughout the year unless greenhouse grown plants are available. Mature embryos are thus an alternative explant for production of calli which can in turn be used as target tissue for genetic transformation studies (Khurana *et al.*, 2002, Patnaik & Khurana, 2003, Chugh & Khurana, 2003b). Mature embryo-derived calli also showed a regeneration response comparable to that achieved from calli initiated from immature embryos. The present work was initiated to develop an efficient method of transformation for Indian cultivars of bread, durum and emmer wheat. Calli derived from mature embryos of the cultivars CPAN1676 of *T. aestivum,* PDW215 of *T. durum,* and the embryogenic basal segment calli of DDK1001 *T. dicoccum* were chosen for genetic transformation experiments based on their regenerating capability (Chugh & Khurana, 2003b). A common protocol was developed for the introduction of marker gene constructs. The regenerating cultures were double bombarded with gold microprojectiles at a target distance of 6 cm. Southern analysis using the *bar* gene as a probe revealed the integration of transgenes in the T_0 transformants. The *bar* gene was active in both T_0 and T_1 generations as evidenced by phosphinothricin leaf paint assay. The transformation frequency obtained was approx. 8% with *T. aestivum* and 10% with *T. durum.* Approximately 30 to 35% of the primary transformants of *T. aestivum* and *T. durum,* respectively, were fertile. The transmission of *bar* gene to T_1 progeny was demonstrated by PCR analysis of germinated seedlings with primers specific to the *bar* gene. Besides introduction of marker genes like *hyg* and *bar*, the optimized protocol was used for the introduction of the barley gene encoding a late embryogenesis abundant protein (*HVA1*) in *T. aestivum* and *T. durum* with the plasmid vector pBY520. The presence of the *HVA1* transgene was confirmed by Southern analysis in the T_0 generation in case of *Triticum aestivum*, and T_0 and T_1 generation in *Triticum durum.* In general, the transformation efficiency was found better with *Agrobacterium* (7.5%) than with particle bombardment (4 %). The success of any transformation process, however, depends largely upon the regeneration capability of the explant employed and the success achieved can be attributed to development of efficient regeneration and an optimized selection system employed.

In addition, novel transformation strategies are also being developed simultaneously. Cellular permeabilization has been developed for direct exogenous DNA uptake in *Brassica* and wheat zygotic embryos, without compromising cell viability and proliferative ability (Parihar *et al.,* 1994, Mahalakshmi *et al.*, 2000). On similar lines, a non-tissue culture approach has been developed for germ line transformation in bread wheat employing *Agrobacterium* and mature-seeds as the target tissues. These techniques are not only simplistic and cost effective in approach but are also relatively free of the intricacies associated with tissue culture practices and are thus applicable to various genotypes. Further refinements in these approaches would undoubtedly have application in areas of functional genomics where gene function needs to be characterized simplistically in a short time span.

Acknowledgements

This work was financially supported by the Department of Biotechnology, Government of India.

REFERENCES

Baron E.P.E.L., Gharyal P.K. & Schindler M. (1988) Pectins as mediators of wall porosity in soybean cells. *Planta.* 175: 389-395.

Baweja K., Khurana J.P. & Gharyal-Khurana P. (1995) Light regulation of somatic embryogenesis in hypocotyls of *Albizzia lebbeck. Curr. Sci.* 68: 544-546.

Bhatnagar S., Das M. & Khurana P. (2002a) Rapid *in vitro* TDZ-mediated micropropagation of *Morus indica* C176 and C776 through axillary buds. *Ind. J Sericulture* 41: 80-83.

Bhatnagar S., Kapur A. & Khurana P. (2001) TDZ-mediated differentiation in commercially valuable Indian mulberry, *Plant Biotechnology.* 18: 61-65.

Bhatnagar S., Kapur A. & Khurana P. (2002b) Evaluation of parameters for high efficiency gene transfer via particle bombardment in Indian mulberry, *Morus indica* cv. K2. *Ind. J. Exp. Biol.* 40: 1387-1392.

Bhatnagar S., Kapur A. & Khurana P. (2003) Evaluation of parameters for high efficiency gene transfer via *Agrobacterium tumefaciens* and production of transformants in Indian mulberry, *Morus indica* cv. K2. *Plant Biotechnology* (In Press).

Bhatnagar S. & Khurana P. (2003) *Agrobacterium tumefaciens*-mediated transformation of Indian mulberry, *Morus indica* cv. K2: A time phased screening strategy. *Plant Cell Reports.* **21:** 669-675.

Chugh A. & Khurana P. (2002) Gene expression during somatic embryogenesis–recent advances. *Current Science.* 83: 715-730.

Chugh A. & Khurana P. (2003a) Herbicide resistant transgenics of bread wheat (*T. aestivum*) and emmer wheat (*T. dicoccum*) by particle bombardment and *Agrobacterium*-mediated approaches. *Current Science.* 84: 78-83.

Chugh A. & Khurana P. (2003b) Regeneration via somatic embryogenesis from leaf basal segments and genetic transformation of bread wheat and emmer wheat by particle bombardment. *Plant Cell Tissue & Organ Culture* (In Press).

Chugh A. & Khurana P. (2003c) A non-tissue culture approach for germ line transformation of Indian bread wheat (*T. aestivum* L.). (Submitted).

Gandhi R. & Khurana P. (1999) Stress-mediated regeneration of mature embryo-derived calli and *Agrobacterium*-mediated gene transfer in *indica* rice. *Ind. J. Exp. Biol.* 37: 332-339.

Gandhi R. & Khurana P. (2001) Regeneration from leaf protoplasts of *Arabidopsis thaliana* ecotype 'Estland'. *Ind. J. Exp. Biol.* 39: 705-709.

Gandhi R., Maheshwari S.C., Khurana J.P. & Khurana P. (2001) Genetic and molecular analysis of *Arabidopsis thaliana* (Ecotype 'Estland') transformed with *Agrobacterium*. *In vitro Cell. Develop. Biol. Plant.* 37: 629-637.

Gandhi R., Maheshwari S.C. & Khurana P. (1999) Transient gene expression and influence of promotors on foreign gene expression in *Arabidopsis thaliana. In vitro Cell. Develop. Biol. Plant.* 35: 232-237.

Gharyal P.K. (1992) Advances and prospects of biotechnology for forest tree improvement. In: *Proc. Symposium on Biotechnology for Forest Tree Improvement*, Biotrop Special Pub. No: 49, pp. 141-148, Bogor, Indonesia.

Gharyal P.K., Ho S.C., Wang J.L. & Schindler M. (1989) *Bradyrhizobium japonicum* lipopolysaccharide inhibits symplastic communication in soybean (*Glycine max*) cells. *J. Biol. Chem.* 264: 12119-12121.

Gharyal P.K. & Maheshwari S.C. (1980) Plantlet formation from callus cultures of a legume, *Lathyrus sativus* cv. L.S.D.-3. *Z. Pflanzenphysiol.* 100: 359-362.

Gharyal P.K. & Maheshwari S.C. (1981) *In vitro* differentiation of somatic embryoids in a leguminous tree-*Albizzia lebbeck* L. *Naturwissenschaften* 69: 379-380.

Gharyal P.K. & Maheshwari S.C. (1982b) Plantlet formation in tissue cultures of the sensitive plant *Mimosa pudica* L. *Z. Pflanzenphysiol.* 105: 179-181.

Gharyal P.K. & Maheshwari S.C. (1982b) *In vitro* differentiation of plantlets from tissue cultures of *Albizzia lebbeck* L. *Plant Cell, Tissue and Organ Culture* 2: 49-53.

Gharyal P.K. & Maheshwari S.C. (1983) Genetic and physiological influences on differentiation in tissue cultures of a legume, *Lathyrus sativus. Theor. Appl. Genet.* 66: 123-126.

Gharyal P.K. & Maheshwari S.C. (1989) Differentiation from mature explants in leguminous trees. *Plant Cell Reports* 8: 550-553.

Gharyal P.K., Rashid A. & Maheshwari S.C. (1983a) Androgenic response from cultured anthers of a leguminous tree, *Cassia siamea Lam. Protoplasma.* 118: 91-93.

Gharyal P.K., Rashid A. & Maheshwari S.C. (1983b) Production of haploid plantlets in anther cultures of *Albizzia lebbeck* L. *Plant Cell Reports.* 2: 308-309.

Kapur A., Bhatnagar S. & Khurana P. (2001) Protocol for efficient *in vitro* regeneration from leaf explants of mature Indian mulberry via organogenesis. *Sericologia* 42: 207-214.

Kapur A., Bhatnagar S. & Khurana P. (2003) Isolation and culture of leaf protoplasts of Indian mulberry (*Morus indica* var. K-2). *Sericologia* (In Press).

Khurana J., Chugh A. & Khurana P. (2002) Regeneration from mature and immature embryos and transient gene expression via *Agrobacterium*-mediated transformation in emmer wheat (*Triticum dicoccum* Schuble). *Ind. J. Exp. Biol.* 40: 1295-1303.

Khurana J. & Khurana P. (2000) Biolistic-mediated DNA delivery into hypocotyls of a leguminous tree-*Albizzia lebbeck* L.: influence of biological and physical parameters. *J. Plant. Biochem. Biotech.* 9: 31-34.

Khurana P., Kapur A. & Bhatnagar S. (2001) Studies on morphogenesis and genetic manipuation of Indian mulberry. *Ind. J. Sericulture.* (Special Issue) (In Press).

Khurana P. & Khurana J. (1999) Application of genetic transformation to tree biotechnology. *Ind. J. Exp. Biol.* 37: 627-638.

Mahalakshmi A., Chugh A. & Khurana P. (2000) Exogenous DNA uptake via cellular permeabilization and expression of a foreign gene in wheat zygotic embryos. *Plant Biotechnology* 17: 235-240.

Mahalakshmi A., Khurana J.P. & Khurana P. 2003. Rapid induction of somatic embryogenesis in leaf explants of wheat (*Triticum aestivum*). *Plant Biotechnology* (In Press).

Mahalakshmi A. & Khurana P. (1995) *Agrobacterium*-mediated transient gene expression in various tissues and genotypes of wheat (*Triticum aestivum* L.) *J. Plant. Biochem. Biotech.* 4: 55-59.

Mahalakshmi A. & Khurana P. (1997) *Agrobacterium*-mediated Cereal Transformation—A Critical Appraisal. *Ind. J. Exp. Biol.* 35: 416-426.

Mahalakshmi A., Maheshwari S.C. & Khurana P. (1993) High frequency divisions in leaf base protoplasts of wheat (*Triticum aestivum* L.). *J. Plant. Biochem. Biotech.* 2: 61-65.

Maheshwari S.C., Gill R., Maheshwari N. & Gharyal P.K. (1986) Isolation and regeneration of protoplasts from higher plants. In: Reinert J., Binding H. (Eds.) *Differentiation of Protoplasts and of Transformed Plant Cells.* pp. 3-36. Springer-Verlag, Berlin, Heidelberg.

Maheshwari S.C., Gupta R. & Gharyal P.K. (1982) Cholinesterases in plants. In: *Recent Developments in Plant Sciences, S.M. Sircar Memorial Volume,* pp. 145-160.

Meiners S., Gharyal P.K. & Schindler M. (1991) Permeabilization of the plasma membrane and wall of soybean root cells to macromolecules. *Planta* 184: 443-447.

Parihar D.S., Maheshwari S.C. & Khurana P. (1994) Uptake of exogenous DNA by *Brassica napus* dry seed embryos via cellular permeabilization and transient gene expression of NPTII activity. *J. Pl. Biochem. Biotech.* 3: 107-111.

Parihar D.S., Maheshwari S.C. & Khurana P. (1995) High frequency plantlet regeneration via somatic embryogenesis in *Brassica napus. Plant Cell Tissue and Organ Culture* 42: 113-115.

Parihar D.S., Maheshwari S.C. & Khurana P. (1997) Influence of heat shock and UV irradiation on PEG-mediated DNA uptake and transient expression of nptII gene in protoplasts of *Brassica napus. Ind. J. Exp. Biol.* 36: 1002-1006.

Patnaik D. & Khurana P. (2001) Wheat Biotechnology: A mini review. Electronic *J. Biotechnology.* (www.ejb.org/content/vol4/issue2/full/4/-).

Patnaik D. & Khurana P. (2003) Genetic transformation of *Triticum aestivum* and *Triticum durum* for herbicide tolerance employing mature embryos as target tissues. (Submitted).

Patnaik D., Vishnudasan D. & Khurana P. (2003) *Agrobacterium*-mediated transformation of mature embryos of *Triticum aestivum* and *T. durum.* (Submitted).

Schindler M., Gharyal P.K. & Jiang L.W. (1988) The dynamic parameter-fluorescence photobleaching as a tool to dissect space in biological system. In: *Fluorescence Methods in Biochemistry and Cell Biology.* Plenum Press.

Tewari A., Bhatnagar S. & Khurana P. (1999) *In vitro* response of commercially valuable cultivars of *Morus* species to thidiazuron and activated charcoal. *Plant Biotechnology* (Japan) 16: 413-417.

CHAPTER

3

Agrobacterium Mediated Transformation of Chickpea (*Cicer arietinum* L.)

S. Tripathi, Anju Arora and H.S. Chawla

ABSTRACT

Two genotypes of chickpea, Pusa256 and PG186 were tested for plant regeneration through multiple shoot induction using embryo axes from mature seeds as explant. The explant gave good response for multiple shoot induction on SR medium containing MS macro salts, 4X MS micro salts, B5 vitamins, 3.0 mg/l BAP, 0.05 mg/l NAA and 3% (w/v) sucrose and a reproducible transformation system was developed for chickpea (*Cicer arietinum* L.). The explants were transformed with *Agrobacterium tumefaciens* strain LBA4404 harbouring binary vector pCAMBIA 2301 containing CaMV35S—neomycin phosphotransferase (*npt*II) gene and CaMV35S-*gus* gene. Various parameters affecting *Agrobacterium*-mediated transformation were standardized. GUS expression in the transformed explants was observed. The presumptive transformants were histochemically stained and found positive for GUS expression. Multiple shoots were regenerated on selection medium containing 50 mg/l kanamycin. The integration of *npt*II gene in the kanamycin-resistant shoots was confirmed through polymerase chain reaction.

Keywords: Chickpea, Transformation, Kanamycin-selection

Abbreviations: BAP-6-benzylaminopurine; NAA-Naphthalene acetic acid; CaMV-cauliflower mosaic virus; GUS-β-glucuronidase; MS-Murashige and Skoog (1962); B5 vitamins-Gamborg *et al.* (1968); RT-room temperature; DMSO-Dimethylsulphoxide

INTRODUCTION

Grain legumes are one of the most valuable sources of proteins for human as well as animal nutrition. Chickpea (*Cicer arietinum* L.) is one of the major pulse crops of tropical and subtropical regions including the Indian subcontinent. Much of its grain productivity is adversely affected by

*Address for correspondence: Plant Biotech. Lab., Department of Genetics and Plant Breeding, Gobind Ballabh Pant Univ. of Agric. & Tech., Pantnagar-263 145, Uttaranchal (India)

damage caused by a a lepidopteran pest, the pod borer (*Helicoverpa armigera*). At farmers field around 30% of yield loss is caused annually due to damage inflicted by pod borer. Raising insect resistant transgenic lines of chickpea is considered a potential improvement strategy.

The prerequisites for the successful transfer of foreign gene into any plant species by genetic engineering are: an efficient gene delivery system, such as *Agrobacterium*-mediated DNA transfer, an effective selectable marker for selection of transformed cells and the ability to regenerate mature, fertile, transgenic plants from transformed tissue under *in vitro* conditions.

A number of factors like genotype, explant, composition of nutrition medium and hormonal balance of the regeneration medium influence plant regeneration (Murthy *et al.*, 1996). *In vitro* plant regeneration in chickpea has been reported through organogenesis from shoot meristem (Kartha *et al.*, 1981), immature cotyledon (Shri and Davis, 1992), hypocotyls cotyledonary node (Chatrath *et al.*, 1996), immature embryo (Singh *et al.*, 1997) shoot apex (Adkins *et al.*, 1995) and mature embryo and cotyledon explants (Arora and Chawla, 2005).

Agrobacterium-mediated transformation of various chickpea explants has been reported. Transformed callus was obtained using wild strains of *Agrobacterium* (Mohapatra and Sharma, 1991; Islam *et al.*, 1994), while hairy roots were obtained by treating chickpea plants with root – inducing strains of *Agrobacterium* (Siejkes-Boer *et al.*, 1995). Transformed chickpea plants were obtained by Fontana *et al.* (1993), Kar *et al.* (1996) and Krishnamurthy *et al.* (2000) from mature seed-derived embryo axes explants. Transformation in chickpea via somatic embryogenesis was achieved by Ramana *et al.* (1996) and Ghanti *et al.* (2002).

In this paper, we report the development of a routine regeneration and transformation system for chickpea.

MATERIALS AND METHODS

Plant Material

Mature seeds of chickpea (*Cicer arietinum* L.) genotypes Pusa256 and PG186 were obtained from G.B. Pant Univ. of Agric. & Tech., Pantnagar. Dried mature seeds were washed with Tween 20 for 20 min followed by vigorous washing with water. The seeds were then submerged in sterile water and soaked overnight (20 h). On the next day, the seeds were surface sterilized in 70 percent alcohol for 10 sec followed by treatment with 0.1% $HgCl_2$ for 8-10 min. Seeds were rinsed 5-6 times with sterile distilled water. The embryos were excised from the seeds and shoot and root meristems were cut. These decapitated embryo axes explants were used for transformation.

Agrobacterium Strain and Plasmid

For inoculation of embryo axes explants, *Agrobacterium* strain LBA4404 containing pCAMBIA2301 was used, which has an *npt*II gene with CaMV35S promoter and 35S terminator and *gus* gene driven by CaMV35S promoter and nopaline synthase terminator.

Plant Regeneration

10-12 explants per 90 mm Petri dish were cultured on shoot regeneration (SR) medium. The medium contained MS (Murashige and Skoog, 1962) macro salts, 4X MS micro salts, B5 vitamins

(Gamborg *et al.*, 1968), 3 mg/l BAP, 0.05 mg/l NAA and 3% sucrose (w/v) and solidified with 0.8% agar. Cultures were placed at 20°C and 16h photoperiod under a light intensity of 140 μ mol $m^{-2}s^{-1}$, provided by cool-white fluorescent tubes. The shoot regeneration medium induced maximum number of shoots. After shoot regeneration, the individual shoots were separated and transferred to root induction medium (MS salts, B5 vitamins, 1.0 mg/l IBA, 0.2% activated charcoal and 2% sucrose). Shoots with roots were transferred to half-strength MS medium without growth regulator.

Transformation and Selection of Transformants

Agrobacterium strain LBA4404 was grown at 28°C overnight (16h) in liquid YEM medium containing 50 mg/l kanamycin and 10 mg/l rifampicin. The bacterial cells were collected by centrifugation at 6000 rpm at RT for 10 min and the pellet resuspended in liquid SR medium to a final OD_{600} of 0.1-0.2. The explants were precultured for 24h on SR medium prior to cocultivation with *Agrobacterium*. 40-50 explants were treated with 10 ml *Agrobacterium* suspension for 10 min with gentle agitation. The explants were then blot dried on sterile filter paper and cocultivated in petri dishes on agar-solidified SR medium for 48h at 20°C under a light intensity of 140 μ mol $m^{-2}s^{-1}$ and 16h photoperiod. After cocultivation, the explants were washed thoroughly 3-4 times with sterile distilled water containing 600 mg/l cefotaxime, blot dried on sterile fiter and cultured on SR medium supplemented with 300 mg/l augmentin. Two weeks later, the explants were transferred to S1 selection media (SR medium containing 300 mg/l augmentin and 50 mg/l kanamycin) for selection of kanamycin-resistant shoots. The surviving green shoots after 3 weeks on S1 selection media were separated and subcultured on S2 selection medium containing 150 mg/l augmentin and 50 mg/l kanamycin. The surviving green shoots were subjected to two more passages of selection by repeated excision of green shoots and their exposure to selection medium. The concentration of augmentin was reduced at each subculture. Green shoots growing in the presence of kanamycin selective agent were transferred to root induction medium without kanamycin.

Tissue Staining for GUS Activity

The histochemical GUS assay was carried out according to Jefferson (1987). After one week of infection, the explants were taken at random and rinsed briefly in 0.2 M phosphate buffer. The explants were then placed in eppendorf tubes containing X-Gluc stain (0.1 M Potassium Ferricyanide, 0.1 M Potassium Ferrocyanide, 0.1% Triton X-100, 0.5 M phosphate buffer and 0.1% X-Gluc). X-Gluc stock (10 mg/ml) was prepared in DMSO. The samples were incubated at 37°C for 36h and subsequently bleached in 95% ethanol for observation under microscope.

DNA Isolation and Polymerase Chain Reaction

The presence and integration of *npt*II gene was analysed by polymerase chain reaction in T_0 shoots. Genomic DNA was isolated from non-transformed plants and from kanamycin-resistant shoots according to the protocol of Doyle & Doyle (1990) with some modifications. For the detection of *npt*II gene coding sequence in the putative transformants, the genomic DNA was subjected to PCR using following primers:

Forward primer (23 mer) 5' ACGTTGTCACTGAAGCGGGAAGG 3' and
Reverse primer (24 mer) 5' GTAAAGCACGAGGAAGCGGTCAGC 3'

The primers were used at a concentration of 50 pM to amplify a 487 bp region of the *npt*II gene. PCR was carried out for 30 cycles (denaturation—1 min, 94°C; annealing—1 min, 55°C; elongation—2 min, 72°C). The final elongation step was carried out at 72°C for 5 min. For detection of the 487 bp fragment, DNA was separated on 1% agarose gel.

RESULTS AND DISCUSSION

Multiple shoots were successfully induced from mature embryo axes explants on SR media containing MS macro salts, 4X MS micro salts, B5 vitamins, 3.0 mg/l BAP, 0.05 mg/l NAA and 3% sucrose (w/v). The number of shoots/explant after one week of culture ranged from 5-9 in both the genotypes. The regeneration capability was strongly related to the amount of apex removed and best results were obtained when less than 1.0 mm sections of shoot apex was removed. After three weeks of culture on SR medium on an average 16-18 shoots/explant could be obtained in the genotype Pusa256. Regenerated shoots were rooted on MS medium with B5 vitamins and supplemented with 1.0 mg/l IBA, 0.2% activated charcoal and 2% (w/v) sucrose. Nearly 50-60% of the shoots rooted within two weeks of culture on this medium. The roots proliferated only when transferred to half-strength MS medium.

Prior to starting gene transfer experiments, chickpea embryo axes explants were tested *in vitro* for kanamycin sensitivity. The optimal selection pressure was determined by culturing injured embryo axes (20-25 per treatment) on SR medium containing 25, 40, 50 and 75 mg/l kanamycin under the conditions mentioned above for cocultivation. The cultures were scored after two weeks of inoculation. There was no inhibition of growth at 25 mg/l kanamycin concentration while at 75 mg/l kanamycin concentration complete inhibition of shoot growth was observed. Significant inhibition of shoot growth from mature embryo axes was observed on medium containing 50 mg/l kanamycin, hence this concentration was used for selection of transformed embryo axes explants (Fig. 1a). Fontana *et al.* (1993) and Kar *et al.* (1996) have also used 50 mg/l kanamycin for selection of transformants. In the present study no escapes were observed using kanamycin as selective agent and hence *npt*II gene can be used as an efficient selectable marker gene in chickpea. This finding does not agree with that of Puonti- Kaerlas *et al.* (1990) and Schroeder *et al.* (1993) who reported that *npt*II gene is not a suitable selectable marker in legumes.

Transformation was carried out using embryo axes explants from two genotypes of chickpea. Two weeks after cocultivation, the embryo axes explants with emerging green shoots were transferred to selection medium (S1) containing 50 mg/l kanamycin (Fig. 1c). After three weeks of culture on S1 selection medium most of the shoots appeared bleached, leaving only a few green shoots, suggesting the elimination of untransformed shoots. These surviving green shoots and chimaeric shoots with partial white branches were separated and transferred to S2 selection medium containing 50 mg/l kanamycin and reduced concentration of augmentin. Green shoots continued to propagate on media containing 50 mg/l kanamycin (Fig. 1d). Table 1 summarizes the results derived from three independent experiments, each consisting of three selection cycles. A total of 13

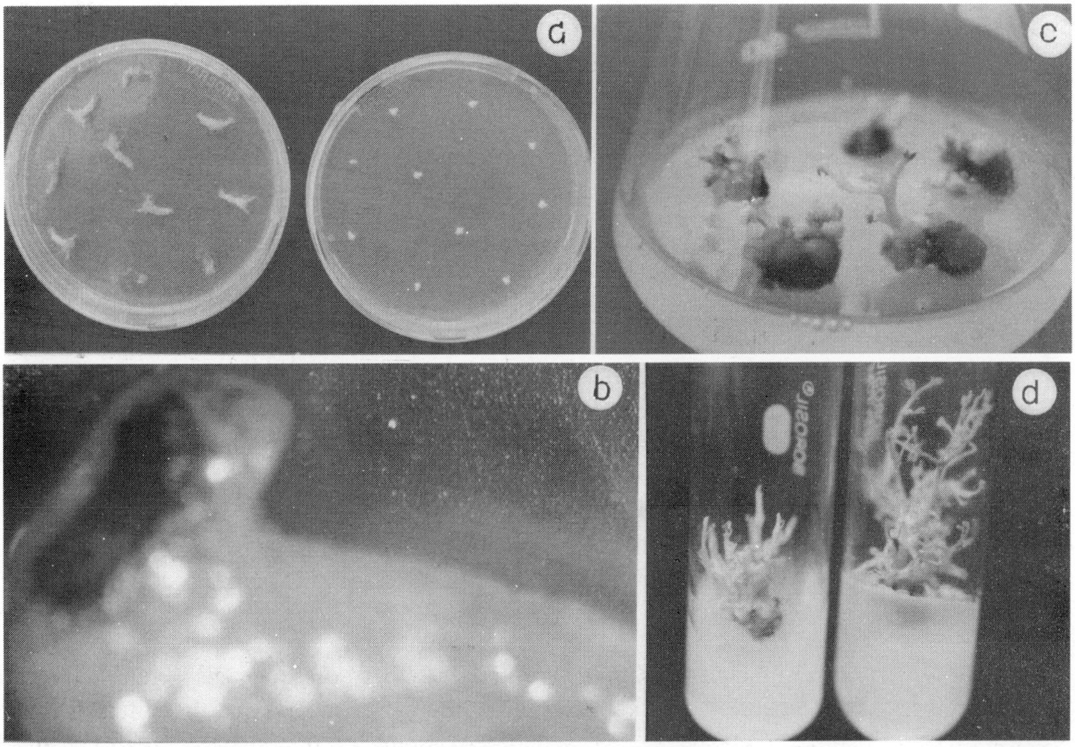

Fig. 1 Transformation in chickpea; (a) Inhibition of shoot growth on SR media containing 50 mg/l kanamycin; (b) GUS staining of transformed explant; (c) Surviving green shoots proliferating on selection media containing 50 mg/l kanamycin; (d) Selected kanamycin resistant shoots proliferatingon selection medium.

Table 1 Frequency of transformation in two cultivars of chickpea

Genotype	*Expt No.*	*No. of explants cocultivated*	*No. of kanamycin resistant shoots*			*Transformation frequency (%) (a/b)*
			S1 selection	*S2 selection*	*S3 selection*	
Pusa256	1	212	37	13	5	
	2	194	31	13	4	
	3	221	40	15	4	
	Total	627 (b)			13 (a)	2.1
PG186	1	317	61	24	6	
	2	303	58	20	7	
	3	287	63	17	4	
	Total	907 (b)			17 (a)	1.8

kanamycin-resistant shoots could be recovered from approximately 650 cocultivated explants of genotype Pusa256 while 17 kanamycin resistant shoots were obtained from approximately 900 cocultivated explants of PG186. At the end of selection procedure the percentage of transformation was 1.8–2.0%. Elimination of all untransformed shoots became possible by repeated separation and subculturing of green shoots on the selection media containing 50 mg/l kanamycin. The green shoots surviving the kanamycin selection were induced to form roots in the rooting medium without kanamycin.

For standardization of various parameters affecting transformation efficiency in chickpea, β-glucuronidase gene was used as reporter gene and histochemical GUS assay was performed. After one week of infection, the explants were picked randomly and stained histochemically for detection of GUS expression. The number of embryo axes explants showing GUS activity was between 7-10% in the two genotypes under study (Fig. 1b). The expression of GUS was localized to the meristematic region as also observed by Fontana *et al.* (1993) and Krishnamurthy *et al.* (2000).

The PCR analysis of genomic DNA from the transformed green shoots surviving on kanamycin selection and untransformed plants was done using *npt*II gene primers. The analysis revealed that most of the shoots were putative transformants (Table 2). The presence of the expected 487 bp fragment in the transformed lines confirmed the integration of *npt*II in chickpea genome (Fig. 2).

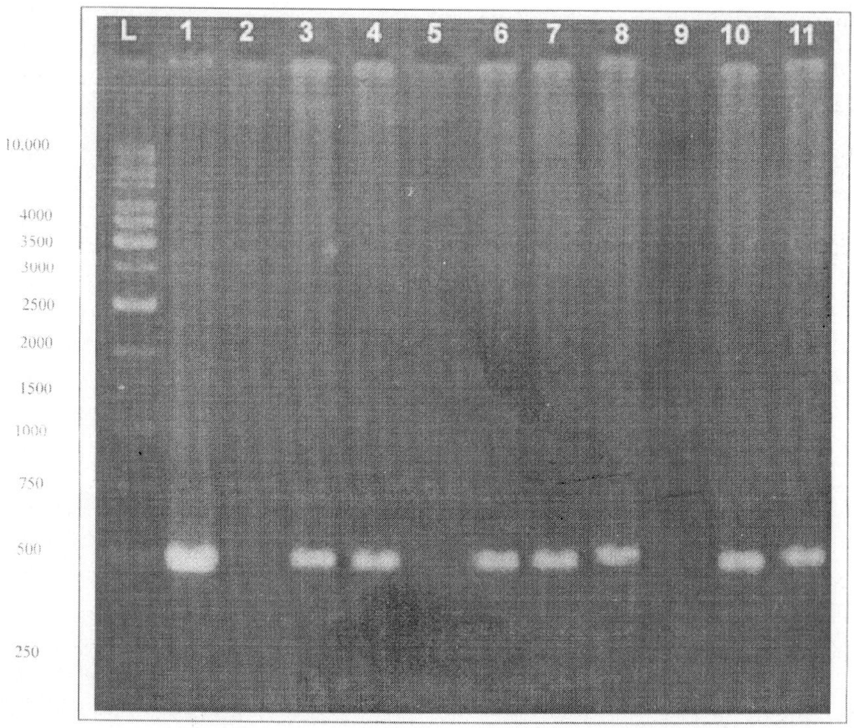

Fig. 2 PCR analysis of kanamycin-resistant shoots with *npt*II gene primers. L-ë 1kb ladder, 1- Positive control, 2- Negative control, 3 to 11- DNA from transformed shoots

Table 2 Frequency of GUS expression in mature embryo axes explants

Genotype	*Treatment*	*No. of explants stained*	*GUS positive*	*Transformation frequency (%)*
Pusa256	Control	10	–	–
	Cocultivated	43	4	9.3
PG186	Control	10	–	–
	Cocultivated	41	3	7.3

This study reveals that multiple shoots can be successfully induced from embryo axes explants of chickpea and *Agrobacterium*-mediated transformation protocol can be used in future to develop insect-resistant transgenic lines of chickpea.

Acknowledgements

S. Tripathi and A. Arora are thankful to CSIR-UGC and UP Council of Agriculture Research for financial assistance received during the course of investigation. Thanks are due to Dr. M.V. Rajam for giving pCAMBIA vector.

REFERENCES

Adkins A.L., Godwin I.D. & Adkins S.W. (1995) An efficient *in vitro* regeneration system for Australian grown chickpea (*Cicer arietinum* L.) cultivars. *Australian Journal of Botany* 43: 491-497.

Chatrath A., Chandra R., Khetrapal S. & Polisetty R. (1996) Changes in nitrate, amino acid and sugar content during growth and differentiation of chickpea explants (*Cicer arietinum* L.). *Indian Journal Plant Physiology* 1: 80-83.

Doyle J.J. & Doyle J.J. (1990) Isolation of plant DNA from fresh tissue. *Focus* 12: 13-15.

Fontana G.S., Santini L., Caretto S., Frugis G. & Mariotti D. (1993) Genetic transformation in the grain legume *Cicer arietinum* L. (chickpea). *Plant Cell Reports.* 12: 194-198.

Gamborg O.L., Miller R.A. & Ojima K. (1968) Nutrient requirement of suspension cultures of soybean root cells. *Experimental Cell Research.* 50: 148-151.

Ghanti S.K., Srinath R., Venugopal R.B., Rao S.R., Kaviraj C.P. & Barad A. (2002) Somatic embryogenesis and *Agrobacterium*-mediated transformation in chickpea (*Cicer arietinum* L.) *Plant Cell Biotechnology and Mol. Biology* 3 (1&2): 51-58.

Islam R., Malik T., Husnain T. & Riazuddin S. (1994) Strain and cultivar specificity in the *Agrobacterium*-chickpea interaction. *Plant Cell Reports.* 13: 561-563.

Jefferson R.A. (1987) Assaying chimaeric genes in plants. The GUS gene fusion system. *Plant Molecular Biology Reporter.* 5: 387-405.

Kar S., Johnson T.M., Nayak P. & Sen S.K. (1996) Efficient transgenic plant regeneration through *Agrobacterium* mediated transformation of chickpea (*Cicer arietinum* L.). *Plant Cell Reports.* 16, 32-37.

Kartha K.K., Pahl K., Leung N.L. & Mroginski L.A. (1981) Plant regeneration from meristems of grain legumes: soybean, cowpea, peanut, chickpea and bean. *Canadian Journal of Botany* 59: 1671-1679.

Krishnamurthy K.V., Suhasini, Sagare A.P., Meixner M., Kathen de A. & Pickardt, Schieder O. (2000) *Agrobacterium* mediated transformation of chickpea (*Cicer arietinum* L.) embryo axes. *Plant Cell Reports.* 19, 235-240.

Mohapatra S.T. & Sharma R.P. (1991) *Agrobacterium*-mediated genetic transformation of chickpea (*Cicer ariretinum* L.) *Indian Journal Exp. Biology.* 29: 758-761.

Murashige T. & Skoog F. (1962) A revised medium for rapid growth and bioassays for tobacco tissue cultures. *Physiol. Plantarum,* 15, 473-497.

Murthy B.N.S., Victor J., Singh R.P., Flectcher R.A. & Saxena P.K. (1996) *In vitro* regeneration of chickpea (*Cicer arietinum* L.): Stimulation of direct organogenesis and somatic embryogenesis by thidiazuron. *Plant Growth Regulation.* 19, 233-240.

Puonti-Kaerlas J., Eriksson T. & Engstrom P. (1990) Production of transgenic pea (*Pisum sativum* L.) plants by *Agrobacterium tumefaciens*-mediated gene transfer. *Theor. Appl. Genet.* 80: 246-252.

Ramana R.V., Venu Ch, Jayasree T. & Sadanadam A. (1996) Direct embryogenesis and transformation in *Cicer arietinum* L. *Indian Journal Exp. Biology.* 34: 716-718.

Schroeder H.E., Schotz A.H., Richardson T.W., Spencer D. & Higgins T.J.V. (1993) Transformation and regeneration of two cultivars of pea (*Pisum sativum* L.) *Plant Physiol.* 101: 751-757.

Shri and Davis, (1992) Zeatin induced shoot regeneration from immature chickpea (*Cicer arietinum* L.) cotyledons. *Plant Cell and Organ Culture* 28: 45-51.

Siefkes-Boer H.J., Noonan M.J., Bullock D.W. & Conner A.J. (1995) Hairy root transformation system in large-seeded grain legumes. *Isr Journal Plant Science.* 43: 1-5.

Singh A., Singh N.P., Asthana A.N. & Singh A. (1997) Callus induction and direct regeneration from immature embryos in chickpea. *International Chickpea and Pigeonpea Newsletter,* 4, 39-40.

CHAPTER

4

Constraints and Relevant Strategies in Genetic Transformation of Sorghum

N. Sai Kishore, B. Ramana Kumari, D. Bala Krishna, S.V. Rao, B. Venkatesh Bhat, N. Seetharama and K.B.R.S. Visarada*

ABSTRACT

Sorghum (*Sorghum bicolor* (L.) is an important dry land crop and is used as food, fodder and feed. The increasing shift in the demand for sorghum from staple food to industrially preferred commodity in poultry, bakery and brewing industries requires large scale production according to the needs of different sectors. Therefore, it is important to develop and deploy novel technologies such as genetic transformation to generate sorghum with highly enhanced level of specific end-products (e.g. quality protein) and resistance to pests that are difficult to breed by conventional means. Poor response of sorghum under *in vitro* culture, and declining demand of the crop in the developed countries are the two reasons for lack of incentive for genetic transformation of this crop. Transgenic plants of sorghum have been produced worldwide using particle bombardment and *Agrobacterium* infection. Among the different explants tested by us immature inflorescences followed by shoot tips were found to be ideal sources for embryogenic callus induction and plant regeneration. However, the regeneration potential was lost quickly after 30-35 days in case of calli from shoot tips while it was retained till 90 days in calli from immature inflorescences. Infection of different explants by *Agrobacterium* showed that immature embryos followed by immature inflorescence derived calli are the most suitable material for transient expression. However, immature embryos had poor callus induction and plant regeneration after infection. Repeated washes with antibiotics to control the overgrowth of *Agrobacterium* after cocultivation lead to total loss of regeneration in immature inflorescence derived calli. Alternate approaches like transformation of shoot apices followed by induction of multiple shoots or direct rooting were found useful. Details of study on such new explant sources, genetic transformation tools that are applicable for improvement of sorghum along with the current worldwide scenario will be presented.

Key Words: *Agrobacterium,* biolistics, multiple shoots, regeneration

Abbreviations: BAP-6-Benzyl Amino Purine, *gus*- ß-glucuronidase, MS-Murashige & Skoog, NAA-Naphthalene acetic acid, TDZ-Thiadiazuron

*Address for correspondence: National Research Center for Sorghum, Rajendranagar, Hyderabad 500030, India
E-mail: visarada@nrcsorghum.res.in; Fax: 091 40 24016378

INTRODUCTION

Ability to introduce foreign genes into crop plants has opened new avenues in increasing the agricultural production and productivity. It has more relevance to improvement of cereals, constituting the staple food of the humans and animals. Genetic improvement of crop plants can be achieved by transferring useful alleles at existing loci through conventional breeding or by adding new loci across diverse sources through genetic transformation (Miflin, 2002). Traditional plant breeding techniques together with classic biotechnological approaches have been widely used to increase crop yields by selecting improved varieties that are more productive and are resistant to diseases and pests.

Sorghum bicolor is a crop of worldwide importance and has the unique ability to produce under a wide array of harsh environmental conditions (House, 1995). It is an important cereal crop that forms the staple diet in much of the third world countries. In India sorghum is cultivated as staple food source by poor farmers in two distinct seasons: the rainy season (*Kharif*) and the winter season (*Rabi*). Rainy season crop is grown on annual rainfall, while the winter season crop is dependent on the residual soil moisture. High vulnerability of *Kharif*-adopted material to biotic stresses and low heterosis for yield in *Rabi* material are the important limitations in increasing the production of Indian sorghum (Annual, 2001). Apart from its multipurpose utilization as food, feed, fodder and fuel, there is an increasing demand for sorghum crop towards the production of alcohol, beer and other sorghum based derivatives like syrup, jaggery and confectionery (Ratnavathi *et al.*, 2003). Though significant improvements have been made, there is a tremendous challenge ahead for sorghum in view of its continued diversification, both locally and internationally. Considering the increasing economic importance of sorghum it is imperative to incorporate the modern biotechnology like genetic transformation for crop improvement (Bennetzen, 1995, Nwanze *et al.*, 1995, Maqbool *et al.*, 2001).

Limitations of Sorghum *in vitro* Response

Transformation systems have been studied in detail for cereals of major agricultural importance including maize, rice, wheat with notable exception in the world being barley and sorghum (Subudhi and Nguyen, 2000). Crop improvement programs require improvement of a variety that is specific to an agro-ecological region for an identified character rather than use of a single or a few genotypes for genetic transformation. Ideal requirements for such a routine genetic engineering technique are

(i) highly efficient genotype independent system of plant regeneration
(ii) efficient protocols for introduction of DNA
(iii) a suitable *in vitro* selection strategy.

Sorghum is considered to be the most recalcitrant species among the cereals for *in vitro* response, thereby for genetic transformation. Declining of demand as staple food and utilization only as feed and fodder in the developed world are other reasons for the limited attention that this crop could earn for genetic transformation. On the other hand the experience gained from other cereals like rice, wheat and corn can be incorporated into sorghum to take full advantage of the opportunity to augment appropriate biotechnological applications. Genetic transformation protocols are dependent on the callus induction and plant regeneration potential of the species. These protocols are tuned to suit the needs of gene transfer, selection and regeneration. Initially critical parameters that contribute to the development of transgenic production in sorghum crop were studied. These factors include: (1) model genotype; (2) ideal explant; (3) sustainability of regeneration; (4) rapid proliferation; (5) genotype independent methods and (6) transient GUS expression.

Model genotypes

Extensive tissue culture studies in other important cereals have identified a few genotypes as high quality callus producers (Nabors *et al.,* 1983, Wang *et al.,* 1987). Identification of model genotypes greatly aid in adoption of new technologies in a new crop system. These model genotypes can serve as initial test genotypes for extending the newly developed biotechnological methods. Likewise, in other cereals, highly responsive genotypes for *in vitro* culture and also highly receptive genotypes for introduction of foreign genes have been identified. In rice the *japonica* type Taipei 309 and *indica* type, IR72, Pusa Basmati and Radon are a few examples of such identified models. These genotypes were used from initial experiments on transformation to the latest introduction of *Agrobacterium* mediated method of gene transfer (Christou, 1994; Ahmad *et al.*, 2002). Similarly in maize A188 and its derivatives, Bobwhite lines in wheat, Golden Promise and Igri in barley were the most responsive genotypes for genetic transformation (Locatelli *et al.*, 1992; Ishida *et al.,* 1996, Jacobsen *et al.,* 1999; Altpeter & Varshney, 2001). We evaluated 26 sorghum germplasm lines along with three cultivated varieties using seed as the explant source to identify a model genotype with high potential for callus induction and plant regeneration. Most of the lines exhibited very low regeneration, (< 44%). The highest regeneration (66%) in the genotypes IS16401 and EP45 was on par with that of a popular *Rabi* sorghum genotype, M35-1 (65%). None of the genotypes showed regeneration potential as high as 80% which is a prerequisite for recovering the transgenic plants after selection. Thus none of the genotypes were found ideal for ready use in transformation. In order to improve the regeneration through manipulation of culture media we tested two callus induction media *viz.,* MS based medium and NBKNB medium for screening the genotypes with seed as the explant source (Murashige & Skoog, 1962; Sivamani *et al.,* 1996). Sivamni *et al.,* (1996) used NBKNB medium for culturing recalcitrant rice genotypes and could recover transgenic plants. This medium is rich in nutrients and thus was found useful for large scale generation of embryogenic callus in *indica* rice genotypes which are less responsive to tissue culture (Visarada *et al.,* 2002). In our study none of the sorghum seed explants responded on NBKNB medium with satisfactory callus induction and regeneration. Kaeppler and Pederson (1997) evaluated 41 genotypes from diverse inbred sorghum lines to identify high quality callus producing genotypes. Pigment less, mucilage-free, relatively fast growing and embryogenic callus producing genotypes were identified for further studies. Highly significant differences among genotypes were observed for callus quality traits. Therefore, the locally cultivated variety M35-1, which showed the highest regeneration was used for optimizing the transformation parameters.

Ideal explant

Callus induction and plant regeneration in sorghum was first reported from the culture of seedling tissues (Masteller and Holden, 1970). Tissue culture response in sorghum is highly genotype dependent and is very low. Though sorghum genotypes had good potential for callus induction, subsequent regeneration was poor. We studied different explant tissues for callus induction and plant regeneration. Seed is the most ideal explant source because it is available round the year and can be stored conveniently for tissue culture experiments. Nevertheless, the response of sorghum seed material was very poor. Immature inflorescences followed by shoot tips showed high callus induction but regeneration was high with immature inflorescences while it was low in shoot tips. Shoot discs were plated after slicing 2-3 mm shoot tips from 5d old germinated seedlings. Though the callus induction in these calli was moderate (60.7 ± 12.7) regeneration was poor (17 ± 7). Other explants such as immature embryos, immature seed and mature seed had low callus induction and plant regeneration. Immature inflorescences and shoot tips were reported to be the best among different explants (Murty *et al.,* 1990a,b, Mythili *et al.,* 1999; Seetharama *et al.,* 2005 Harshavardhan *et al.,* 2002). Most of the successful genetic transformation experiments in other cereals used immature embryos as the target tissues. *In vitro* response in terms of both callus induction and plant regeneration was very limited in case of immature embryos in sorghum. Also isolation of immature embryos is tedious, labour-intensive and has concomitant contamination problems. In sorghum they are available for only a limited period of time because of photoperiodic sensitivity of Indian sorghum genotypes. Genotypes of sorghum cultivated in *Kharif* and *Rabi* are exclusive to season. Immature inflorescences are available seasonally though the isolation and culture are easier to adopt. Cai and Buttler (1990) evaluated plant regeneration in eight high tannin sorghum genotypes from immature inflorescences. No critical differences were found in potential for callusing and regeneration. But genotype variability was observed for shoot differentiation and maintenance of regeneration capacity.

Sustainability of regeneration

Transformation experiments require good quality calli having high regeneration capacity. Protocols that facilitate at least 70-80% of regeneration and callus tissues that can give rise to a number of plants from unit culture, can only serve as the target material. Since the regeneration capacity will be drastically reduced due to the stress given during the process of transformation this necessitates the development of initial regeneration system, which can induce at least 70-80% of the calli for the development of shoots. We feel this is more so in the case of *Agrobacterium* mediated method, where the calli are cocultured with the bacterium for a recommended period. Later several washes with antibiotic solutions are required to remove the bacterial cells, which further exert stress on the calli. Selection of the transformed calli on marker substrates *viz.,* hygromycin or phosphinothricin further decreases the regeneration. Thus, to recover at least a few fertile transgenic plants it is necessary to start with a tissue that has a good regeneration potential. Also selection on marker gene for at least 2-3 cycles of 15-20d is required to eliminate the non-transformed tissues and to recover only the transformed tissues. Regeneration of the putative transformed tissues is done after 6-7 weeks of callus induction. Thus regeneration potential of the calli has to be sustained till that period to recover a few plants. Initial regeneration protocols should support this requirement. We tested the retention of regeneration potential up to 90d in calli from different explants. In sorghum such sustainability is found only in case of calli derived from immature inflorescences and it was retained till 90d after callus induction. Elkonin and Pakhomova (2000) could maintain the regeneration ability till 100 to 125d by altering the concentration of nitrogen or phosphorous in the medium.

Rapid proliferation

Callus cultures having high rate of proliferation are recommended as target tissues for transformation. Transformed tissues multiply rapidly and can be regenerated earlier. We observed that calli developed from immature inflorescences grow faster while shoot tips are the next. Other explants have calli that multiply slowly.

Transient GUS expression

Transformation of sorghum was reported through particle bombardment as well as *Agrobacterium* infection (Casas *et al.,* 1993; Krishnaveni *et al.,* 1999; Pratibha Devi *et al.,* 1999; Tadesse and Jacobs, 1999). Initially transient GUS expression studies were conducted to identify the optimum parameters. Five different explant sources *viz.,* calli derived from immature inflorescence, calli derived from shoot tips, immature embryos, freshly isolated shoot tips and mature embryos were infected with *Agrobacterium* strain AGL-1 carrying the plasmid pDM805 (*courtesy:* Richard Brettel, CSIRO, Australia) having *gus* as the reporter gene and *bar* as the plant selection marker. Among the different explant sources the highest GUS expression (100%) was observed in freshly isolated immature embryos. Freshly isolated shoot apices were the second best explant tissues that showed 93% GUS expression, while calli induced from these explant tissues did not show GUS expression. We tested mature embryos using the construct pDM805 and 20 ± 8.2% GUS expression was observed. Calli derived from immature inflorescences showed moderate expression (60%). Thus there is an immediate need to devise protocols to adopt immature embryos for genetic transformation through *Agrobacterium.*

Genotype independent methods

Like many other cereals the *in vitro* response in sorghum is genotype dependent. Gray *et al.* (1999) developed less laborious and less manipulative organogenic pathway in sorghum which resulted in tenfold increase in the regeneration frequency, high survival rate after transplantation, greater maintenance of regeneration capacity(>ten months). They could extend this method to genetic transformation through biolistics. In our studies we used shoot apices from 3d old germinated seedlings, since seed is the most suited material for laboratory studies. *In vitro* response from direct use of seed was poor, so shoot apices were infected directly after isolation from three-day old seedlings. These explants could be generated round the year, prone to less contamination and genotype independent. Transient GUS expression was observed in shoot apices after five days of cocultivation. Intense GUS expression was seen at the cut ends in the outer whorls of leaves while it was feeble in the inner whorls. Central core of leaves showed faint GUS staining indicating the transformation of meristematic zones. Induction of multiple shoots from shoot apices isolated from three-day old germinated seedlings was tested with different concentrations of Benzyl Amino Purine(BAP) and Thiadiazuron (TDZ) in three elite sorghum genotypes (data unpublished). *In vitro* rooting of shoot apices and /or induction of multiple shoots is easy, which is also a cost effective and time saving procedure for transformation studies. It eliminates the problems of regeneration and seasonal availability of the explant and is genotype independent. In a similar study Gould *et al.,* (1991) successfully used shoot apices isolated from three-day old seedlings for transformation in maize using *Agrobacterium.* The current procedure facilitates the direct generation of plants after *Agrobacterium* infection without callus phase using seed based explant sources. It can be applied to other crop species, which have low regeneration capacity. Production of transgenic plants using particle bombardment technique in sorghum was reported by a few groups but subsequent developments were slow (Hagio *et al.,* 1991; Zhu *et al.,* 1998; Jeoung *et al.*, 1999; Zhao *et al.*, 2000). Immature embryos were used as starting material for biolistics (Hagio, 1991; Zhu *et al.*, 1998). Zhao *et al.* (2000) could produce fertile transgenic plants in sorghum through double agro infection. They increased the density of the bacterial culture in the second step and obtained transformation frequency of 2.1%. Jeoung *et al.*, (1999) used three genotypes for *Agrobacterium* transformation with chitinase gene construct and could recover plants only from one genotype. These studies indicate the possibility of extending the *Agrobacterium* technology to sorghum genotypes and immature embryos seem to be the best source. Nevertheless development of genotype independent methods like multiple shoot induction and direct rooting can be explored further for incorporation of useful genes into sorghum and to facilitate sorghum improvement.

Acknowledgements

The authors acknowledge National Agricultural Technology Project (# RNPS 20) and Andhra Pradesh Netherlands Biotechnology (APNL) program for support through financial grant. We thank Mrs. A.Annapurna for her technical support during the period of investigation.

REFERENCES

Ahmad A, Maqbool S.B., Riazuddin S. & Sticklen M.B. (2002) Expression of synthetic *cry* 1 Ab and *cry* 1 Ac genes in Basmati rice (*Oryza sativa* L.) variety 370 via *Agrobacterium*-mediated transformation for the control of the European corn borer (*Ostrinia nubilalis*), *In Vitro Cell. Dev. Biol. Plant.* 38(9): 213-220.

Altpeter F. & Varshney A. (2001) Stable transformation and tissue culture response in current European Winter Wheats (*Triticum aestivum* L.), *International Plant and Animal Genome Conference,* San Diego, CA, Jan 13-17.

Annual Report (2001) National Research Centre for Sorghum, Rajendra Nagar, Hyderabad-30, India.

Bennetzen J.L. (1995) Biotechnology for sorghum improvement, *African Crop Science Journal* 3: 161-170.

Cai T. & Buttler L.G. (1990) Plant regeneration from embryogenic callus initiated from immature inflorescences of several high-tannin sorghums, *Plant Cell Tiss Org Cult.* 20: 101-110.

Casas A.M., Kononowicz A.K., Zehr U.B., Tomes D.T., Axtell J.D., Butler L.G., Bressan R.A. & Hasegawa P.M. (1993) Transgenic sorghum plants via micro projectile bombardment. (En). *PNAS*, USA 90: 212-216.

Christou P. (1994) *Rice biotechnology and genetic engineering.* Technomic Publishing Co. Inc. USA.

Elkonin L.A. & Pakhomova N.V. (2000) Influence of nitrogen and phosphorus on induction embryogenic callus of sorghum, *Plant Cell. Tiss. Org. Cult.* 61: 115-123.

Gould J.H., Devey M., Hasegawa O., Ulian E.C., Peterson G. & Smith (1991) Transformation of *Zea mays* L., using *Agrobacterium tumefaciens* and the shoot apex, *Plant Physiol.* 95: 426-434.

Gray S.J., Zhang S., Rathus C., Lemaux P.G. & Godwin I.D. (1999) Development of sorghum transformation: Organogenic regeneration and gene transfer methods, *Sorghum tissue culture, transformation and genetic engineering workshop* ICRISAT 8-9 Mar.

Hagio T., Blowers A.D. & Earle E.D. (1991) Stable transformation of sorghum cell cultures after bombardment with DNA-coated microprojectiles. (En), *Plant Cell. Rep.* 10: 260-264.

Harshavardhan D., Rani T.S., Ulganathan K. & Seetharama N. (2002) An improved protocol for regeneration of *Sorghum bicolor* from isolated shoot apices, *Plant Biotechnol.* 19(3): 163-171.

House L.R. (1995) Sorghum: One of the world's great cereals, *African Crop Science Journal.* 3: 135-142.

Ishida Y., Saito H., Ohta S., Hiei Y., Komari T. & Kumashiro T. (1996) High efficiency transformation of maize (*Zea mays* L.) mediated by *Agrobacterium tumefaciens. Nat. Biotechnol.* 14: 745-750.

Jacobsen J., Matthews P., Abbott D., Wang M.B., Waterhouse P., Buza L., Thornton S. & Gubler F. (1999) Barley transformation breeding: further progress and remaining problems, *Proceedings of the 10th Australian Barley Technical Symposium*, Australia.

Jeoung J.M., Krishnaveni S., Jayaray J., Haiqing Yi, Trick H.N., Muthukrishnan S. & Liang G.H. (1999) *Agrobacterium*-mediated transformation of grain sorghum, *Sorghum tissue culture transformation and genetic engineering workshop* ICRISAT 8-9 Mar.

Kaeppler H.F. & Pederson J.F. (1997) Evaluation of 41 elite and exotic inbred sorghum genotypes for high quality callus production, *Plant. Cell. Tiss. Org. Cult.* 48: 71-75.

Krishnaveni S., Jeoung J.M., Zhu H., Muthukrishnan S. & Liang G.H. (1999). Biolistic transformation of sorghum and influence of a transgenic chitinase gene, *Sorghum tissue culture transformation and genetic engineering workshop* ICRISAT 8-9 Mar.

Locatelli F., Bertolini M. & Lupotto E. (1992) *In vitro* response to somatic embryogenesis of inbred lines of maize, *MNL* 66: 17-18.

Maqbool S.B., Prathiba Devi & Mariam B. Sticklen (2001) Biotechnology: Genetic improvement of sorghum (*Sorghum bicolor* (L.) Moench), *In vitro Cell. Dev. Biol. Plant.* 37: 504-515.

Masteller V.J. & Holden D.J. (1970) The growth of an organ formation from callus tissues of sorghum, *Plant Physiol.* 45: 362-364.

Miflin (2002) Crop improvement in the 21st century, *J. Exp. Bot.* 51: 1-8.

Murashige T. & Skoog F.A. (1962) Revised medium for rapid growth and bioassays with tobacco tissue cultures, *Physiol. Plant.* 15: 473-479.

Murty U.R., Visarada, Annapurna A. & Bharathi M. (1990a) Developing tissue culture system for Sorghum, *Sorghum bicolor* (L.) Moench. I. Embryogenic callus induction from elite genotypes, *Cereal Research Communications.* 18: 257-262.

Murty U.R., Visarada, Annapurna A. & Bharathi M. (1990b) Developing tissue culture system for Sorghum, *Sorghum bicolor* (L.) Moench. II Plant regeneration from embryogenic callus, *Cereal Research Communications.* 18: 355-358.

Mythili P.K., Sairam R.V., Shyamala T., Reddy V.D. & Seetharama N. (1999) Sorghum tissue culture and transformation research at ICRISAT, *Sorghum tissue culture transformation and genetic engineering workshop* ICRISAT 8-9 Mar.

Nabors M.W., Heyser J.M., Dykes T.A. & Demott K.J. (1983) Long duration, high frequency plants regeneration from cereal tissue cultures, *Planta* 157: 385-391.

Nwanze K.F., Seetharama N., Sharma H.C. & Stenhouse J.W. (1995) Biotechnology in pest management: Improving resistance in sorghum to insect pests, *African Crop Science Journal* 3: 209-215.

Prathibha Devi, Heng Zhong & Mariam Sticklen (1999) Production of transgenic sorghum plants with related *HVAI* gene, *Sorghum tissue culture transformation and genetic engineering workshop* ICRISAT 8-9 Mar.

Ratnavathi C.V., Dayakar Rao B. & Seetharama N. (2003) Sweet sorghum stalks: A suitable raw material for fuel alcohol production, p. 8 *NATP Research Bulletin*, National Research Centre for Sorghum, Rajendranagar, Hyderabad-30.

Seetharama N., Sairam R.V. & Rani T.S. (2000) Regeneration of sorghum from shoot tip cultures and field performance of the progeny, *Plant Cell. Tiss. Org. Cult.* 61: 169-173.

Sivamani E., Shen P., Opalka N., Beachy R.N. & Fauquet C.M. (1996) Selection of large quantities of embryogenic calli from *indica* rice seeds for production of fertile transgenic plants using biolistic method, *Plant Cell. Rep.* 15: 322-327.

Subudhi P.K. & Nguyen H.T. (2000) New horizons in Biotechnology, In: *Sorghum: Origin, History, Technology and Production.* Smith CW & RA Frederiksen (Eds.) (John Wiley & Sons, NY, USA).

Tadesse Y. & Jacobs M. (1999) Nutritional quality improvement of sorghum through genetic transformation, *Sorghum tissue culture, transformation and genetic engineering workshop* ICRISAT 8-9 Mar.

Visarada K.B.R.S., Sailaja M. & Sarma N.P. (2002) Effect of callus induction media on morphology of embryogenic calli in rice genotypes, *Biologia plantarum* 45: 495-502.

Wang M.S., Zapata E.J. & Decastro D.C. (1987) Plant regeneration through somatic embryogenesis from mature seed and young inflorescence of wild rice (*Oryza perennis* Moench), *Plant. Cell. Rep.* 6: 294-296.

Zhao Z., Cai T., Tagliani L., Miller M., Wang N., Pang H., Rudent H., Schoeder S., Hondred D., Seltzer J. & Pierce D. (2000) *Agrobacterium* mediated sorghum transformation, *Plant. Mol. Biol.* 44: 789-798.

Zhu H., Muthukrishnan S., Krishnaveni S., Wilde G., Jeoung J.M. & Liang G.H. (1998) Biolistic transformation of sorghum using a rice chitinase gene, *J. Genet. & Breed.* 52: 243-252.

CHAPTER

5

Selection of Transgenic Plants Using Positive Selectable Markers and Other non-Antibiotic Marker Genes

P. Suprasanna & V.A. Bapat

ABSTRACT

In recent years, considerable progress has been made in the genetic manipulation of plants owing to the fine-tuning of gene transfer methods and selection of transformed cells, and designing the control of transgene expression. Ever increasing number of transgenic plants do possess marker genes that are often antibiotic resistance or herbicide tolerance based. Beyond their usage under laboratory circumstances, generally these markers have no marked role and their presence in crops and food has raised much concerns. This has led to either removal of the marker genes or use of alternative marker genes, which have often become the choice. During the past few years, advances have been made in the development of alternative marker systems that are based on the concept of positive selection and in several important crop plants, these have been successfully employed in transgenesis.

INTRODUCTION

Plant genetic transformation technology began in the early'80s with the revelation that *Agrobacterium tumefaciens*, the causative agent of crown gall disease, could be a prime candidate for the introduction of defined pieces of DNA into plant cells (Fraley *et al.*, 1983). Subsequently several other techniques have been developed for genetic transformation (Hansen and Wright 1999, Newell, 2000). With the advancement in the techniques of transformation, appropriate tissue culture methods, selection systems and vector cassettes to host desired genes, genetic manipulation of plants has become useful for diverse applications in crop improvement (Rao and Suprasanna, 1999; Kempken, 2001; Job, 2002).

*Address for correspondence: Plant Cell Culture Technology Section, Nuclear Agriculture and Biotechnology Division, Bhabha Atomic Research Centre, Trombay, Mumbai 400 085.
E-mail: prasanna@magnum.barc.ernet.in; Tel.: 022 2559 3276, Fax: 022 25505151.

Plant transformation includes the delivery, integration and expression of foreign genes into plant cells. The transformed cells are then regenerated into transgenic plants. Despite the development of several gene transfer techniques, efficiency of transformation is often not very high with only a fraction of the cells become transformed and integrate alien DNA into their genomes. It becomes then significant to select these transformed cells, tissues or organisms from the bulk of non-transformed ones and regenerate genetically transformed plants (Ziemienowicz, 2001). The prime characteristic features of a selectable marker gene are: (i) a minimum of non-transformed cells or tissue should escape the selection, (ii) selection should result in a large number of independent transformation events, (iii) it should be applicable to different plant species, and (iv) should have an assay that is easily available to confirm the presence of the marker gene. The selection systems that have been in practice are mostly based on the marker genes encoding resistance to antibiotics or tolerance to herbicides (Table 1). Among these, the most extensively used is the neomycin phosphotransferase gene that confers resistance to aminoglycoside-type antibiotics such as kanamycin, neomycin and G-418 (geneticin). Other selection systems include those conferring resistance to bleomycin, bromoxynil, chlorsulfuron, 2,4-dichlorophenoxy-acetic acid, glyphosate, hygromycin or phosphinothricin (Bowen, 1993).

Table 1 Some of the selectable marker genes used in plant transformation

Marker gene	*Encoded product*	*Selective agent*
nptII	Neomycin phosphotransferase II	Kanamycin
hpt	Hygromycin phosphotransferase	Hygromycin
SPT	Streptomycin phosphotransferase	Streptomycin
dhfr	Dihydrofolate reductase (methotrexate-insensitive)	Methotrexate
ALS	Acetolactate synthase	Chlorsulfuron
aro A (epsps)	5-enolpyruvyl-shikimate-3-phosphate synthase	
bar** and **pat	Phosphinothricin-N-acetyltransferase	Phosphinothricin, bialaphos
CP4 + GOX	Glyphosate tolerant 5-enol pyruvylshikimate-3-phosphate	Glyphosate
nptII with an intron	Neomycin phosphotransferase II	Kanamycin
ipt	Isopentenyl transferase	None
hemL	GR6 glutamata-1-semialdehyde aminotransferase	Gabaculine

Generally, both the desired gene of interest and the selective gene are introduced together into plant cells. The corresponding resistance genes are in most cases not relevant to the transgenic trait of interest and thus may be undesirable once the selection process has been accomplished and transgenic plants generated. Moreover, although transformation procedures have been established for a wide range of plant species, some of these species are naturally resistant to one or more of the above selective agents that may give rise to an unacceptable proportion of non-transgenic escapes. Therefore, alternative selective agents and more efficient selection methods are essential.

There is a growing debate on the safety of the marker genes (antibiotic and other marker genes) in genetically modified food organisms as they might present an unpredictable hazard to the human health and ecosystem (ACNFP, 1994). Despite an increasing scientific knowledge about the horizontal transfer of these genes and safety aspect of human consumption, different opinions have been expressed. There are disadvantages associated with antibiotic or herbicide resistance genes in a selection system, such as potential toxicity or allergenicity of the gene product and interference with antibiotic treatment (Flavell *et al.*, 1992; Fuchs *et al.*, 1992) and possible higher herbicide applications. Based on these critical considerations, there has been rising interest on the removal of selectable marker genes (Puchta, 2000). As there is a need to introduce more gene combinations, either one-step introduction of gene fusions or artificial operons or consecutive transformation steps are required. During the consecutive transformation, a selectable marker gene should be used that is different from the one introduced together with the first gene of interest. This necessitates that a wide range of alternative selection systems is available. Most of the selection systems currently used are based on the principle of negative selection: the selective agent effectively kills untransformed cells. It is understood that during such selection regime, most transformed cells may not regenerate into plants with desired efficiency as the dying non-transformed cells release growth inhibitors and toxic substances (Lindsey and Gallois, 1990) and may compromise the uptake of essential minerals and vitamins from the culture medium.

During the past few years, advances have been made in the development of alternate marker systems that are based on the concept of positive selection (Suprasanna *et al.*, 2002). In this selection method, transformed cells benefit from a metabolic advantage over the untransformed cells that are starved rather than killed. There is a need for these and other novel selectable marker systems, to increase transformation frequencies by improved selection, to allow efficient selection of transgenic plants, and to minimize the use of antibiotic and herbicide selectable marker genes. In this article, some examples on the use of non-antibiotic selectable marker genes in plant transformation, are presented.

POSITIVE SELECTABLE MARKERS

GUS (β-Glucuronidase) Gene

This gene encodes for the enzyme β-glucuronidase which is mostly used as a reporter gene in plant transformation. Joersbo and Okkels (1996) devised a selection strategy wherein transgenic tissues harboring the GUS (β-glucuronidase) gene hydrolyze benzyladenine-glucuronide into active cytokinin and hence are stimulated to grow whereas non-transformed cells are arrested in their development as cytokinin is in its inactive form. In this study, transformed tobacco leaf disks selected using benzyladenine-glucuronide as the selective agent yielded high transformation frequencies, 1.7 to 2.9-fold higher than with kanamycin selection. Interestingly, this selection method can be applied both for selection as well as a reporter gene system. However this is the only report demonstrating the success of this selection method.

manA Gene

This gene codes for the enzyme Phosphomannose isomerase (PMI) catalyzing the conversion of mannose-6-phosphate to fructose-6-phosphate. Most plant species cannot metabolize mannose (or mannose-6-phosphate). It has been demonstrated that mannose-6-phosphate toxicity in plants cells is responsible for the apoptosis or programmed cell death via the induction of DNA laddering (Stein and Hansen, 1999). The transgenic cells with the *manA* gene attain a metabolic advantage over the non-transgenic ones, as the latter starve in the presence of mannose in the culture medium. The *manA* gene of *E. coli*, has been successfully used in transformation experiments in cassava, maize, rice, sugar beet, wheat and Arabidopsis (Table 2).

Table 2 Different positive selection systems used in plant transformation

Selectable marker gene	*Gene source*	*Transgenic crop*	*Selective agent*
β–Glucuronidase (*gusA*)	*Escherichia coli*	Tobacco	Benzyladenine-glucuronide
Phosphomannose isomerase (*manA*)	*Escherichia coli*	Cassava Maize Rice Sugar beet Wheat Arabidopsis	Mannose
Xylose isomerase (*xylA*)	*Streptomyces rubiginosus* *Thermoanaerobacterium thermosulfurogenes*	Potato Potato; Tobacco Tomato	Xylose

In sugar beet, Joersbo *et al.* (1998) selected transformed cells using 1.0-1.5% mannose with a tenfold increase in transformation frequency and 20-30% shoot forming efficiency. A stepwise increase in the selection pressure also ensured the growth of the non-transgenic shoots and rooting problems of the transgenic plants were not encountered as observed with kanamycin selection.

In maize, transgenic plants were obtained by polyethylene glycol-mediated protoplast transformation using selection with 2% mannose (Wang *et al.*, 2000). The transgenic calluses grew on the mannose medium and the T2 generation gave a 3:1 segregation ratio indicating that a single copy of the *manA* gene was stably integrated into the maize genome. Further, Negrotto *et al.* (2000) tested the potential use of PMI via *Agrobacterium*-mediated transformation of maize immature embryos. The recovery of transgenic lines was found to be as high as 30% in about 12 weeks after transformation.

In cassava, Zhang and Pounti-Kaerlas (2000) and Zhang *et al.* (2000) compared the efficiency of positive selection (with mannose) over negative selection (hygromycin). The mannose concentration required for efficient selection in cassava was ten times higher (2%) than that in sugar beet (0.3%, Joersbo *et al.*, 1998). Complete inhibition of rooting of control shoots on 10 g/l mannose was observed compared to normal rooting in the transgenic shoots, indicating that mannose may be useful for a rapid screening of primary transformants.

In wheat, fertile transgenic wheat plants were produced via Biolistics™ following mannose selection (Wright *et al.*, 2001). Average transformation frequency was 20% (the highest 45%) compared to 12% with Basta selection. The authors also employed chlorophenol red (CPR) assay with leaf material of transgenic plants and found 46% expression of the *pmi* gene. In rice, this marker system has been employed successfully with a transformation frequency up to 41% (Lucca *et al.*, 2001). Transformed rice embryogenic calluses were exposed to stepwise selection with 3% mannose initially, increasing to 5% in the last two weeks of selection. Mannose tolerance of transgenic roots was also evaluated using chlorophenol-red assay (Kramer *et al.*, 1993) as an indirect measure of the PMI expression. In a recent study, Zhengquan *et al.*, (2003) evaluated the use of *pmi* system using *Agrobacterium* mediated immature embryo transformation.

In *Arabidopsis*, Todd and Tague (2001) evaluated the selection system using mannose during germ-line transformation. A significant advantage of PMI selection was observed in its adaptability to both soil (non-sterile) grown and tissue cultured (sterile) plants. The authors compared the relative costs of four selective agents in sterile plates and in soil, and found that compared to high costs associated with Kanamycin, hygromycin and phosphinothricin, mannose required moderate to low application costs.

xylA

Xylose isomerase is involved in the isomerisation of D-xylose to D-xylulose in the xylose metabolic pathway of microorganisms. Like mannose, xylose cannot be metabolized by most plant species, while D-xylulose can be metabolized. Haldrup *et al.* (1998a,b) successfully employed xylose isomerase gene (*xylA*) as a selectable marker for transformation experiments with potato, tobacco and tomato. This selection gave an eightfold increase in transformation frequency compared to that with kanamycin selection. On the contrary, the transformation frequency with xylose selection was slightly less (13%) than that of kanamycin (18%) in tobacco, while transformation frequencies were comparable (7% and 9%) in tomato. These observations realize the application of xylose isomerase as a positive selectable marker gene. The enzyme is generally considered to be safe since it has been routinely used in the starch industry and for food processing.

Safety Considerations

Phosphomannose isomerase (PMI) is common in nature across bacteria, yeast and humans. However in plant kingdom, its occurrence is not widespread. In humans, lack of the PMI enzyme is associated with carbohydrate-deficient glycoprotein syndrome resulting in hereditary fructose intolerance, galactosemia and hyperinsulinemic hypoglycemia.

The *pmi* marker system has proven to be successful for transformation in many plant species (as outlined above). Privalle *et al.* (1999) conducted an assessment of different safety parameters viz., adverse effects on growth, allergenicity, effects on agronomical traits and nutritional composition of the transgenic plants. The preliminary results of this study showed that,

(i) in a public database search, *man*A gene product had no significant homology to any known toxin or allergen,

(ii) *in vitro* digestibility experiments proved that PMI protein was readily digestible like any conventional protein under typical mammalian gastric conditions,

(iii) glycoprotein profiles of the PMI expressing and PMI non-expressing showed no differences suggesting no unintended effects on the process of glycosylation,

(vi) Acute oral mouse toxicity tests showed that PMI administered mice did not show any growth abnormalities,

(v) comparison of agronomical features of transformed maize plants (with controls) indicated no adverse growth effects, and

(vi) grain composition (moisture, fiber, fat and protein) of transgenic maize was indistinguishable from their non-transgenic control plants.

OTHER SELECTABLE MARKER GENES

ipt (isopentenyl transferase)

In this selection system, the oncogene of *Agrobacterium tumefaciens* is used as the selectable marker gene to morphologically select the transgenic tissue without the use of antibiotics or herbicides. Ebinuama *et al.* (1997) developed the MAT vector system with the *ipt* gene as the marker and the maize transposable element *Ac* for removing the *ipt* gene. Endo *et al.* (2001) compared the efficiencies of the *ipt* based and kanamycin based selection. Over-expression of the former gene resulted in 2.7 fold higher transformation frequency and also favoured plant regeneration from transformed cells. The authors believe that the use of this system may reduce the use of antibiotic or herbicide resistance marker genes, thereby reducing the concerns associated with the use of these later selection systems.

GSA-AT (Glutamate-1-semialdehyde aminotransferase)

Gough *et al.* (2001) developed an enzyme based system involving a mutant form of cyanobacterial glutamate-1-semialdehyde aminotransferase (GSA-AT) that confers resistance to the phytotoxin gabaculine (3-amino-2,3-dihydrobenzoicacid). The *hem*L gene of the gabaculine resistant cyanobacterium Synechococcus PCC6301 strain GR6 was expressed in tobacco following *Agrobacterium* mediated leaf-disc transformation encodes the enzyme. Selection of the transformants on gabaculine containing medium produced green and white phenotypes; Mendelian segregation and PCR analysis concluded that greens corresponded to the presence of GR6 *hem*L gene. This selection system could also be an ideal candidate for chloroplast transformation with advantages such as gene containment and increased transgene expression (Gough *et al.*, 2001).

Intron containing *nptII* Gene

Most of the first generation transgenic crops contain *nptII* gene as the selectable marker and this antibiotic has been best studied with respect to safety aspects of allergencity, interference with orally administered antibiotics and horizontal gene flow both to intestinal and soil microorganisms (U.S Food and Drug Administration, 1994). Insertion of intron in coding sequences can be used to prevent translation of the interrupted genes in prokaryotes. Using this strategy, Libiakova *et al.* (2001) introduced an intron (IV2 from the potato gene ST-LS1) in the center of the *npt*II –coding sequence where only correct splicing will yield an active NPTII enzyme. Transformation of potato and tobacco with the intron containing constructs was done and the analysis of DNA and RNA showed that the gene was present in the transgenic plants and that the intron was processed correctly. The study suggests that intron + *npt*II selection system can be effective in plant genetic transformation.

In summary, the above described marker systems offer alternatives to antibiotic or herbicide resistance selection, which have often been associated with certain disadvantages. Although there is no scientific evidence against the use of antibiotic marker gene(s), it may be desirable to remove selectable markers from transgenic plants. Several strategies have been developed for the removal of these genes including multi-auto-transformation (MAT) system, co-transformation, site-specific recombination, targeted gene replacement, intragenomic relocation of transgenes via transposable elements (Yoder and Goldsbrough, 1994; Puchta, 2000, Brasileiro and Aragao, 2001, Hare and Chua, 2002). All that remains at this point of time in the era of plant transformation is the fact that there are several possibilities to explore and use alternative selectable markers and strategies for marker gene removal. We need more examples on their use in genetic transformation in economically important crop plants.

In author's laboratory, research is currently underway on employing positive selectable & non-antibiotic marker gene strategy in plant genetic transformation. Studies have yielded positive results with phosphomannose isomerase (PMI) gene in rice and xylose isomerase (*xylA*) gene in banana, tobacco and potato. Transgenic lines have been developed and molecular analysis is being done. Ebmeier *et al.* (Planta 218:751-758, 2004) considered the use of *E. coli* threonine deaminase gene a selectable marker gene and L-O-methylthreonine as a selection agent for tobacco transformation. Tewari-Singh et al. (Plant Cell Rep. 22:576-583, 2004) investigated a desensitized aspartate kinase (AK) gene as a non-antibiotic marker for the production of chickpea transgenic plants using selection with lysine plus threonine. Erikson *et al.* (Nature Biotechnology March 2004) used a conditional marker gene, *daol*(D-aminoacid oxidase) that catalyzes the oxidative deamination of a range of amninoacids. The marker system can be used for either positive or negative selection depending on the substrate (D-alanine, D-serine, D-isoleucine and D-valine). The authors have used this marker successfully with *Arabidopsis* with unambiguous results allowing selection immediately after germination.

REFERENCES

ACNFP, Report, Department of Health and Ministry of Agriculture, Fisheries and Food, London. 1994.

Bowen B.A. (1993) Markers for plant gene transfer. In: *Transgenic Plants*, eds SD Kung & R Wu. Vol. 1, Academic Press, San Diego. 89-95.

Brasileiro A.C.M. & Aragao F.J.L. (2001) Marker genes for *in vitro* selection of transgenic plants. *J. Plant Biotechnol.* 3(3): 113-121.

Ebinuama H., Sugita K., Matsunaga E. & Ymakado M. (1997) Selection of marker free transgenic plants using *isopentenyl transferase* gene. *Proc. Natl. Acad. Sci.* (USA) 94: 2117-2121.

Endo S., Kasahara T., Sugita K., Matsunaga E. & Ebinuama H. (2001) The iso[pentenyl transferase gene is effective as a selectable marker gene for plant transformation in tobacco (*Nicotiana tobaccum* cv, Petite Havana SRI). *Plant Cell Rep.* 20: 60-66.

Flavell R.B., Dart E., Fuchs R.L. & Fraley R.T. (1992) Selectable marker genes: Safe for plants? *Bio/Technol.* 10: 141-144.

Fraley, R.T., Rogers, S.G., Horsch R.B., Sanders P.R., Fick J.S., Adams S.P., Bittners M.L., Brand L.A., Fink C.L., Fry Y.S., Galluppi G.R., Goldberg S.B., Hoffmann N.L., Woo S.C. (1983) Expression of bacterial genes in plant cells. *Proc. Natl. Acad. Sci.* USA, 80: 4803-4807.

Fuchs R.L., Ream J.E., Gammond B.G., Naylor M.W., Leimgruber R.M. & Berberich S.A. (1992) Safety assessment of neomycin phosphotransferase II (NPT II) protein. *Bio/Technol.* 11: 1543-1547.

Gough K.C., Hawes W.S., Kilpatrick J. & Whitelam G.C. (2001) Cyanobacterial GR6 glutamate-1-semialdehyde aminotransferase: a novel enzyme based selectable marker for plant transformation. *Plant Cell Rep.* 20: 296-300.

Haldrup A., Petersen S.G. & Okkels F.T. (1998a) Positive selection: a plant selection principle based on xylose isomerase, an enzyme used in the food industry. *Plant Cell Rep.* 18: 76-81.

Haldrup A., Petersen S.G. & Okkels F.T. (1998b) The xylose isomerase gene from *Thermoanaerobacterium thermosulfurogenes* allows effective selections of transgenic plant cells using D-xylose as the selection agent. *Plant Mol. Biol.* 37: 287-296.

Hansen G. & Wright M.S. (1999) Recent advances in the transformation of plants. *Trends Plant Sci.* 4: 226-231.

Hare P.D. & Chua N.H. (2002) Excision of selectable marker genes from transgenic plants. *Nature Biotechnol.* 20: 577-580.

Job D. (2002) Plant biotechnology in agriculture. *Biochemie* 84: 1105-1110.

Joersbo M. & Okkels F.T. (1996) A novel principle for the selection of transgenic plant cells: positive selection. *Plant Cell Rep,* 16: 219-221.

Joersbo M., Donaldson I., Kreiberg J., Petersen S.G., Brunstedt J. & Okkels F.T. (1998) Analysis of mannose selection used for transformation of sugar beet. *Mol. Breed.* 4: 111-117.

Kempken F. (2001) Plant biotechnology: transgenic crops for the third millennium. *Progr. Bot.* 62: 114-139.

Kramer C., DiMaio J., Carswell G.K. & Shillito R.D. (1993) Selection of transformed protoplast derived *Zea mays* L. colonies with phosphinothricin and a novel assay using the pH indicator chlorophenol red. *Planta.* 190: 454-458.

Libiakova G., Jorgensen B., Palmgren G., Ulskov P. & Johansen E. (2001) Efficacy of an intron-containing kanamycin resistance gene as a selectable marker for plant transformation. *Plant Cell Rep.* 20: 610-615.

Lindsey K. & Gallois P. (1990) Transformation of sugarbeet (*Beta vulgaris* L.) by *Agrobacterium tumefaciens*. *J. Exp. Bot.* 41: 529-536.

Lucca P., Ye X. & Potrykus I. (2001) Effective selection and regeneration of transgenic rice plants with mannose as selective agent. *Mol. Breed.* 7: 43-49.

Negrotto D., Jolley M., Beer S., Wenck A.R. & Hansen G. (2000) The use of phosphomannose-isomerase as a selectable marker to recover transgenic maize plants (*Zea mays* L.) via *Agrobacterium* transformation. *Plant Cell Rep.* 19: 798-803

Newell C.A. (2000) Plant transformation technology. *Mol. Biotechnol.* 16: 53-65.

Privalle L.S., Wright M., Reed J., Hansen G., Dawson J., Dunder E.M., Chang Y.F., Powell L. & Meghji M. (1999) Phosphomannose isomerase, a novel selectable plant selection system: mode of action and safety assessment. *Proc. 6th Int. Symp.* On The Biosafety of genetically modified organisms. Pp. 171-178, eds Fairbairn C., Scoles G. & McHughen A., Saskatoon, Canada, (University Extension Press, Univ. Saskatchewan, Canada).

Puchta H. (2000) Removing selectable marker genes: taking the short cut. Trends *Plant Sci.* 5: 273-274.

Rao P.S. & Suprasanna P. (1999) Augmenting plant productivity through plant tissue culture and genetic engineering. *Jour. Plant Biol.* 26(2): 119-127.

Stein J.C. & Hansen G. (1999) Mannose induces an endonuclease responsible for DNA laddering in plant cell. *Plant Physiol.* 121: 1-9.

Suprasanna P., Sagi L. & Swennen R. (2002) Positive selectable markers for routine plant transformation. *In Vitro* Cell Develp. *Biol.–Plant.* 38(2): 125-128.

Todd R. & Tague B.W. (2001) Phosphomannose isomerase: a versatile selectable marker for *Arabidopsis thaliana* germ-line transformation. *Plant Mol. Biol. Rep.* 19: 307-319.

U.S. Food and Drug Administration (1994). Secondary direct food additives permitted in food for human consumption; food additives permitted in feed and drinking water of animals; aminoglycoside 3' phosphotransferase II. Federal Register 23 May 1994. Vol. 59, pp 26700-26711.

Wang A.S., Evans R.A., Altendorf P.R., Hanten J.A., Doyle M.C. & Rosichan J.L. (2000) A mannose selection system for production of fertile transgenic maize plants from protoplasts. *Plant Cell Rep.* 19: 654-660.

Wright M., Dawson J., Dunder E., Suttie J., Reed J., Kramer C., Chang Y., Novitzky R., Wang H. & Artim-Moore L. (2001) Efficient biolistic transformation of maize and wheat using the phosphomannose isomerase gene, *pmi* gene as the selectable marker. *Plant Cell Rep.* 20: 429-436.

Yoder J.I. & Goldsbrough A.P. (1994) Transformation systems for generating marker-free transgenic plants. *Bioetch.* 12: 263-267.

Zhang P. & Pounti-Kaerlas J. (2000) PIG-mediated cassava transformation using positive and negative selection. *Plant Cell Rep.* 19: 1041-1048.

Zhang P., Potrykus I. & Puonti-Kaerlas J. (2000) Efficient production of transgenic cassava using negative and positive selection. *Transgenic Res.* 9: 405-415.

Zhengquan H., Fu Y., Si H., Hu G., Zhang S., Yu Y. & Sun Z. (2003) Phosphomannose isomerase (*pmi*) gene as a selectable marker for rice tansformation via *Agrobacterium. Plant Sci.* (In Press)

Ziemienowicz A. (2001) Plant selectable markers and reporter genes. *Acta Physiol. Plant.* 23(3): 363-374.

CHAPTER

6

Production of Hybrid Plants in Thompson Seedless Grapes (*Vitis vinifera*) through Breeding and *in ovulo* Embryo Rescue Methods

P.V. Bharathy[1], D.D. Kulkarni[1], A.B. Biradar[2], G.S. Karibasappa[2], A.U. Solanke[3], S.G. Patil[3] and D.C. Agrawal[1]*

ABSTRACT

The present investigations, a part of study to introgress downy mildew resistance in Thompson Seedless through breeding and *in ovulo* embryo rescue methods, was carried out to study the effect of eight male parents on the cross compatibility, percentage of embryo recovery, germination percentage and development of hybrid plants. Eight cross-combinations consisted of Thompson Seedless, susceptible to downy mildew used as female parent crossed with 8 resistant cultivars viz., Lake Emerald, Concord, Catawaba, Fruhroter Veltliner, Seyve Villard (SV 18402), *Vitis tilifolia, Vitis candicans* and St. George as male parents. All the 8 cross-combinations developed viable embryos. However, percentage of embryos recovered varied among the crosses. The percentage of embryo recovery, germination of embryo and development of plants was higher in crosses TS × SV 18402, TS x Concord and TS x Catawba as compared to the other crosses. Plants too were healthier in these crosses. The results obtained in the present study suggest a higher compatibility of cultivars SV 18402, Concord and Catawaba with Thompson Seedless.

KEY WORDS: Breeding, downy mildew, cultivars, germination.

ABBREVIATIONS: BAP-Benzylaminopurine

INTRODUCTION

Thompson Seedless (*Vitis vinifera*) is one of the most widely cultivated table grape varieties in the world. Taxonomically, grapes are divided into two sub-genera, *Euvitis* Planch. (2n=38) and

*Address for correspondence: [1]Plant Tissue Culture Division, National Chemical Laboratory, Pune – 411 008, INDIA
Email: dc.agrawal@ncl.res.in; Fax: -91-20-2589 3438
[2]National Research Centre for Grapes, Manjri, Solapur Road, Pune-412 307
[3]Agharkar Research Institute (MACS), GG Agharkar Road, Pune-411 004

Muscadinia Planch. (2n=40) (Einset and Pratt, 1975). Most commercial grapes belong to sub-genus *Euvitis* with *Vitis vinifera* being the most widely grown species. In the year 2000, the world production of grapes was to the tune of 62, 409 million tonnes (Dutruc-Rossett, 2001). In India, grape is grown on an area of 34,000 ha with a production of about 1,00,000 tonnes, 85% of which comprises mainly of table grapes. Seedless table and raisin grapes are preferred by consumers all over the world. Seedlessness, which is the most important marketable trait in Thompson Seedless is due to stenospermocarpy in which pollination and fertilization occur but the embryos get aborted. Hence the seed appears flattened and gives the fruit its seedless quality.

Thompson Seedless is highly susceptible to downy mildew (*Plasmopara viticola* Berl. et de Toni), one of the most devastating fungal diseases affecting the grapevine worldwide. In India, it mainly affects grapevines in the tropical region of the country. Intergeneric and interspecific hybridization to combine the fruit quality of *Euvitis* and the hardiness of *Muscadine* grapes forms the basis of development of new cultivars of good quality. However, success in terms of useful cultivars is yet to be realized, mainly due to the differences in chromosome number between the two genera and embryo abortion. *In ovulo* embryo rescue thus provides the solution not only for obtaining good interspecific crosses, but also allows the use of seedless vines as female parents (Cain *et al.*, 1983; Emershad and Ramming, 1984; Spiegel-Roy *et al.*, 1985; Gray *et al.*, 1987).

The present study was taken up mainly to introgress downy mildew resistance in Thompson Seedless using breeding and *in ovulo* embryo rescue methods. It is the first instance in India where a seedless variety has been used as female parent. The other objective of the present investigations was to study the effect of eight male parents on (i) cross-compatibility, (ii) embryo recovery, (iii) germination and (iv) development of hybrid plants.

MATERIALS AND METHODS

Breeding Work

Breeding work was initiated by identification of both male and female parental plants in the germplasm maintained at the vineyard of National Research Centre for Grapes, Pune. Thompson Seedless was selected as the female parent. Eight male parental lines (seeded) which had shown field tolerance to the downy mildew disease were as follows: Lake Emerald, Concord, Catawaba, (all belonging to *Vitis labruska*), Fruhroter Veltliner (*Vitis vinifera*) Seyve Villard (S.V. 18402, *Vitis* spp.), *Vitis tilifolia*, *Vitis candicans* and St. George (*Vitis rupestris* var. du Lot). Defoliation of both male and female parental plants was done a week prior to their pruning to induce flowering.

After about 25 to 30 days of pruning, flowering commenced in male parental plants. Bagging of panicles in male plants was done to collect pollen grains. Emasculation in female plants and immediate bagging of emasculated panicles and subsequent hand cross pollination using hair brush with the designated male pollen was carried out for each cross-combination.

Ovule Culture

Immature berries (Fig. 1A) from the crosses were collected 40 days after pollination. These were washed with liquid soap for 10 min; surface sterilized with mercuric chloride (0.1% w/v) solution for 10 min and were rinsed 3 times with sterile distilled water. Ovules from the berries were

Fig. 1 (A) Immature berries, (B) Ovules, (C) Germination of embryos, (D) Early stages of plantlets, (E) Hybrid plantlet in test tube, (F) Hybrid plants.

aseptically excised and cultured in petri dishes (15-20 ovules per dish) (Fig. 1B) containing Emershad and Ramming medium (ER) (Emershad and Ramming, 1984) supplemented with sucrose (6%) and activated charcoal (0.3%). The pH of medium was adjusted to 6.2 before autoclaving. The total number of ovules inoculated per cross was recorded. The petri dishes were kept under diffused light (irradiance of 6.1 μmol m^{-2} s^{-1}) with a 24 hr photoperiod at 25 ± 2°C for 8 weeks to allow the development of embryo within the ovule. The experiment was set in a completely randomized design, replicates being either single ovules or embryos.

Embryo Culture and Plantlet Development

After 8 weeks, the ovules were dissected aseptically under microscope and embryos were excised from the micropylar end of the ovule. The embryos were transferred to petri dishes containing Woody Plant Medium (WPM) (Lloyd and McCown, 1981), supplemented with BA (1 μM), sucrose (1.5 %), activated charcoal (0.3 %) and pH of medium was adjusted to 6.2. Embryos were incubated at 25 ± 2°C under fluorescent light (12.2 μmol m^{-2} s^{-1}) with a 24 hr photoperiod. The germinated embryos showing the first pair of leaves (Fig. 1D) were transferred to test tubes (**Fig. 1E**) containing WP medium, supplemented with BAP (1 μM) and kept under similar light and temperature conditions in order to develop into plants.

Hardening of Plantlets

After 4 weeks, the plantlets were transferred to plastic cups containing a mixture of soil + sand (1:1). Initially these cups were covered with thin and transparent polythene bags and kept under high light intensity (24.4 μmol m^{-2} s^{-1}) with a 24 hr photoperiod at 25 ± 2°C for hardening. Thereafter, hardened plants (Fig. 1F) were transferred to greenhouse.

RESULTS AND DISCUSSION

Berry Growth

Berries were obtained in all the eight cross-combinations although its number varied depending upon the male parent type. Among the crosses, only marginal difference was observed in the size and weight of the berries. Almost all the crosses had an average berry weight between 0.37 to 0.47 g. The difference in the average weight of berries was not significant. The berries were firm in all the cross-combinations.

Ovule Growth

Berry to ovule ratio

A total of 4038 berries were used for embryo excision in Thompson Seedless, from which 5800 ovules were excised and cultured (Table 1). The berry to ovule ratio signifies the number of ovules recovered per berry. The average berry to ovule ratio was 1.44. There was variation among the different cross-combinations as regards the berry to ovule ratio. The cross TS × St. George had the highest berry to ovule ratio (1.77), followed by TS × *V. candicans* (1.58) and TS × Catawba (1.56). The least berry to ovule ratio was observed in TS × Concord (1.27). The variation in the berry to ovule ratio in different cross-combinations may be due to male parent type and/or levels of

Table 1 Influence of male parent types on recovery of embryos, germination and development of hybrid plants in Thompson Seedless grapes (*Vitis vinifera*).

Sr. No.	*TS x Male parent*	*No. of berries used*	*Average wt. of berries (gm)*	*No. of ovules excised*	*Berry to ovule ratio*	*Av. Wt of ovules (mg)*	*No. of embryos excised*	*Response/ovule (%)*		
								Embryos recovered	*Embryos germinated*	*Plantlets developed*
1.	Lake Emerald	745	0.37	1041	1.40	9.81	79	7.59	6.34	4.80
2.	St. George	249	0.46	441	1.77	9.33	39	8.84	8.16	6.80
3.	SV 18402	1237	0.38	1724	1.39	8.14	221	12.82	9.69	8.29
4.	*V. tilifolia*	300	0.39	408	1.36	10.77	10	2.45	2.21	2.21
5.	*V. candicans*	299	0.38	471	1.58	10.20	56	11.89	7.86	7.86
6.	Concord	325	0.45	413	1.27	9.63	116	28.09	20.34	18.40
7.	Catawba	480	0.47	749	1.56	11.02	88	11.75	10.15	9.08
8.	Fruhroter Veltliner	403	0.39	553	1.37	9.61	45	8.14	4.52	4.52
	Total	4038	0.41	5800	1.44	9.81	654	11.28	8.62	7.55
	S.E. ±	41.29	0.10	67.77	0.38	2.36	8.85	7.14	5.00	4.80
	C.D.	117.95	0.28	193.59	1.09	6.75	25.29	20.39	14.27	13.72

endogenous hormones. A brown coloured, papery integument surrounded the embryo or the endosperm, if present. The ovules were mostly hollow without endosperm. Nitsch *et al.* (1960) and Barritt (1970) have earlier reported that embryo/endosperm breakdown in stenospermic grapes occur about 3 weeks after bloom.

The size of the ovules increased during the incubation period. Ovules in some cases turned brown in spite of the addition of charcoal to the medium. The size of green ovules was more as compared to brown ovules at the end of the incubation period and was observed in almost all the crosses. This may be associated with veraison and berry softening. Also, since it was difficult to extract the ovules from the softened berries, as small portion of the pulp remained around the ovule which may have caused oxidation and subsequent browning of the ovules during culture. Similar observations have been reported by Gray *et al.* (1990). In the present study, though browning occurred, the size of ovules increased during culture.

Callus formation was observed in a few ovules in some crosses. Callus was mostly white in colour with some anthocyanin pigmentation. Callus formation however, did not affect the embryo development. Similar results were obtained in Thompson Seedless by Emershad and Ramming (1984) and Cain *et al.* (1983).

Average weight of ovules

Variation in the average weight of ovules was observed among the different cross-combinations (Table 1). Among the crosses, TS × Catawba had the highest ovule weight (11.02 mg), followed by TS × *V. tilifolia* (10.77 mg). Lowest weight was observed in TS × SV 18402 (8.14 mg). Weight of the ovules was not directly proportional to the weight of berries, which may be due to genotypic characteristics, parental compatibility and nutritional factors.

Embryo Growth and Development

Effect of male parent type on embryo development

Embryo development was influenced by the male parent type among the cross-combinations (Table 1). The highest percentage of embryo recovery was recorded in the cross TS × Concord (28.09), followed by TS × SV 18402 (12.82) (Table 1), while the lowest percentage of embryo recovery was observed in TS × *V. tilifolia* (2.45). In rest of the crosses, it ranged between 7.59 and 11.75. An average percentage of embryo recovery for all the crosses was 11.28. In an earlier study, Gray *et al.* (1990) made crosses using Orlando Seedless as female parent with other seedless varieties as pollen parents, including Thompson Seedless and obtained 17.1% embryo recovery from 70 ovules cultured at 40 days after pollination. In our studies, some embryos pushed out of the ovule into the ovule culture media and were better developed than those which had to be excised. The embryos obtained from ovules with brown seed coat were mostly globular and smaller in size than those from ovules with green seed coat. The endosperm in most of the ovules was dried up, but in some cases it was fleshy and surrounded by brown and papery integument. The data presented in Table 1 shows that the berry weight and ovule weight did not affect the embryo recovery. The embryos were white, glossy and situated at the micropylar end of the ovule. The embryos were bigger in size in case of TS × SV 18402, Ts × Concord and TS × Catawba as compared to the other crosses. Goldy and

Amborn (1987) showed differences in embryo recovery between open pollinated ovules of several seedless cultivars. Hence it is reasonable to assume that the vast difference in the number of embryos obtained among the crosses may be due to the different combining abilities of the parents which would lead to variation in embryo recovery.

Percentage embryo germination

Most of the embryos germinated (Fig. 1C) on the WP medium (Table 1) within 5 days, although globular embryos took a longer time and showed very less germination. Size of embryo did not affect the germination. The germination percentage varied among the crosses and ranged between 2.21 to 20.34. From an overall embryo recovery of 11.28%, the percentage of germinated embryos was 8.62. Abnormal embryos were found, too, which either had enlarged, giant cotyledons and a small root system, an elongated hypocotyl with roots but without cotyledons or stunted embryos without cotyledons and profuse rooting. There was a variation in the number of cotyledons too, ranging from 1 to 4 cotyledons. Two cotyledonary embryos were most frequently observed while one and three cotyledons were more frequent than four cotyledons. Multiple cotyledonary embryos (cabbage shaped) were also observed. The cotyledons expanded normally and the embryos had a tap root system with elongated hypocotyls. Multiple cotyledonary embryos took almost a month to develop the first pair of leaves.

Plantlet development

The highest percentage of plantlet development was obtained in the cross TS × Concord (18.40) followed by TS × SV 18402 (8.29) (Table 1). All the embryos that germinated in the crosses TS × *V. tilifolia*, TS × *V. candicans* and TS × Fruhroter Veltliner developed into plantlets. During hardening, variation was observed in the growth and vigour of the plants from the same cross. TS × SV 18402, TS × Concord and TS × Catawba produced healthier plants as compared to the other crosses. Morphological variations like difference in stem colour and thickness, leaf shape and colour were observed in the plantlets obtained in the same cross.

From the present investigations, it may be inferred that crosses TS × SV. 18402, TS × Concord and TS × Catawba seem to have a better compatibility, hence resulting in a higher recovery of embryos, percentage germination and development of hybrid plants, which makes them promising varieties to be used in future cross-breeding programmes for downy mildew resistance.

Acknowledgements

The financial assistance in the form of research grant by the National Agriculture Technology Programme (NATP), ICAR, New Delhi, India is gratefully acknowledged.

REFERENCES

Barritt B.H. (1970) Ovule development in seeded and seedless grapes, *Vitis* 9: 7-14.

Cain D.W., Emershad R.L. & Tarailo R.E. (1983) *In-ovulo* embryo culture and seedling development of seeded and seedling development of seeded and seedless grapes (*Vitis vinifera L.*), *Vitis* 22: 9-14.

Dutruc-Rossett G. (2001) Extract of the Report on World Vitiviniculture (World Congress of the OIV in Adelaide, 12 October 2001).

Einset J. & Pratt C. (1975) Grapes, In: *Advances in Fruit Breeding,* pp. 130-153, eds Janick J. & Moore J.N. (Purdue Univ. Press, West Lafayette, Ind.).

Emershad R.L. & Ramming D.W. (1984) In-ovulo embryo culture of *Vitis vinifera* L. cv. Thompson Seedless, *Amer. J. Bot.* 71: 837-877.

Goldy R.G. & Amborn U. (1987) *In vitro* culturability of ovules from 10 seedless grape clones, *Hort. Science* 22(5): 952

Gray D.J., Fischer L.C. & Mortensen J.A. (1987) Comparison of methodologies for *in ovulo* embryo rescue of seedless grapes, *Hort. Science* 22: 1334-1335.

Gray D.J., Mortensen J.A., Benton C.M., Durham R.E. & Moore G.A. (1990) Ovule culture to obtain progeny from hybrid seedless bunch grapes, *J. Amer. Soc. Hort. Sci.* 115: 1019-1024.

Llyod G. & McCown B. (1981). Commercially feasible micropropagation of mountain laurel Kalmia latifolia by use of shoot tip culture, *Proc. Int. Plant Prop. Soc.* 30: 421-427.

Nitsch J.P., Pratt C., Nitsch C. & Shaulis N.J. (1960) Natural growth substances in Concord and Concord Seedless grapes in relation to berry development, *Amer. J. Bot.* 47: 566-576.

Spiegel-Roy P., Sahar N., Baron J. & Lavi U. (1985) *In vitro* culture and plant formation from grape cultivars with abortive ovules and seeds, *J. Amer. Soc. Hort. Sci.* 34: 416-420.

CHAPTER

7

Somatic Embryogenesis in Chickpea (*Cicer arietinum* L.)

Anju Arora, Anuj Choudhary and H.S. Chawla

ABSTRACT

In vitro regeneration in chickpea (*Cicer arietinum* L.) was achieved via somatic embryogenesis in genotypes Pusa256 and PG186 from two explants viz. mature embryo axes and cotyledon explants of mature seeds. For embryo axes explant embryo was excised from mature seed and decapitated by discarding root meristem and the shoot apex. The cotyledon explants were slightly cut at three sides leaving the attachment site of embryo intact. Both the explants were cultured on MS media supplemented with different concentrations of 2,4-D (0.5, 1.0, 1.5 and 2.0 mg/l) with or without 1 mg/l BAP. Embryogenic callus was induced from both the explants but cotyledon derived cultures showed good embyrogenesis and regeneration efficiency. Medium supplemented with 2 mg/l 2,4-D and 1 mg/l BAP yielded maximum embryos during dark incubation of cultures with least or no differentiation of embryos. However, medium supplemented with 0.5 mg/l 2,4-D alone or in combination with BAP was found suitable for embryo induction followed by subsequent maturation of embryos on MS medium containing 1.0 mg/l BAP and 0.05 mg/l NAA. When 0.5 mg/l 2,4-D containing MS medium was fortified with amino acids (100 mg/l glutamine and 50 mg/l proline) and 100 mg/l casein hydrolysate, resultant cultures showed stimulatory effect on embryogenic callus induction and embryo germination. The well formed embryos germinated on MS medium with B5 vitamins. Genotypic differences existed for frequency of embryo formation and germination of somatic embryos. In general, PG186 responded better than Pusa256 for regeneration via embyrogenesis.

Key Words: Chickpea, Regeneration, Somatic embryogenesis

Abbreviations: 2,4-D - 2,4-Dichlorophenoxy acetic acid; BAP-6-benzylaminopurine; NAA-Naphthalene acetic acid

INTRODUCTION

Chickpea (*Cicer arietinum* L.) is an important leguminous crop of Indian subcontinent as it represents one of the valuable sources of proteins for human and animal nutrition. In the world,

*Address for correspondence: Plant Biotech. Lab., Department of Genetics and Plant Breeding, Gobind Ballabh Pant Univ. of Agric. & Tech., Pantnagar-263 145, Uttaranchal (India).

India covers maximum area under pulses out of which chickpea is grown on approximately 6.9 mha area with 780 kg/ha productivity. This low productivity is due to the attack of major pathogens. Therefore, there is need to develop resistant lines by incorporating foreign genes through genetic engineering. For developing transgenic line in any crop, an efficient *in vitro* plant regeneration protocol is the prerequisite. *In vitro* plant regeneration can be achieved either through organogenesis or somatic embryogenesis approach. Since rooting in regenerated shoots is a problem, many of the workers attempted regeneration from various explants via somatic embryogenesis approach. Plant regeneration via somatic embryogenesis has been reported from various explants (Rao and Chopra, 1989; Barna and Wakhlu, 1993; Nalini *et al.*, 1993; Sagare *et al.*, 1993, 1995; Suhasini *et al.*, 1994, 1997; Islam, 1994; Kumar *et al.*, 1994, 1995; Adkins *et al.*, 1995; Vani and Reddy, 1996; Hita *et al.*, 1997). 2,4-D, a synthetic auxin is the most frequently used growth regulator to initiate embryogenesis in chickpea either alone or in combination with other growth regulator. Some workers have used cytokinin (BAP, zeatin or TDZ) with auxin IAA or amino acid proline (Nalini *et al.*, 1993; Murthy *et al.*, 1996; Adkins *et al.*, 1995; Hita *et al.*, 1997). With the aim of developing protocol at a pilot level, mature seed explants were taken for obtaining complete plantlets via somatic embryogenesis in chickpea. The effect of 2,4-D and 2,4-D with BAP was tested for embryogenic callus induction frequency as well as interaction of embryogenic callus induction media with regeneration media. The role of amino acids in combination with growth regulators was also tested for embryogenic callus induction.

MATERIALS AND METHODS

Explants from two genotypes viz. Pusa256 and PG186 were taken. The seeds were washed with detergent and soaked for 20 hrs prior to explant excision. The seeds were surface sterilized with 0.2% $HgCl_2$ solution for 8 min and washed with several changes of sterile distilled water. Then embryos and cotyledons were removed from seeds. The excised embryos were decapitated by discarding the root and shoot meristems. The cotyledon explants were slightly cut at three ends leaving the attachment site of embryo intact. For embryogenic callus induction, MS medium (Murashige and Skoog, 1962) was used that contained 0.5 to 2.0 mg/l 2,4 Dichlorophenoxy acetic acid (2,4-D) with or without 1 mg/l Benzylamino purine (BAP) and a control. The media were designated as ECM_0, $ECM_{0.5D}$, $ECM_{0.5D1B}$, ECM_{1D}, ECM_{1D1B}, $ECM_{1.5D}$, $ECM_{1.5D1B}$, ECM_{2D} and ECM_{2D1B}. The values in suffix indicate the concentration of growth regulator in mg/l and letters D and B stands for growth regulators 2,4-D and BAP, respectively. The calluses were once sub-cultured on the same media after four weeks of callus initiation. In another experiment, the callusing media supplemented with 0.5 mg/l 2,4-D with or without 1 mg/l BAP was further fortified with amino acids with or without protein. In addition to the growth regulators, media comprised of 50 mg/l proline (P) and 100 mg/l glutamine (G) with or without 100 mg/l casein hydrolysate (C). The media were designated as $ECM_{0.5DPG}$, $ECM_{0.5DPGC}$, $ECM_{0.5D1BPG}$ and $ECM_{0.5D1BPGC}$. The calluses after eight weeks of induction were transferred to media for embryo germination. Modified MS medium with MS basal salts and B_5 vitamins (Gamborg *et al.*, 1968) was used for germination of embryos. The embryo germination medium was supplemented with 1, 2 and 3 mg/l BAP with or without 0.05 mg/l NAA and a control. The media were designated as R_0, R_{1B}, $R_{1B0.05N}$, R_{2B},

$R_{2B0.05N}$, R_{3B} and $R_{3B0.05N}$. The letters B and N stand for growth regulators BAP and NAA, respectively. The optimum culture conditions were 20±2°C temperature with complete darkness during callus initiation and 16/8h photoperiod during embryo maturation and germination. Eight explants were cultured in one flask/petri dish. The data was recorded on each explant and mean response from one flask/petri dish constituted one replication. The data was analyzed statistically for testing the significance of main and interaction effects of genotype, media and explant. The experimental designs for embryogenic callus induction frequency and number of somatic embryos per embryogenic callus were multi-factorial with four replications.

RESULTS AND DISCUSSION

Analysis of variance for embryogenic callus induction frequency showed significant differences for callusing media, explants and interactions involving genotype × callusing media, callusing media × explant and genotype × explant (Table 1). Embryogenic callus induction frequency from cotyledon and embryo axes explants of two genotypes on nine different media containing 2,4-D alone and with BAP has been presented in Table 2. When medium was not supplemented with any growth regulator there was no callus induction from embryo axes explants while 7% callus induction frequency was seen from cotyledon explants. On different media supplemented with increasing concentrations of 2,4-D from 0.5 to 2.0 mg/1 with or without BAP (1.0 mg/1), embryogenic callus induction frequency varied from 65 to 100% for embryo axes explants and 5-100% for cotyledon explants. Embryogenic callus induction frequency decreased for cotyledon explants from 100 to 10% when cytokinin was not added to the auxin containing media. In both the genotypes, maximum embryogenic callus induction frequency of 95-100% was obtained at higher auxin level in the callusing media (Fig. 3a). The combination of auxin 2,4-D with cytokinin BAP had no influence on embryogenic callus induction frequency for embryo axes explants while it was stimulatory for cotyledon explants. Low frequency of embryogenic callus induction was seen from cotyledon explants on medium with auxin alone because it formed roots (Vani and Reddy, 1996). For cotyledon explants when media were supplemented with 2,4-D only, 5-7% embryogenic callus induction frequency was observed but when auxin was supplemented with BAP, embryogenic callus induction frequency increased with increasing concentration of 2,4-D. A maximum of 100% embryogenic callus induction frequency was observed on callusing media supplemented with 2 mg/

Table 1 Analysis of variance for embryogenic callus induction frequency

Source of variation	*d.f.*	*Embryogenic callus induction frequency*
Genotype	1	1.85ns
Callusing media	8	429.94**
Explant	3	246.82**
Genotype × callusing media	8	3.03**
Callusing media x explant	24	66.08**
Genotype × explant	3	5.44**
Genotype × callusing media x explant	24	0.82ns

** Significant at 1% level of significance

Table 2 Embryogenic callus induction frequency from different seed explants

Callusing	*Pusa256*				*PG186*			
Media	*Embryo axes*		*Cotyledon*		*Embryo axes*		*Cotyledon*	
No.	*Total no. of explants*	*Embryogenic calluses (%)*	*Total no. of explants*	*Embryogenic calluses (%)*	*Total no. of explants*	*Embryogenic calluses (%)*	*Total no. of explants*	*Embryogenic calluses (%)*
ECM_0	36	– (0)	39	3 (7.7)	40	– (0)	41	3 (7.3)
$ECM_{0.5D}$	37	26 (70.3)	42	28 (66.7)	42	37 (88.1)	41	25 (61.0)
$ECM_{0.5D1B}$	40	26 (65.0)	40	35 (87.5)	41	35 (85.4)	41	35 (85.4)
ECM_{1D}	42	36 (85.7)	39	7 (18.0)	41	37 (90.2)	40	6 (15.0)
ECM_{1D1B}	39	32 (82.1)	39	35 (89.7)	40	35 (87.5)	38	34 (89.5)
$ECM_{1.5D}$	40	39 (97.5)	40	6 (15.0)	39	37 (94.9)	39	4 (10.3)
$ECM_{1.5D1B}$	38	36 (94.7)	41	40 (97.6)	41	39 (95.1)	42	41 (97.6)
ECM_{2D}	41	41 (100)	42	4 (9.5)	42	41 (97.6)	40	2 (5.0)
ECM_{2D1B}	40	40 (100)	41	41 (100)	38	36 (94.7)	42	41 (97.6)

l 2,4-D with or without 1 mg/l BAP. A maximum of 65% embryogenic calluses from cotyledon explants was obtained on the media containing 2 mg/l 2,4-D and 1 mg/l BAP. Considering explants over genotypes and callusing media, cotyledon explants showed 0 to 20% higher embryogenic callus induction frequency as compared to embryo axes explants when media contained auxin and cytokinin. In general, for embryo axes explants callus induction frequency increased with increase in 2,4-D concentration in the medium irrespective of BAP concentration while for cotyledon explants it increased on callusing media containing increasing concentrations of 2,4-D with BAP but decreased on callusing media with 2,4-D auxin alone. Different callusing media tested showed significant variation in callus induction frequency. Callusing media containing 2,4-D at higher concentration induced somatic embyrogenesis with a maximum frequency of 100% for embryo axes explants on media containing 2,4-D with or without BAP. For cotyledon explants, maximum callus induction frequency was observed on media containing 2,4-D at higher concentration in combination with BAP 1 mg/l. In general medium with auxin alone tend to induce roots from the developing embryos.

The embryo axes and cotyledon explants of both the genotypes showed variable frequencies when embryogenesis was induced in the presence of amino acids with or without casein hydrolysate. Fig 1 shows the effect of proline and glutamine with or without 100 mg/l casein hydrolysate in combination with growth regulators. Addition of amino acids to the callusing media showed 1-5% increase in embryogenic callus induction frequency (Fig. 3b). Further the addition of 100 mg/l casein hydrolysate had a stimulatory effect on embryogenic callus induction frequency. Though the increase in embryogenic callus induction frequency was not significant but the addition of growth supplements may have a positive role on embryo maturation and germination. The trend of increase in embryogenic callus induction frequency was similar in cotyledon and embryo axes explants of both the genotypes. Murthy *et al.* (1996) and Vani and Reddy (1996) showed the positive effect of amino acids proline and glutamine respectively, on embryogenesis and subsequent germination of embryos.

The quality of callus varied on different callusing media. The cotyledon and embryo axes explants induced different morphology of callus on same callusing media. The calluses induced on callusing media with 2,4-D alone were yellowish and compact while on media containing auxin with cytokinin, calluses were creamish yellow, compact and nodular.

After four weeks of embryogenic callus induction, the calluses were cut into pieces and kept on same media for proliferation. After 8 weeks, the calluses were transferred on different regeneration media containing varying levels of BAP (1, 2 and 3 mg/l) with or without 0.05 mg/l NAA and a control. The cultures were placed in 16/8 h photoperiod for germination of embryos. The calluses induced on a particular callusing media responded differently for maturation and germination on various regeneration media. Somatic embryogenesis induced from embryo axes and cotyledon explants showed embryoid structures on regeneration media. The data was recorded on number of somatic embryos/explant and percent germination of somatic embryos.

The effect of interaction of callusing media with regeneration media on somatic embryos/ explant has been shown in Fig. 2. It was observed that callusing media without growth regulator did not induce somatic embyrogenesis while a maximum of 16-18 somatic embryos/ explant were

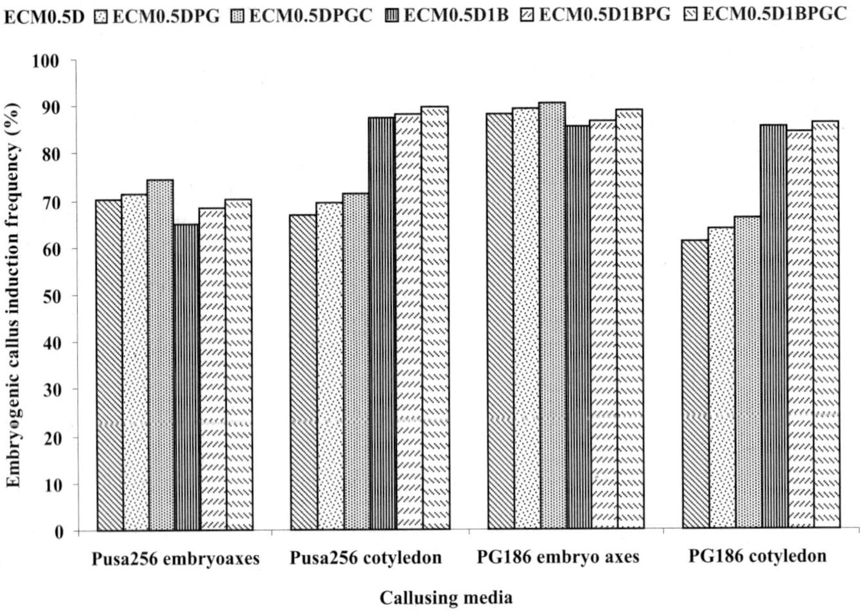

Fig. 1 Effect of amino acids in callusing media on embryogenic callus induction frequency

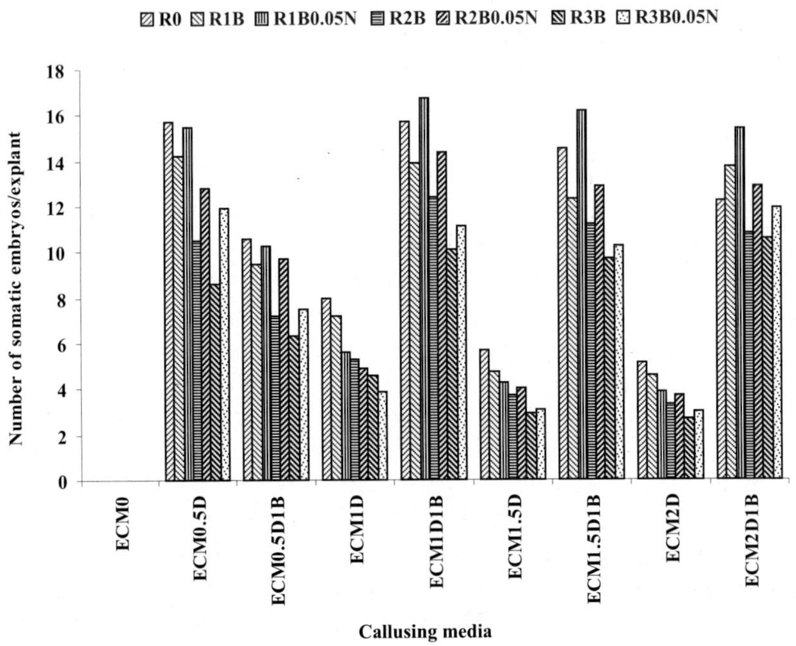

Fig. 2 Effect of callusing media and regeneration media on somatic embryos/explant

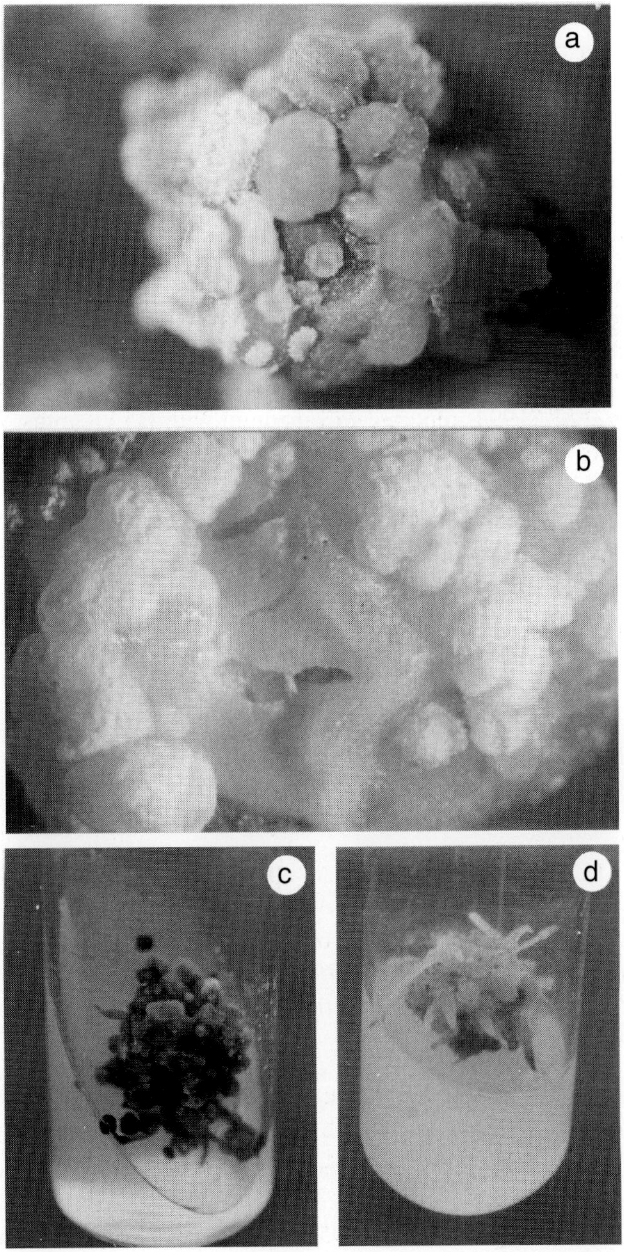

Fig. 3 Embryogenesis in chickpea; **(a)** Embryogenic callus induced on callusing media containing 0.5 mg/l 2,4-D and 1 mg/l BAP; **(b)** Embryogenic callus induced on callusing media with 2,4-D, BAP and growth supplements (proline, glutamine, casein hydrolysate); **(c)** Maturation of somatic embryos on regeneration media containing 1 mg/l BAP and 0.05 mg/l NAA; **(d)** Embryo germination on regeneration media without growth regulators.

obtained when embryogenesis was induced on callusing media containing 0.5 mg/l 2,4-D alone or with cytokinin at higher auxin level (Fig. 3c). It was seen that callusing media containing lower auxin level (0.5 or 1 mg/l) resulted in higher frequency of embryo maturation and germination as compared to higher auxin level (1.5 or 2 mg/l) in the callusing media. On transfer of somatic embryos to different regeneration media, embryos differentiated with comparatively higher frequency on media without growth regulator than media supplemented with growth regulator (Fig. 3d). The influence of regeneration media showed that number of somatic embryos/explant decreased with increase in cytokinin concentration in the regeneration media. The addition of auxin to cytokinin containing regeneration media had stimulatory effect on embryo maturation and germination. In general regeneration media without growth regulators or with lower level of cytokinin along with auxin showed maximum conversion of developing embryos into well developed embryos. Whereas higher levels of cytokinin with or without auxin were comparatively less effective. Kumar *et al.* (1994) reported MS medium with B5 vitamins, 2 mg/l BAP and 0.05 mg/l IBA as the best for maturation of embryos induced from immature leaf explants. Barna and Wakhlu (1993) obtained better embryo maturation and germination on hormone-free medium. Though the present study shows that maximum conversion of embryoids to globular structures took place on 1 mg/l BAP with auxin while the germination frequency was higher on growth regulator free medium. It is possible that this difference is due to the growth regulator used in the callusing medium.

Acknowledgements

A. Arora and A. Choudhary are thankful to UP Council of Agriculture Research for financial assistance received during the course of investigation.

REFERENCES

Adkins A.L., Godwin I.D. & Adkins S.W. (1995) An efficient *in vitro* regeneration system for Australian grown chickpea (*Cicer arietinum* L.) cultivars. *Australian Journal of Botany.* 43: 491-497.

Barna K.S. & Wakhlu A.K. (1993) Somatic embryogenesis and plant regeneration from callus cultures of chickpea (*Cicer arietinum* L.). *Plant Cell Reports.* 12: 521-524.

Gamborg O.L., Miller R.A. & Ojima K. (1968) Nutrient requirement of suspension cultures of soybean root cells. *Experimental Cell Research.* 50: 148-151.

Hita O., Lafarga C. & Guerra H. (1997) Somatic embryogenesis from chickpea (*Cicer arietinum* L.) immature cotyledons; the effect of zeatin, gibberellic acid and indole-3-butyric acid. *Acta Physiologia Plantarum.* 19: 333-338.

Islam R. (1994) Somatic embryogenesis from immature cotyledons of chickpea (*Cicer arietinum* L.). *Pakistan Journal of Botany.* 1: 197-199.

Kumar P.A., Bisaria S., Rai R.A. & Sharma R.P. (1995) Comparative shoot regeneration in different genotypes of chickpea, *Cicer arietinum* L. *Indian Journal of Experimental Biology.* 33: 77-78.

Kumar V.D., Kirti P.B., Sachan J.K.S. & Chopra V.L. (1994) Plant regeneration via somatic embryogenesis in chickpea (*Cicer arietinum* L.). *Plant Cell Reports.* 13: 486-472.

Murashige T. & Skoog F. (1962) A revised medium for rapid growth and bioassays for tobacco tissue cultures. *Physiol. Plantarum.* 15: 473-497.

Murthy B.N.S., Victor J., Singh R.P., Flectcher R.A. & Saxena P.K. (1996) *In vitro* regeneration of chickpea (*Cicer arietinum* L.). Stimulation of direct organogenesis and somatic embryogenesis by thidiazuron. *Plant Growth Regulation.* 19: 233-240.

Nalini M., Moss J.P. & Mallikarjuna N. (1993) Regeneration from elite chickpea genotypes. *International Chickpea Newsletter.* 29: 4-6.

Rao B.G. & Chopra V.L. (1989) Regeneration in chickpea (*Cicer arietinum* L.) through somatic embryogenesis. *Plant Physiology.* 134: 637-638.

Sagare A.P., Suhasini K. & Krishnamurthy K.V. (1993) Plant regeneration via somatic embryogenesis in chickpea (*Cicer arietinum* L.). *Plant Cell Reports.* 12: 652.

Sagare A.P., Suhasini K. & Krishnamurthy K.V. (1995) Histology of somatic embryo initiation and development in chickpea (*Cicer arietinum* L.). *Plant Science.* 109: 87-93.

Suhasini K., Sagare A.P. & Krishnamurthy K.V. (1994) Direct somatic embryogenesis from mature embryo axes in chickpea (*Cicer arietinum* L.). *Plant Science.* 102: 189-194.

Suhasini K., Sagare A.P., Shankar S.R. & Krishnamurthy K.N. (1997) Comparative study of the development of zygotic and somatic embryos of chickpea (*Cicer arietinum* L.). *Plant Science Limerick.* 128: 207-216.

Vani A.K.S. & Reddy V.D. (1996) Morphogenesis from callus cultures of chickpea *Cicer arietinum. Indian J. of Exp. Bio*. 34: 285-287.

CHAPTER

8

Polyamines and Long-Term Plant Regeneration

M.V. Rajam

ABSTRACT

The efficient regeneration is important for plant propagation and for producing transgenic crops. Although regeneration procedures are available for several plants, regeneration from long-term cultures is still a problem in many plants, especially cereals which exhibit loss of regeneration after few sub-cultures. Several reasons have been given for regeneration failure, including change in osmotic potential and ploidy level or related chromosomal aberrations. Regeneration from such long-term cultures could be revived by modulating the osmotic potential and polyamine metabolism as well as the culture medium. The underlying mechanism for loss of regeneration appears to be complex as it is the manifestation of changes at the cellular and molecular level. It is suggested that failure of regeneration could be due to the silencing of plant regeneration associated genes as a result of DNA methylation and such genes may be getting re-activated by the changes in polyamine pools. Thus, genetic modification of polyamine metabolism in such long-term cultures may be a novel approach for the restoration of regeneration.

INTRODUCTION

Efficient plant regeneration via organogenesis or somatic embryogenesis is very important for plant multiplication as well as for the development of transgenic crops. In fact, good regeneration procedures have been developed for a large number of plant species. However, regeneration from long-term cell and tissue cultures is still a problem in many plants, although regeneration was achieved in such long-term cultures of a few plant systems (Freeling *et al.*, 1976; Stavarek *et al.*, 1980; Heyser *et al.*, 1983; Nabor *et al.*, 1983; Hussey & Gunn, 1984; Raghava Ram & Nabors, 1984; Vasil *et al.*, 1984, Kamo & Hodges, 1986; Kavi Kishor & Reddy, 1986a; Qiao *et al.*, 1992; Perl *et al.*, 1995; Cailloux *et al.*, 1996; Kothari & Varshney, 1998; Chakravarty & Goswami, 1999; Rubluo *et al.*, 2002). There are various reasons for loss of regeneration in such long-term cultures.

*Address for correspondence: Plant Polyamine and Transgenic Research Laboratory, Department of Genetics, University of Delhi, South Campus, Benito Juarez Road, New Delhi 110021, India
E-mail: mv_rajam@hotmail.com; Fax: 091-011-24112437.

This review article was planned to discuss the state-of-art of the long-term plant regeneration with special emphasis on the role of polyamines.

WHY LONG-TERM CULTURES LOSE PLANT REGENERATION CAPACITY?

A rapid decrease in morphogenetic ability with age in culture still remains a serious limitation for regeneration in several plants, especially cereals (Bajaj & Rajam, 1996). Callus or cell suspension cultures often show the loss of morphogenetic potential after a few sub-cultures (Bajaj & Rajam 1995, 1996). Loss of regeneration capacity has been reported in a number of plants like rice, maize, Oats, Sorghum, Arabidopsis, *Heavea brasilensis*, Brassica, etc. (Negrutiu & Jacobs, 1978; Chandler & Vasil 1984; Binh & Heszky 1990; Beaumont & Widholm 1993; Pius *et al.*, 1993, Asano *et al.*, 1994; Bajaj & Rajam, 1995; 1996; Gasper, 1995; Bajaj, 1996; Shoeb, 1999), Several explanations have been given for regeneration loss in such long-term cultures. For instance, the dominance of non-embryogenic cells with increasing sub-cultures (Vasil, 1987), *in vitro* cultured induced variation or mutation, which increase with the increase in time in culture (Muller *et al.*, 1990), abnormal growth rate (Gasper, 1995; Shoeb, 1999), disappearance of a specific factor contained in the primary explant (Gautheret 1966), aneuploidy (Moyne *et al.*, 1993, Henry *et al.*, 1996), polyploidization or related chromosomal abnormalities (Sheridan, *et al.*, 1975, Beaumont and Widholm 1993, Moyne *et al.*, 1993, Fukuoka *et al.*, 1994), polyamine accumulation and the high putrescine to spermidine ratio (Bajaj & Rajam, 1995, 1996; Bajaj, 1996; Shoeb, 1999), altered ethylene levels (Shoeb, 1999), gene silencing (Shoeb, 1999) and neoplastic features (Hasler *et al.*, 2003). There might be some other physiological factors, which are unknown (Brisbie *et al.*, 1994). Chromosomal localization of genes controlling plant regeneration in long-term cultures in wheat has also been reported (Hendry *et al.*, 1994).

Rice long-term (72- and 84-month old) callus cultures exhibited abnormal high growth rate, hyperhydricity and hormone habituation (Shoeb, 1999). Previously, such observations were also made in some dicots (Gasper, 1995). Genetic differences between fresh and long-term rice callus cultures were also examined by checking polypeptide and RAPD profiles. A clear difference between protein profiles of fresh and long-term cultures were obtained. Some proteins (~ 100 and 36 kDa) were completely lost and some proteins (~ 38 kDa) were over-expressed in long-term rice callus cultures as compared to fresh cultures (Shoeb, 1999). Moreover, RAPD analysis also exhibited some polymorphism between fresh and long-term rice callus cultures (Shoeb, 1999).

Restoration of Plant Regeneration in Long-Term Cultures

The revival of plant regeneration in long-term cultures has been reported by using various strategies such as changing the osmolarity of the culture medium (Kavi Kishor & Reddy, 1986b; Kavi Kishor 1987; Asano *et al.*, 1994), reduction of sub-culture interval and modification of growth and regeneration medium (Heyser *et al.*, 1983; Siriwardana & Nabors, 1983; Ozawa & Komamine, 1989, Ghosh-Biswas & Zapata, 1992; Brisbie *et al.*, 1994; Chaturvedi & Jain, 1994; Binh *et al.*, 1995, Etienne *et al.*, 1997, Lambe *et al.*, 1999), salt pre-treatment (Binh & Heszky, 1990, Binh *et al.*, 1992), the addition of polyamine, spermidine or specific inhibitor of polyamine biosynthesis, difluoromethylarginine (Bajaj & Rajam 1995, 1996), addition of Ca^{2+} ($CaCl_2$) (Etienne *et al.* 1997) and treatment with plant growth regulators (i.e. 2,3,5-triiodobenzoic acid in combination with benzyladenine or thidiazuron) (Nakano *et al.* 2000). In spite of these studies, the biochemical and molecular mechanisms of the loss of regeneration potential in long-term cultures is poorly understood.

Revival of Plant Regeneration in Long-Term Rice Callus Cultures by Modulation of Polyamine Metabolism

Polyamines, putrescine, spermidine and spermine are naturally occurring polycationic compounds that are present ubiquitously in all living cells. These small molecules play a vital role in a variety of biological processes, including the regulation of the function of macromolecules (DNA, RNA and proteins), membrane stability, growth and development, biotic and abiotic stress responses and *in vitro* morphogenetic responses (e.g. organogenesis and somatic embryogenesis) (Rajam, 1997; Rajam *et al.,* 1998a, b). It has also been suggested that these new growth regulators (Galston & Kaur-Sawhney, 1990) may be involved in the signal transduction pathways, which may be one of the key mechanisms of mode of polyamine actions in several cellular functions (Kumar, 2003).

Since polyamines have been involved in plant regeneration (Minocha & Minocha, 1995; Yadav & Rajam, 1998; Rajam *et al.,* 1998b) and ageing (Galston and Kaur-Sawhney, 1990; Rajam, 1997), the possible association of polyamine metabolism in the loss of morphogenetic potential in long-term rice (*Oryza sativa* L., cv. TN-1) callus cultures has been studied (Bajaj & Rajam, 1995, 1996; Bajaj, 1996; Shoeb, 1999).

In rice, the regeneration frequency as well as number of plantlets/embryogenic callus culture decreases gradually up to 3 years, and then it results in the complete loss of regeneration potential (Asano *et al.,* 1994; Bajaj & Rajam, 1995, 1996). Interestingly, a massive accumulation of polyamines, particularly the diamine putrescine (5-fold in 12-month old callus cultures and 26-fold in 20 month old callus cultures) was observed in long-term callus cultures of rice, resulting in a change of putrescine to spermidine ratio, which seems to be important for maintaining the morphogenetic response (Faure *et al.,* 1991; Shoeb *et al.,* 2001). Further, the exogenous application of spermidine to long-term rice callus cultures showed increased levels of spermidine, and restored the putrescine/spermidine ratio comparable to that found in freshly induced callus cultures, concomitantly promoting the plant regeneration in 12-month old cultures (Bajaj & Rajam, 1995), and restoration of regeneration in long-term (20- and 36-month old) callus cultures of rice (Bajaj & Rajam, 1996). The accumulation of polyamines in long-term rice cultures was due to the increase in putrescine synthesis enzyme arginine decarboxylase (ADC). The blockage of polyamine accumulation due to the reduction in ADC activity by a specific and irreversible inhibitor of ADC difluoromethylarginine (DFMA), completely restored plant regeneration capacity in these long-term rice callus cultures (Bajaj & Rajam, 1996). Similar observations were also found in 72- and 84-month old callus cultures of rice (Shoeb, 1999).

Since ethylene and polyamine biosynthesis share S-adenomethionine (SAM) as a common intermediate, an attempt was also made to examine the effect of treatment of such long-term rice callus cultures with $AgNO_3$ for one month, which resulted initially in the green dot formation and then complete plantlets upon sub-culture, while $AgNO_3$ treatment caused a drastic reduction in regeneration or lethal effects in fresh callus cultures (Shoeb, 1999). This observation suggests that long-term cultures may be producing more ethylene, which might be responsible for significant reduction or loss of regeneration as high ethylene levels can trigger senescence, and since $AgNO_3$ interfer with ethylene action, it must have resulted in the revival of regeneration capacity. Further, the restoration of regeneration by spermidine may be due to its involvement in the feedback inhibition of ethylene forming enzymes (Mehta *et al.,* 2002).

The effect of azacytidine (DNA methylation inhibitor) on the long-term regeneration in rice was also tested with the idea that the loss of regeneration could be due to the silencing of some genes that are associated with *in vitro* morphogenesis, and also the involvement of polyamines in the modulation of DNA methylation (Heby, 1995). Indeed, azacytidine treatment caused the restoration of plant regeneration in 84-month old callus cultures, supporting for the notion of gene silencing in long-term cultures due to DNA methylation (Shoeb, 1999).

Since long-term rice callus cultures exhibited polyamine accumulation and as polyamine accumulation is closely associated with stress responses (Rajam, 1997; Rajam *et al.,* 1998a, b), the long-term (72- and 84-month old) rice callus cultures were utilized to test tolerance against some abiotic stresses. In fact, such long-term callus cultures did show increased tolerance to salinity, PEG-mediated drough and heavy-metal cadmium (which was also accompanied by the restoration of regeneration). In addition, these cultures also showed tolerance to heat and cold stresses. These results substantiate the importance of polyamine accumulation for abiotic stress tolerance in transgenic tobacco over-expressing polyamine biosynthesis genes ornithine decarboxylase (Kumria & Rajam, 2002) and SAM decarboxylase (Waie & Rajam, 2003).

Concluding Remarks

It has been observed that plant regeneration in many plants, especially cereals is lost after few transfers, and the underlying mechanism of such regeneration failure appears to be complex as evidenced in the literature and it is due to the manifestation of changes at the physiological, biochemical and molecular level. One of the main reasons for loss of regeneration in long-term cultures seems to be the dramatic alteration of osmotic potential and that is the reason that osmolytes like mannitol, sorbitol, polyamines, etc. can restore regeneration from such long-term cultures. Further, polyamines seem to be playing a crucial role in the regulation of long-term plant regeneration as they are involved in the regulation of a variety of cellular and molecular processes, including DNA methylation and gene expression. At the molecular level, failure of regeneration appears mainly due to the silencing of plant regeneration related genes (may be because of altered DNA methylation in such cultures) and such genes might be getting re-activated by the adjustment of cellular polyamine pools and putrescine to spermidine ratio in long-term cultures. Although the regeneration could be restored in long-term cultures by a variety of approaches, an elucidation of cellular and molecular mechanism is important for the improvement of regeneration ability as it facilitates the transfer of useful genes using long-term culture systems.

Expression of polyamine biosynthesis genes [i.e. sense *SAMDC* or antisense *ADC* or sense diamine oxidase (putrescine catabolic enzyme) gene, which would optimise the altered ratio between putrescine and spermidine in the transformed long-term cultures or other genes associated with plant regeneration would be a novel approach for achieving regeneration from long-term cultures. In fact, the enhancement of regeneration of rice calli was achieved by the integration of the gene (*Os22A*) involved in regeneration ability of the callus. Interestingly, this *Os22A* gene has high homology to a barley gene coding for glucose dehydrogenase (Ozawa *et al.*, 2003), substantiating the idea that osmolytes play an important role in the improvement/restoration of regeneration in the long-term cultures.

Acknowledgements

Research work in my laboratory has been generously supported by grants from the Department of Biotechnololgy and Department of Science and Technology, New Delhi. I thank Dr. S. Bajaj, Dr. F. Shoeb and Mr. S. V. Kumar for their help.

REFERENCES

Asano Y., ITO Y., Ohara M., Sugiura K. & Fujiie A. (1994) Improved regeneration response of creeping bentgrass and japonica rice by maltose and lactose, *Plant Cell Tissu. Org. Cult.* 39: 101-103.

Bajaj S. (1996) Role of polyamines in *in vitro* plant regeneration and stress responses in rice (*Oryza sativa* L.), Ph. D Thesis, University of Delhi, Delhi.

Bajaj S. & Rajam M.V. (1995) Efficient plant regeneration from long-term callus cultures of rice by spermidine, *Plant Cell Rep.* 14: 717-720.

Bajaj S. & Rajam M.V. (1996) Polyamine accumulation and near loss of morphogenesis in long-term callus cultures of rice: restoration of plant regeneration by manipulation of cellular polyamine levels, *Plant Physiol.* 112: 1343-1348.

Beaumont V.H. & Widholm J.M. (1993) Ploidy variation of pronamide-treated maize calli during long term culture, *Plant Cell Rep.* 12: 648-651.

Binh D.Q. & Heszky L.E. (1990) Restoration of the regeneration potential of long-term cell culture in rice (*Oryza sativa* L.) by salt pre-treatment, *Plant Physiol.* 136: 336-340.

Binh D.Q., Heszky L.E., Gyulai G. & Csillag A. (1992) Plant regeneration of NaCl-pretreated cells from long-term suspension culture of rice (*Oryza sativa* L.) in high saline conditions, *Plant Cell Tissu. Org. Cul.* 29: 75-82

Binh D.Q., Kiss P., Heszky L.E., Kiss E. & Asboth B. (1995) Some specific responses and requirements during growth and plant redifferentiation of long-term salt adapted rice cells. *Plant Physiol.* 14: 1-10.

Brisbie E.A., Miyake H., Taniguchi T. & Maeda E. (1994) Regulation of somatic embryogenesis in long-term callus cultures of sugarcane (*Saccharum officinarum* L.), *New Phytol.* 126: 301-307.

Cailloux F., Guerrier J.J., Linossier L. & Coudret A. (1996) Long-term somatic embryogenesis and maturation of somatic embryo in *Hevea brasiliensis, Plant Sci.* 120: 185-196.

Chakravarty B. & Goswami B.C. (1999) Plantlet regeneration from long-term callus cultures of *Citrus acida* Roxb. and the uniformity of regenerated plants, *Sci. Horti.* 82: 159-169.

Chandler S.F. & Vasil I.K. (1984) Optimization of plant regeneration from long-term embryogenic callus cultures of *Pennisetum purpureum* Schum (Napier grass), *J. Plant Physiol.* 117: 147-156.

Chaturvedi H.C. & Jain M. (1994) Restoration of regeneration potential in prolonged culture of *Digitalis purpurea. Plant Cell Tissu. Org. Cul.* 38: 73-75.

Etienne H., Lartaud M., Carron M.P. & Michaux F.N. (1997) Use of calcium to optimise long-term proliferation of friable embryogenic calluses and plant regeneration in *Hevea brasiliensis* (Mull. Arg.), *J. Exp. Bot.* 48: 129-137.

Faure O., Mengoli M., Nougarede A. & Bagni N. (1991) Polyamine pattern and biosynthesis in zygotic and somatic embryo stages of *Vitis vinifera, J. Plant Physiol.* 138: 389-549.

Freeling M., Woodman J.C. & Cheng D.S.K. (1976) Developmental potentials of maize tissue cultures, *Maydica.* 21: 97-112.

Fukuoka H., Kawata M. & Takaiwa F. (1994) Molecular changes of organelle DNA sequences in rice through dedifferentiation, long-term culture, or the morphogenesis process, *Plant Mol. Biol.* 26: 899-907.

Galston A.W. & Kaur-Sawhney R. (1990) Polyamines in plant physiology, *Plant Physiol.* 94: 406-410.

Gasper T.H. (1995) The concept of cancer in *in vitro* plant cultures and the implication of habituation to hormones and hyperhydricity, *Plant Tissu. Cul. Biotech.* 1: 126-136.

Gautheret R.J. (1966) In *Cell Differentiation and Morphogenesis* (ed. Beerman W.), North Holland, Amsterdam, pp 55-95.

Ghosh Biswas G.C. & Zapata F.J. (1992) Plant regeneration from long-term cell suspension cultures of Indica Rice (*Oryza sativa* L. cv. IR43), *J. Plant Physiol.* 139: 523-527.

Hasler J., Wuest J., Gaspar T. & Crevecoeur M. (2003) Long term *in vitro*-cultured plant cells show typical neoplastic features at the cytological level, *Biol. Cell.* 95: 357-364.

Heby O. (1995) DNA methylation and polyamines in embryonic development and cancer, *Int. J. Dev. Biol.* 39: 737-757.

Henry Y., Marcotte J.L. & De Burser J. (1994) Chromosomal location of genes controlling short-term and long-term somatic embryogenesis in wheat revealed by immature embryo culture of aneuploid lines, *Theor. Appl. Genet.* 89: 344-350.

Henry Y., Marcotte J.-L. & Buyser J. (1996) The effect of aneuploidy on karyotype abnormalities in wheat plants regenerated from short and long term somatic embryogenesis, *Plant Sci.* 114: 101-109.

Heyser J.W., Dykes T.A., Demott K.J. & Nabors M.W. (1983) High frequency long term regeneration of rice from callus culture, *Plant Sci. Lett.* 29: 175-182.

Hussey G. & Gunn H.V. (1984) Plant production in pea (*Pisum sativum* L. cvs. Puget and Upton) from long-term callus with superficial meristems, *Plant Sci. Lett.* 37: 143-148.

Kamo K.K. & Hodges T.K. (1986) Establishment and characterization of long-term embryogenic maize callus and cell suspension cultures, *Plant Sci.* 45: 111-117.

Kavi Kishor P.B. (1987) Energy and osmotic requirement for high frequency regeneration of rice plants from long-term cultures, *Plant Sci.* 48: 189-194.

Kavi Kishor P.B. & Reddy G.M. (1986a) Regeneration of plants from long-term cultures of *Oryza sativa* L, *Plant Cell Rep.* 5: 391-393.

Kavi Kishor P.B. & Reddy G.M. (1986b) Retention and revival of regenerating ability by osmotic adjustment in long-term cultures of four varieties of rice, *J. Plant Physiol.* 126: 49-54.

Kothari S.L. & Varshney A. (1998) Morphogenesis in long-term maintained immature embryo-derived callus of wheat (*Triticum aestivum* L)—Histological evidence for both somatic embryogenesis and organogenesis, *J. Plant Biochem. & Biotech.* 7: 93-98.

Kumar S.V. (2003) Transgenic manipulation of polyamine biosynthesis in *Solanum melongena* and *Chlamydomonas reinhardtii*. Ph. D Thesis, University of Delhi, Delhi.

Kumria R. & Rajam M.V. (2002) Ornithine decarboxylase transgene in tobacco affects polyamines, *in vitro*-morphogenesis and response to salt stress, *J. Plant Physiol.* 159: 983-990.

Lambe P., Nutambel H.S.N., Deltour R. & Dinant M. (1999) Somatic embryogenesis in pearl millet (*Peunisetium glaucum*): strategies to reduce genotype limitation and to maintain long-term totipotency, *Plant Cell Tissu. Org. Cul.* 55: 23-29.

Mehta R.K., Cassol T., Li N., Ali N., Handa A.K. & Mattoo A.K. (2002) Engineered polyamine accumulation in tomato enhances phytonutrient content, juice quality, and vine life, *Nature Biotech.* 20: 613-618.

Minocha S.C. & Minocha R. (1995) Role of polyamines in somatic embryogenesis, In: *Biotechnology in Agriculture and Forestry*, Vol 30, *Somatic Embryogenesis and Synthetic Seeds* I (Ed Bajaj Y.P.S.), Springer-Verlag, Berlin, pp 53-70.

Moyne A.L., Souq F., Yean L.H., Brown S.C., Boulay M. & Sangwan-Norreel B.S. (1993) Relationship between cell ploidy and regeneration capacity of long-term *Rosa hybrida* cultures, *Plant Sci.* 93: 159-168.

Muller E., Brown P.T.A., Hartke S. & Lorz H. (1990) DNA variation in tissue culture derived rice plants, *Theor. Appl. Genet.* 80: 673-679.

Nabor M.V., Heyser J.W., Dykes T.A. & Demott K.J. (1983) Long duration, high frequency plant regeneration from cereal tissue cultures. *Planta.* 157: 385-391.

Nakano M., Sakakibara T., Suzuki S. & Saito H. (2000) Decrease in the regeneration potential of long-term cell suspension cultures of *Lilium formosanum* Wallace and its restoration by the auxin transport inhibitor, 2,3,5-triiodobenzoic acid, *Plant Sci.* 158: 129-137.

Negrutiu I. & Jacobs M. (1978) Restoration of the morphogenetic capacity in long-term callus cultures of *Arabidopsis thaliana. Z. Pflanzenphysiol. Bd.* 90: 431-441.

Ozawa K. & Komamine A. (1989) Establishment of a system of high-frequency embryogenesis from long-term cell suspension cultures of rice (*Oryza sativa* L.), *Theor. Appl. Genet.* 77: 205-211.

Ozawa K., Kawahigashi H., Kayano T. & Ohkawa Y. (2003) Enhancement of regeneration of rice (*Oryza sativa* L.) calli by integration of the gene involved in regeneration ability of the callus, *Plant Sci.* 165: 395-402.

Perl A., Saad A., Sahar N. & Holland D. (1995) Establishment of long-term embryogenic cultures of seedless *Vitis vinifera* cultivars—a synergistic effect of auxins and the role of abscisic acid, *Plant Sci.* 104: 193-200.

Pius J., George L., Eapen S. & Rao P.S. (1993) Enhanced plant regeneration in pearl millet (*Pennisetum americanum*) by ethylene inhibitors and cefotaxime, *Plant Cell Tissu. Org. Cult.* 32: 91-96.

Qiao Y.-M., Cattaneo M., Coatelli F. & Lupotto E. (1992) Plant regeneration from long term suspension culture-derived protoplasts of hexaploid wheat (*Triticum aestivum* L.), *Plant Cell Rep.* 11: 262-265.

Raghava Ram N.V. & Nabors M.W. (1984) Cytokinin mediated long-term, high-frequency plant regeneration in rice tissue cultures, *Z. Pflanzenphysiol. Bd.* 113: 315-323.

Rajam M.V. (1997) Polyamines, In: *Plant Ecophysiology* (ed. Prasad M.N.V.), Academic Press, New Jersey, pp 343-374.

Rajam M.V., Dagar S., Waie B., Yadav J.S., Kumar P.A., Shoeb F. & Kumria R. (1998a) Genetic engineering of polyamine and carbohydrate metabolism for osmotic stress tolerance in higher plants, *J. Biosci.* 4: 473-482.

Rajam M.V., Shoeb F. & Yadav J.S. (1998b) Polyamines as modulators of plant regneration in tissue cultures, In: *Plant Tissue Culture and Molecular Biology: Applications and Prospects* (ed. Srivastava P.S.), Narosa Publishing House, New Delhi, pp 620-641.

Rubluo A., Marin-Hernandez T., Duval K., Vargas A. & Marquez-Guzman J. (2002) Auxin induced morphogenetic responses in long-term *in vitro* subcultured *Mammillaria san-angelensis* Sanchez-Mejorada (Cactaceae), *Sci. Horti.* 95: 341-349.

Sheridan W.F. (1975) Plant regeneration and chromosome stability in tissue culture, In: *Genetic manipulation with plant material* (Ed. Leldoux L.), Plenum Press, New York, p 263.

Shoeb F. (1999) Regulation of plant regeneration by modulating cellular polyamine levels in fresh and long-term callus cultures of indica rice (*Oryza sativa* L.), Ph. D Thesis, University of Delhi, Delhi.

Shoeb F., Yadav J.S., Bajaj S. & Rajam M.V. (2001) Polyamine as biomarkers for plant regeneration capacity: improvement of regeneration by modulation of polyamine metabolism in different genotypes of indica rice, *Plant Sci.* 160: 1229-1235.

Siriwardana S. & Nabors M.W. (1983) Tryptophan enhancement of somatic embryogenesis in rice, *Plant Physiol.* 73: 142-146.

Stavarek S.J., Croughan T.P. & Rains D.W. (1980) Regeneration of plants from long-term cultures of alfalfa cells, *Plant Sci. Lett.* 19: 253-261.

Vasil I.K. (1987) Developing cell and tissue culture systems for the improvement of cereals and crops, *J. Plant Physiol.* 128: 193-218.

Vasil V., Vasil I.K. & Lu C. (1984) Somatic embryogenesis in long-term callus cultures of *Zea mays* L. (Gramineae), *Amer. J. Bot.* 71: 158-161.

Waie B. & Rajam M.V. (2003) Effect of increased polyamine biosynthesis on stress responses in transgenic tobacco by introduction of human S-adenosylmethionine gene, *Plant Sci.* 164: 727-734.

Yadav J.S. & Rajam M.V. (1998) Temporal regulation of somatic embryogenesis by adjusting cellular polyamine content in eggplant, *Plant Physiol.* 116: 617-625.

CHAPTER

9

Somatic Embryogenesis in Chir Pine (*Pinus roxburghii* Sarg.)

Sarita Arya, Rajwant K. Kalia and I.D. Arya

INTRODUCTION

Pines constitute one of the most divergent and economically important group of species, which provides valuable natural resources and contribute significantly to the local and industrial economy of the country. Pines are the pioneer species and are active colonizers of degraded sites. They also protect the water shades, which sustain and regulate the water supply for the millions who inhabit the Himalayan river basin.

Pinus roxburghii Sarg. commonly known as long leaf pine or chir pine is one of the most important species of India and component of Himalayan ecosystem covering an area of 6728 sq. km. It is found in a fairly continuous belt approximately between the longitude 72°E to 95°E, Pakistan to Arunachal Pradesh including Nepal and Bhutan, and latitude 27°N to 35°N between 450-2300 m above sea level, which covers the ridges of Shivalik hill to outer ranges of Himalayas. It is found on the hill slopes of both Eastern and Western Himalayas.

Chir pine is an important source of timber and resin of high commercial value. All parts of the tree are valuable and are used in one form or the other. It has an important role in hill economy and is used for various purposes including, as rafters, poles and post, doors and windows boards and railway sleepers. Chir pine yields a good quality oleoresin, which on steam distillation generates two industrial important products viz. Turpentine oil (17%) and rosin (70%). Rosin is extensively used in many industries like soap, paper, paints, varnishes, pinolium, sealing, wax, oilcloth, inks and disinfectants. Turpentine is chiefly used in preparation of paints and varnishes, polishes, chemical and pharmaceuticals.

There is tremendous burden on the forests due to population explosion, industrialization and expanding cities. Forests and forestlands are shrinking at an alarming rate. Similarly, chir pine

*Address for correspondence: Plant Tissue Culture Lab., Division of Genetics & Tree Propagation, Forest Research Institute, Dehradun-248006 (Uttaranchal).

resources have dwindled considerably in their natural habit. Many pine forests are shrinking due to minimum conservation and reforestation practices. The currently practiced propagation programs are not adequate to fulfill the demands of planting stock material for afforestation and severe shortage of chir pine is expected shortly. To get supply of chir, it is necessary to raise them on mass scale in plantation. Vegetative propagation of chir pine is beset with many problems. The major conventional methods of pine propagation include seeds and rooting of cutting. The seeds are to be collected from trees but it is beset with problem like height of trees, seasonal dependence of cone opening and air dispersal of seeds. The seeds are met with other problems like loss of seed viability during extraction and storage, seed sterility, poor seed set and consumption of seeds by rodents, birds and animals. Additionally cone crops are often irregular and most pine species require at least 10-20 years to produce a substantial seed yield. Propagation through cutting is the other method used for raising new plantation but regeneration through this method is again beset with many hurdles. The major problem, which limits the development of applied rooted cutting programs, is the reduced growth and rooting potential resulting from maturation of stock plant. It is reported that with increase in mother tree age, the ability of cutting to root is reduced.

There is need to switch over to non-conventional methods of propagation, which hold potential of mass production in a short period of time to achieve the afforestation targets.

Somatic embryogenesis is the potential for rapid *in vitro* multiplication of conifers. Somatic embryos are also an important source of totipotent protoplasts useful for genetic transformation. Somatic embryos are useful for long-term germplasm storage eg. using artificial seeds, cold storage and cryopreservation (Chen and Kartha, 1987; Tautorus *et al.*, 1990). Some work has been done with respect to somatic embryogenesis in chir pine (Arya *et al.*, 2000; Mathur *et al.*, 2000). In the present investigation, the detailed study for condition necessary for induction of somatic embryogenesis and multiplication has been carried out in chir pine.

MATERIALS AND METHODS

Immature Zygotic Embryos

Sterilization of explants

The green cones were cut to dissect out the immature seeds. These seeds were surface sterilized with 0.1% $HgCl_2$ for 20 min. The $HgCl_2$ treatment was succeeded by 3-4 washes with distilled water to remove all traces of sterilizing agent. The megagametophyte containing immature zygotic embryos with 0.5-1.2 mm embryonal heads were cultured in all the experiments.

Mature zygotic embryos

The mature seeds of chir pine were rubbed between hands to remove the wings. These seeds were treated with 0.1% $HgCl_2$ for 20 min followed by rinsing with sterile distilled water. H_2O_2 was used at 3% (w/v) strength for 10min to further sterilize the seeds followed by rinsing with autoclaved distilled water. The megagametophytes were dissected out from seeds and further sterilized with 0.1% $HgCl_2$ for 5 min followed by washing with autoclaved distilled water. Finally the embryos were dissected out and cultured.

Incubation of cultures

The cultures raised were incubated in dark for induction and multiplication at 25°C ± 2°C. For maturation of somatic embryos, the embryogenic callus was exposed to 16h light photoperiod provided by cool fluorescent tubes.

Induction of callus

To assess the effect of auxin concentration and auxin-cytokinin combinations on embryogenic callus induction frequency, pre-cotyledonary zygotic embryo and mature zygotic embryo were cultured on DCR medium supplemented with 2,4-D or NAA (5, 10, 15 or 25 µM) alone or in combination with 2.5, 5.0, and 10 µM BAP.

Four basal media viz. DCR (Gupta and Grob, 1995), MS (Murashige and Skoog, 1962), SH (Schenk and Hildebrandt, 1972) B_5 (Gamborg *et al.*, 1968) supplemented with 10 µM 2,4-D and 5 µM BAP were used to study the response of media composition on embryogenic callus induction frequency from pre-cotyledonary zygotic embryos.

Immature embryos collected from 5 trees selected in Nahan (HP) area were cultured on DCR medium containing 5-25 µM 2,4-D and 5 µM BAP to assess the effect of tree genotype on induction frequency.

Multiplication of embryogenic callus

Two embryogenic lines L1 and L2 were used for the experiments pertaining to multiplication of embryogenic callus. For multiplication experiments the callus were maintained on DCR medium supplemented with 10 µM 2,4-D and 5 µM BAP was used.

Four basal media viz. DCR, MS, SH and LP supplemented with 10 µM 2,4-D/NAA and 5 µM BAP were study the effect of basal medium composition on the rate of embryogenic callus proliferation.

To study the effect of auxin-cytokinin interaction on proliferation of embryogenic callus, 500 mg callus was subcultured to DCR medium supplemented with 5-25 µM 2,4-D/NAA and 5 µM BAP. The embryogenic callus originated from two individuals seeds, maintained on DCR medium supplemented with 10 µM 2,4-D and 5 µM BAP, was routinely examined microscopically to adjudge the effect of genotype on callus morphology and anatomy.

Maturation of somatic embryos

ABA was supplemented in DCR medium at 5, 10, 15, 20, 25 and 50 µM concentration to assess its effect on maturation of chir pine somatic embryos.

For germination, the cotyledonary stage somatic embryos were cultured on DCR medium devoid of growth regulators.

Observation and Data Regarding

In the experiments pertaining to induction, data was recorded on factors viz. percent embryogenic cultures induced after one month and those surviving after three months of subculturing. In the multiplication experiments, an initial amount of 500 mg embryogenic callus was cultured on various media and increase in amount (fresh weight basis) in 3 weeks was recorded. For microscopic examination, callus was viewed under inverted microscope (Leica). The number and size of somatic embryos were recorded along with the nature of cells in callus. All the experiments were repeated twice with a minimum of fifty explants per treatment. Variability in the treatment was calculated using standard deviation.

RESULTS

Magagametophytes containing immature zygotic embryos with 0.5-1.2 mm embryonal head were used as explants for culturing. The fertilized second year female cones were collected at different developmental stages. Embryos 10-12 weeks after fertilization were found best for initiation of embryogeny in *P. roxburghii*. The callus was produced within 4-5 weeks from cut micropylar end of megagametophytes (Fig. 1, 2). The callus were initiated from suspensor cells. The size varied from 2 mm to 4 mm in diameter, which increased in the following 2 weeks to 5-8 mm in diameter. Mature zygotic embryos, when cultured produced non-zygotic embryogenic callus.

In 3 months, culture established, the somatic embryos in the culture consisted of an embryogenic region with cells containing dense cytoplasm and elongated suspensor with high vacuolated cells (Fig. 3). The somatic embryos found in mucilaginous callus closely resembled the somatic embryos of other gymnosperm. In established embryogenic cultures different developmental stages of somatic embryos were observed under stereozoom microscope.

Effect of Phytohormones

Embryos with 0.5-1.2 mm embryonal head, produced embryogenic callus (6.00-19.33%) at all the concentrations of 2,4-D but only 1.33-5.33% embryogenic cultures survived subculturing for 3-4 months. Embryogenic callus induction frequency of 5.33% was recorded on medium containing 10 μM after 3 months of culturing (Table 1). The embryogenic callus produced was transparent to translucent white, mucilaginous and glossy in nature.

A similar trend of embryogenic callus production was observed when the megagametophytes containing zygotic embryos were cultured on DCR medium supplemented with 5-25 μM NAA. Maximum percentage of embryogenic callus (4.67%) was achieved on 10 μM NAA supplemented medium.

Addition of 5-25 μM BAP in DCR medium supplemented with 5-25 μM 2,4-D/NAA increased the percent embryogenic response. Highest callus induction frequency of 9.33% was recorded on DCR medium containing 10 μM 2,4-D in association with 5 μM BAP (Table 1). It also had a beneficial effect on the rate of multiplication of induced embryogenic callus induction. A similar trend was recorded with NAA where percent embryogenic callus induction frequency showed an increase from 4.66% to 6.00% at 10mM strength. Increase in BAP concentration beyond 5 μM led to a drastic decrease in embryogenic response.

Effect of Basal Media

The megagametophytes with zygotic embryos were cultured on four basal media viz. DCR, MS, SH and B_5 supplemented with 10 μM 2,4-D and 5 μM BAP to see the effect of basal medium

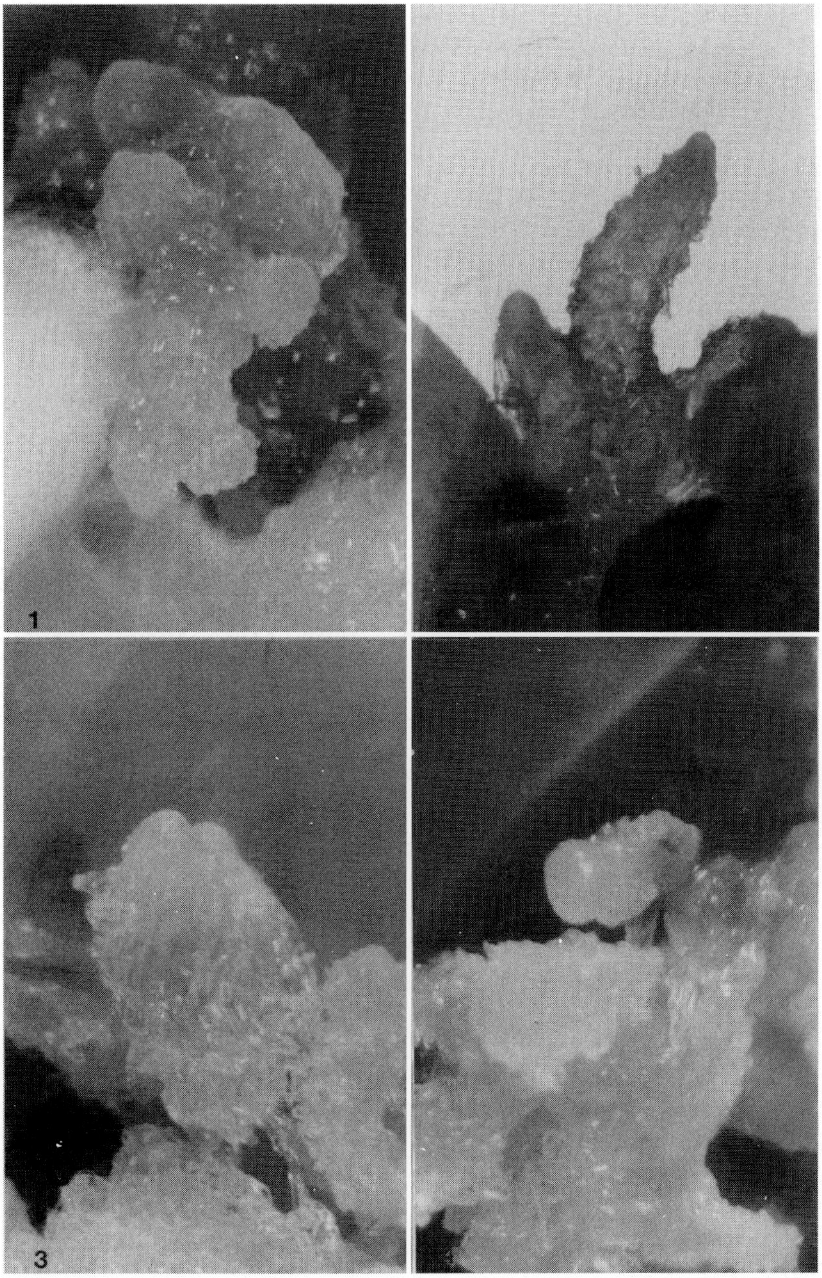

Fig. 1-4 Somatic embryogenesis in chir pine: (1) Embryogenic callus protruding out from micropylar end of megagametophyte after 2 week culture on DCR medium. (2) Somatic embryos developing in the callus protruding out from megagametophyte. (3) Somatic embryos with embryonal head and suspensor in white, translucent mucilaginous embryogenic callus. (4) Stage II somatic embryo with a distinct smooth head.

Table 1 Effect of phytohormones supplemented in DCR medium on induction of embryogenic callus from stage 'c' zygotic embryos of *P. roxburghii.*

S.No.	*Hormones*	*Auxin Conc.(μM)*	*Number of Explants cultured*	*Embryogenic callus induced After 1 month(%)*	*Embryogenic Callus after 3 months (%)*
1	2,4-D	5	150	6.67 ± 1.15	4.00 ± 2.00
2		10	150	19.33 ± 3.06	5.33 ± 1.15
3		15	150	10.00 ± 2.00	4.67 ± 1.15
4		20	150	8.67 ± 3.06	3.33 ± 2.31
5		25	150	6.00 ± 0.00	1.33 ± 1.15
6	NAA	5	150	14.67 ± 2.31	4.00 ± 0.00
7		10	150	17.33 ± 3.06	4.67 ± 1.15
8		15	150	16.00 ± 5.29	4.00 ± 2.00
9		20	150	14.67 ± 4.16	3.33 ± 3.06
10		25	150	11.33 ± 3.06	2.00 ± 0.00
11	2,4-D+BAP*	5	150	34.00 ± 7.21	8.67 ± 2.31
12		10	150	43.33 ± 43.33	9.33 ± 3.06
13		15	150	24.00 ± 4.00	8.67 ± 3.05
14		20	150	41.33 ± 6.43	8.00 ± 2.00
15		25	150	45.33 ± 7.02	6.00 ± 2.00
16	NAA+BAP*	5	150	12.67 ± 6.43	3.33 ± 2.31
17		10	150	25.33 ± 6.11	6.00 ± 4.00
18		15	150	16.67 ± 4.16	4.67 ± 3.06
19		20	150	16.67 ± 2.31	2.67 ± 1.15
20		25	150	29.33 ± 4.16	2.00 ± 0.00

- BAP was added at 5mM concentration.

± Values represent the standard deviation.

constitution on embryogenic callus induction frequency (Table 2). It was found that embryos produced cultures on all the media tried with a maximum frequency of 8.0% on DCR medium followed by 5.33% on MS medium and only 2.0% in SH and B_5 media.

Table 2 Effect of different media supplemented with 10 μM 2,4-D and 5 μM BAP on induction of embryogenic callus from stage 'c' zygotic embryos of *P. roxburghii.*

S.No.	*Basal medium*	*Number of Explants cultured*	*Embryogenic callus induced after 1 month (%)*	*Embryogenic callus after 3 month (%)*
1	DCR	150	27.33 ± 4.16	8.00 ± 2.00
2	B_5	150	11.33 ± 3.06	2.00 ± 0.00
3	MS	150	29.33 ± 4.16	5.33 ± 1.15
4	SH	150	24.00 ± 4.00	2.00 ± 0.00

± Values represent the standard deviation

Effect of Tree Genotypes

The embryogenic callus induction frequency from each tree sample showed wide variation. Moreover, the embryogenic calli derived from different genotypes varied in their texture, microscopic nature as well as their cultural requirements. The megagametophytes collected from five trees when cultured on DCR medium supplemented with different concentration of 2,4-D (5-25 µM) alongwith 5.0 µM BAP resulted in a variable response. Tree A2 produced a maximum of 18.75% embryogenesis on 20 µM 2,4-D supplemented medium while tree A5 and A4 responded most at 5 µM (15.79%) and 10 µM 2,4-D (10%) concentration respectively. Tree A2 was found to produce embryogenic callus at all the five concentration tested while trees A5 and A4 produced embryogenesis at only three and two concentrations of 2,4-D respectively (Table 3). Trees A1 and A3 produced no embryogenic cultures, which could be maintained beyond three months in culture.

Table 3 Effect of tree genotype on induction of embryogenic callus from stage 'c' zygotic embryos of *P. roxburghii* when cultured on DCR medium supplemented with 2,4-D and 5 µM BAP.

S.No.	*2,4-D conc. (mM)*	*Tree genotype*	*Embryogenic callus induced after 1 month (%)*	*Embryogenic callus surviving after 3 months (%)*
1.	5.0	A1	0.00	0.00
2.		A2	62.50	12.50
3.		A3	0.00	0.00
4.		A4	8.33	0.00
5.		A5	31.58	15.79
6.	10.0	A1	0.00	0.00
7.		A2	87.50	12.50
8.		A3	11.11	0.00
9.		A4	16.671	0.00
10.		A5	22.22	0.00
11.	15.0	A1	16.67	0.00
12.		A2	43.75	6.25
13.		A3	0.00	0.00
14.		A4	0.00	0.00
15.		A5	27.78	0.00
16.	20.0	A1	0.00	0.00
17.		A2	68.75	18.75
18.		A3	0.00	0.00
19.		A4	8.33	8.33
20.		A5	55.56	5.56
21.	25.0	A1	0.00	0.00
22.		A2	43.75	12.50
23.		A3	0.00	0.00
24.		A4	0.00	0.00
25.		A5	26.32	5.26

Multiplication of Embryogenic Callus

Two embryogenic lines, one raised from A2 (line L1) and another from A5 (Line L2) were selected for callus multiplication studies. These lines were selected due to their potential to proliferate at a faster rate and presence of more somatic embryos per gram fresh weight compared to other callus lines. Effect of different basal media viz. DCR, MS, SH and LP supplemented with 10 μM 2,4-D/NAA and 5 μM BAP was studied on multiplication of embryogenic callus. The results are presented in Table 4. It was found that line L1 had maximum proliferation of 4.06-fold on SH medium supplemented with 2,4-D with BAP followed by DCR medium (3.47-fold). In line L2, 3.23-fold multiplication was obtained on DCR medium supplemented with 10 μM 2,4-D with 5mM BAP. It was observed that continuous cultures on these media led to variability in the multiplication rate of callus lines. The cultures maintained on DCR medium were vigorous, healthy and maintained a consistent rate of multiplication. Prolonged culturing on SH medium on the other hand reduced the multiplication rate of callus lines and also initiated the conversion of glossy callus into sticky one. So, DCR medium was used for further studies.

Table 4 Effect of different media supplemented with 10mM 2,4-D/NAA and 5mM BAP on multiplication of embryogenic callus of *P. roxburghii*. Data recorded after 25 days.

S.No	*Auxin*	*Basal medium*	*Final Weight of callus (g)*	*Fold increase*	*Final Weight of callus(g)*	*Fold increase*
			LINE L1		**LINE L2**	
1	2,4-D	MS	1.36 ± 0.19	2.72	0.94 ± 0.22	1.88
2		SH	1.69 ± 0.23	3.39	1.09 ± 0.17	2.18
3		LP	1.37 ± 0.25	2.74	1.06 ± 0.23	2.14
4		DCR	1.73 ± 0.27	3.47	1.61 ± 0.15	3.23
5	NAA	MS	1.17 ± 0.20	2.35	1.12 ± 0.16	2.25
6		SH	2.03 ± 0.25	4.06	1.08 ± 0.23	2.18
7		LP	1.66 ± 0.26	3.32	1.13 ± 0.27	2.27
8		DCR	1.67 ± 0.26	3.35	1.15 ± 0.20	2.31

± Values represent the standard deviation.
Initial amount of 0.50g was cultured on all the media.

Effect of Auxin-Cytokinin Interaction

The embryogenic lines L1 and L2 were cultured on DCR medium supplemented with 5-25 μM 2,4-D/NAA and 5 μM BAP for proliferation. Line L1 produced 4.56-fold increase in callus amount in 25 days on medium (Table 5) supplemented with 10 μM 2,4-D. Increase or decrease in 2,4-D content in the medium led to a decline in rate of multiplication (Table). In line L2 maximum proliferation rate of 4.00-fold was achieved on 5 μM 2,4-D supplemented medium followed by 10 μM 2,4-D (3.66-fold). Both the lines were found to produce better response on 2,4-D supplemented DCR medium. During proliferation, line L1 was shiny white mucilaginous while line L2 was off-white.

Table 5 Effect of 2,4-D/NAA with 5mM BAP supplemented in DCR medium on multiplication of embryogenic callus of *P. roxburghii*. Data recorded after 25 days.

S.No	*Auxin*	*Auxin Conc. (μM)*	*Final Weight of Callus(g)*	*Fold increase*	*Final Weight of Callus(g)*	*Fold increase*
			LINE L1		**LINE L2**	
1	2,4-D	5.0	1.96 ± 0.28	3.94	2.00 ± 0.28	4.00
2		10.0	2.28 ± 0.24	4.56	1.83 ± 0.24	3.66
3		15.0	1.22 ± 0.21	2.44	1.51 ± 0.10	3.02
4		20.0	1.89 ± 0.39	3.77	1.30 ± 0.11	2.60
5		25.0	1.39 ± 0.19	2.79	1.26 ± 0.16	2.52
6	NAA	5.0	1.27 ± 0.29	2.54	1.21 ± 0.15	2.44
7		10.0	1.46 ± 0.19	2.92	1.72 ± 0.22	3.45
8		15.0	1.42 ± 0.13	2.84	1.68 ± 0.29	3.36
9		20.0	0.97 ± 0.14	1.94	1.24 ± 0.15	2.49
10		25.0	0.83 ± 0.16	1.66	0.91 ± 0.20	1.82

± Values represent the standard deviation.
Initial amount of 0.50 g was cultured on al the media.

Effect of Genotype

The two embryogenic lines raised from immature seeds of different trees were observed under inverted microscope to study their nature. Both the lines were different in their morphological appearance as well as microscopic nature. Line L1 was shiny white, mucilaginous consisted of pro-embryos and somatic embryos of different size. Line 2 was off-white, mucilaginous and consisted of somatic embryos intermingled with clusters of spherical and elongated cells.

In order to promote maturation of somatic embryos, ABA was added to the medium. Lower concentration of ABA (5-15 μM) predominantly supported multiplication of callus with slight increase in embryo size. Concentrations of 20 and 25 μM ABA neither supported proliferation nor maturation of somatic embryos. Further increase in ABA concentration to 50 μM led to development of few stage II somatic embryos in the callus. Stage II embryos consist of an embryonic region that becomes more opaque with a smooth glossy surface and is attached to a long translucent suspensor (Fig. 4).

These embryos become green when transferred in light condition (Figs. 5, 6). On hormone-free medium the cotyledonary stage embryos germinated into plantlets. These plants were transferred in soilrite and kept in mist chamber and shade house for hardening and acclimatization of the plant (Fig. 7).

DISCUSSION

Somatic embryogenesis provides an ideal experimental system for investigating the process of plant differentiation as well as the mechanisms of expression of totipotency in plant cells. It is a process analogous to zygotic embryogenesis, but one in which a single cell or a small group of vegetative

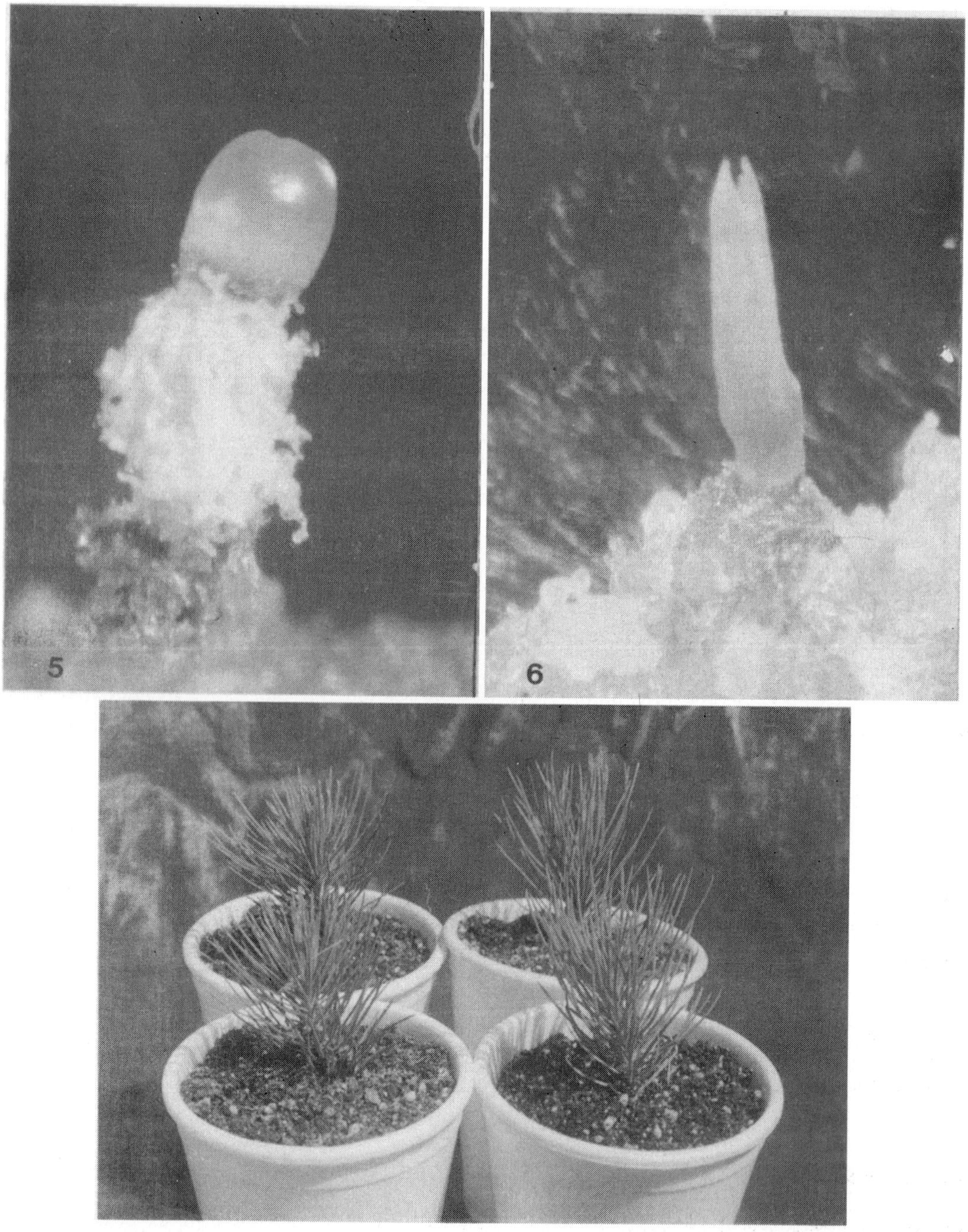

Fig. 5-7 Maturation of somatic embryos: (5) Stage II somatic embryo turned green. (6) Differentiation of cotyledons in developing in somatic embryo. (7) *In-vitro* regenerated plantlets.

i.e. somatic cells are the precursors of embryos. Dunstan (1988), Wann (1988) and Gupta *et al.* (1991) have emphasized the potential of somatic embryogenesis as a method for rapid *in vitro* multiplication of conifers. Somatic embryos are also important source of totipotent protoplasts useful for genetic transformation. Somatic embryos are useful for long term germplasm storage and cryopreservation (Chen and Kartha, 1987; Tautorus *et al.*, 1990). Understanding of the stimuli and conditions necessary for induction of somatic embryogenesis is limited in general and especially for *Pinus roxburghii.* So, the present investigation was carried out to study the same for chir pine. Proper explant selection is critical to achieve successful induction of somatic embryogenesis in conifers. This is important because various tissue of the same plant or tissues at various developmental stages can differ in their response when cultured *in vitro* (Robert *et al.*, 1989; Arya *et al.*, 2000). It was found that only immature zygotic embryos responded towards somatic embryogenesis while all other explant could not produce any embryogenic response. The mature zygotic embryos produced non-embryogenic callus. Thus, age is a critical factor on which activity of the explant depends considerably. The loss of competence of explant has been attributed to anatomical changes, which takes place with maturity of the explants. This decrease is due to progressive specialization of the tissue, which reduces its plasticity and capability of the cells to differentiate.

The auxins, 2,4-D and NAA are generally preferred for initiation of embryogenic calli in most of conifer species (Gupta *et al.*, 1991; Tautorus *et al.*, 1990). However, no difference has yet been reported in embryogenic callus proliferation or development in conifers with 2,4-D as sole auxin source. Study of effect of auxin and auxin-cytokinin combinations supplemented in the medium indicated lower (10-15 μM) auxin strength was sufficient for induction of higher percentages of embryogenesis compared to higher (25 μM) ones in chir pine. The balance between auxin and cytokinin was more beneficial than the absolute concentration of auxin. Earlier reports of Becwar *et al.* (1990) also validate the observations where 2,4-D at 3.0 mg/l level was reported superior to 10 mg/l with the same BA level for initiation of embryogenic cultures using DCR medium.

In the present study comparison of four basal media viz. DCR, MS, SH and B5 supplemented with 10 μM 2,4-D and 5 μM BAP indicated that DCR medium supported maximum induction of embryogenic calli with lesser number of megagametophyte browning.

The study and comparison of induction frequencies from explant collected from different trees demonstrated that the genotype had a bearing on the induction frequency. Embryos from tree A2 were most responsive towards embryogenesis, producing embryogenic calli at all the concentrations used while tree A5 and A4 were responsive to a lesser extent possibly a result of open-pollination and hence genotypic as also reported by Jones and van Staden (1995). The embryogenic calli derived from different genotypes differ in their requirements for proliferation. Variation in performance of half-sib material indicated that small genetic changes could lead to enormous change in embryogenic potential. Jain *et al.* (1989) also reported the existence of genotypic variation in *Pinus elliottii.* Becwar *et al.* (1990) also reported differences in the induction frequency from immature zygotic embryo explants of 10 clones of loblolly pine

The nutritional requirement of different genotypes vary when maintained for multiplication. In our studies line L1 showed optimum multiplication on SH medium while line L2 had maximum multiplication on DCR medium. It was found that cultures maintained on SH medium for longer durations had retarded rates of multiplication but those maintained on DCR medium maintained a consistent rate of multiplication. This difference can be attributed to higher concentrations of nitrate in SH medium, which accumulates over a period of time and hampers the callus growth.

Differences in the hormonal requirements were observed among the embryogenic lines originating from different seeds. Anatomically also these lines were different. Laine and David (1990) reported similar differences in nutritional requirements and anatomical structure of 9 different embryogenic callus line of *Pinus caribaea.* The basal salt medium composition supporting optimal callus proliferation also varies according to the genotype of the embryogenic line. Similar differences were also indicated by Jain *et al.* (1989) when they tested 4 genotypes of *Pinus elliottii.*

Development, maturation and germination of somatic embryos into emblings is by and large a problem and remains the limiting factor in regeneration through somatic embryogenesis (Gupta and Grob, 1995). For development and maturation of embryos, ABA play an important role. It is suggested that favorable effect of ABA may be due to increase in storage reserves, such as storage proteins (Gupta and Pullman, 1991).

In this study, initiation and establishment of embryogenic cultures and in- vitro plantlet development of *Pinus roxburghii* from pre-cotyledonary stage embryos has been achieved.

REFERENCES

Arya S., Kalia R.K. & Arya I.D. (2000) Induction of somatic embryogenesis in *Pinus roxburghii* Sarg. *Plant Cell Rep.* 19: 775-780.

Becwar M.R., Nagmani R. & Wann S.R. (1990) Initiation of embyrogenic culture and somatic embryo development in loblolly pine (*Pinus taeda*). *Can. J. For. Res.* 20: 810-817.

Chen T.H.H. & Kartha K.K. (1987). Cryopreservation of woody species. In: Bonga J.M. and Durzan D.J. (Eds.). *Cell and tissue culture in forestry.* Vol. 2. Martinus Nijhoff Publishers, Dordrecht, The Netherlands. pp. 305-319.

Dustan D.I. (1988) Prospects and progress in conifer biotechnology. *Can. J. For. Res.* 18: 1497-1506.

Gamborg O.L., Miller R.A. & Ojima K. (1968). Nutrient requirements of suspension cultures of soyabean root cells. *Res.* 50: 151-158.

Gupta P.K. & Grob J.A. (1995). Somatic embryogenesis in conifers. In: Jain S., Gupta P.K. & Newton R.J. (Eds.). *Somatic embryogenesis in woody plants.* Vol. I. Kluwer Academic Publishers, Dordrecht, Netherlands. pp. 81-98.

Gupta P.K. & Pullman G.S. (1991). Method of reproducing coniferous plants by somatic embryogenesis using abscisic acid and osmotic potential variation. U.S. Patent No. 5, 036, 007.

Gupta P.K., Timmis R., Pullman G.S. Yancey M., Krettinger M., Carlson W. & Carpenter C. (1991) Development of an embryogenesis system for automated propagation of forest trees. In: Vasil I.K. (Ed). *Scale up and automation of plant propagation.* Academic Press, New York. pp 76-90.

Jain S.M., Dong N. & Newton R.J. (1989) Somatic embryogenesis in slash pine (*Pinus elliottii*) from immature embryos cultured *in vitro. Plant Sci.* 65: 233-243.

Jones N.B. & van Staden J. (1995) Plantlet production from somatic embryos of *Pinus patula. J. Plant Physiol.* 145: 519-525.

Laine E. & David A. (1990) Somatic embryogenesis in immature embryos and protoplast of *Pinus caribaea*. *Plant Sci.* 69: 215-224.

Mathur G., von Arnold S., and Nadgauda R. (2000) Studies on somatic embryogenesis from immature zygotic embryos of chir pine (*Pinus roxburghii* Sarg.). *Curr. Sci.* 79: 999-1004.

Murashige T. and Skoog R. (1962) A revised medium for rapid growth and bioassay with tobacco tissue cultures. *Physiol. Plant.* 15: 473-497.

Rooberts D.R., Flinn B.S., Webb D.T., Webster F.B. & Sutton, B.C.S. (1989) Characterization of immature embryos of interior spruce by SDS-PAGE and microscopy in relation to their competence for somatic embryogenesis. *Plant Cell Rep.* 8: 285-288.

Schenk R.U. & Hildebrandt A.C. (1972) Medium and techniques for induction and growth of monocotyledonous and dicotyledonous plant cell cultures. *Can. J. Bot.* 50: 199-204.

Tautorus T.E., Attree S.M., Fowke L.C. & Dunstan D. (1990) Somatic embryogenesis from immature and mature zygotic embryos and embryo regeneration from protoplast in black spruce (*Picea mariana* Mill.). *Plant Sci.* 67: 115-124.

Wann S.R. (1998) Somatic embryogenesis in woody species. *Hort. Rev.* 19: 153-181.

CHAPTER

10

Molecular Diversity of Elite WA-CMS Maintainer Lines with Eui Gene for Hybrid Breeding in Rice

Mayank Rai, M.P. Pandey and Pankaj Sharma

ABSTRACT

Genetic diversity and agronomic performance of parental lines in addition to combining ability would determine the higher expression of heterosis. Random amplified polymorphic DNA (RAPD) analysis was therefore, employed with 32 rice genotypes including 29 agronomically superior maintainer lines with *eui* gene for good panicle exsertion. Out of the 70 RAPD primers initially screened the 12 reproducible primers generated 102 amplification products, of which 69 were polymorphic. All the primers produced polymorphic amplification products. The extent of polymorphism varied with the primer used. A cluster dendrogram was generated using Jaccard's similarity coefficient. The values for similarity coefficients ranged between 0.54 to 0.92. Cluster dendrogram analysis revealed clear-cut separation of genotypes into two main clusters. Cluster I comprised two genotypes only, both derived from the same cross. Genotypes in cluster II were redistributed into a number of subclusters at different similarity coefficient levels indicating different levels of genetic relationship. The hybrids involving parents with maximum dissimilarity coefficients displayed non-significant estimates of heterosis for grain yield per plant. Study suggested that only optimal level of diversity is needed for heterosis expression.

INTRODUCTION

Exploitation of heterosis is the most pragmatic approaches for increasing the rice yield potential. However, the success is determined not only by the *per se* performance of the parents but also by the combining ability (Dwivedi *et al.*, 1999) and divergence of the parents involved (Yuan *et al.*, 1994 and Gopal and Minocha, 1997). Recent studies on genetic distance analysis of rice cultivars have focussed on molecular genetic markers (Sun *et al.*, 1996 and Cao & Oard, 1997). The level of correlation between molecular marker based distances and hybrid performance is dependent on the

*Address for correspondence: Department of Genetics and Plant Breeding, G.B. Pant University of Agriculture & Technology, Pantnagar-263 145 (Uttaranchal), India.

germplasm (Sanghai *et al.*, 1997). The RAPD analysis (Williams *et al.*, 1990) being relatively simple and based on the use of commercially available random primers provides useful information about the molecular distance within and among varieties.

Since most of the three-line hybrids either already released or in the advanced trials since start of the programme in 1994 utilize IR 58025 A as the sole CMS line with WA cytoplasm, therefore, the extent of exploitable heterosis is limited to a point. Correlation of molecular diversity with heterosis reflects usefulness of RAPD in predicting heterosis based on RAPD marker polymorphism (Xiao *et al.,* 1996). The present study therefore, aimed at assessing the genetic divergence among the maintainer lines specifically bred with improved plant type, duration, yield potential and grain quality at G B Pant University of Agriculture & Technology, Pantnagar and some adapted varieties by employing RAPD (Random Amplified Polymorphic DNA) markers. This will aid to find out new vistas for the greater exploitation of heterosis in hybrid rice improvement programme.

MATERIALS AND METHODS

Plant Material

Thirty-two rice genotypes (Table 1) including 29 improved maintainer lines with good yielding ability and other agronomic characters and bred specifically for the development of superior cytoplasmic male sterile lines. The elite lines were also selected for better panicle exsertion as one of the parents in the cross involved '*eui*', the single recessive gene for elongated uppermost internode. The material also included three widely adapted superior varieties viz., Pant Dhan 4, Ajaya and UPRI 93-287, which have shown fertility restoration capability to wild abortive (WA) based cytoplasmic male sterility.

Raising the Material in Field/Glass House

Seeds of all the 29 maintainer lines were sown in the glass house during 1998-1999 off season and crossed to UPR 93-287 and Pant Dhan 4 in line × tester fashion with hand pollination after emasculation of the maintainers. All the necessary steps were followed and seeds of 58 crosses were successfully harvested at maturity.

To understand the effect of diversity on the extent of heterosis in crosses, the complete set of hybrids and parents were evaluated in the randomized complete block design (RCBD) using two replications and under two different environments based on fertility. The environments were created using two fertilizer doses viz., optimum (120N: 40P: 40K kg/ha) and high (180N: 80 P: 40K kg/ha) fertility, referred as E_1 and E_2, respectively. Data on yield related characters were obtained to calculate heterosis and other genetic parameters.

DNA Extraction and PCR Amplification

For each genotype, 100 mg of fresh leaves from four-week old seedlings were weighed for DNA extraction using CTAB miniprep DNA extraction protocol (Saghai *et al.*, 1984). The amplification of DNA was done in a volume of 25 µl polymerase chain reaction mixture containing 8 ng of DNA

Table 1 Parentage and code of the maintainer lines and improved cultivars

S. No.	*Designation*	*Code used*	*Parentage*
1.	UPR 2899-78-4-2B	L_1	UPRI 95-140 / IR Basmati
2.	UPR 2899-78-4-3B	L_2	-do-
3.	UPR 2899-78-5-1B	L_3	-do-
4.	UPR 2899-78-6-2B	L_4	-do-
5.	UPR 2900-80-1-1B	L_5	Haryana Basmati 1 / UPRI 95-139
6.	UPR 2906-254-2-6-5B	L_6	Govind/UPRI 95-139
7.	UPR 2906-257-2-8-1B	L_7	Haryana Basmati 1 / UPRI 95-139
8.	UPR 2911-271-2-4B	L_8	UPRI 95-139 / IR 58025B
9.	UPR 2911-271-2-6B	L_9	-do-
10.	UPR 2911-271-3-2B	L_{10}	-do-
11.	UPR 2911-271-7-1B	L_{11}	-do-
12.	UPR 2911-271-8-15B	L_{12}	-do-
13.	UPR 2911-271-8-3B	L_{13}	-do-
14.	UPR 2911-271-8-4B	L_{14}	-do-
15.	UPR 2911-271-8-9B	L_{15}	-do-
16.	UPR 2911-271-12-2B	L_{16}	-do-
17.	UPR 2912-273-8-1B	L_{17}	UPRI 95-141 / IRBB 21
18.	UPR 2900-80-1-4B	L_{18}	Haryana Basmati 1 / UPRI 95-139
19.	UPR 2900-80-9-2-1B	L_{19}	-do-
20.	UPR 2900-80-9-2-2B	L_{20}	-do-
21.	UPR 2903-87-1-1B	L_{21}	Pusa Basmati 1 / IR 58025B
22.	UPR 2903-87-1-2B	L_{22}	-do-
23.	UPR 2904-207-1-3B	L_{23}	BG 1321 / UPRI 95-141
24.	UPR 2906-254-2-6-2B	L_{24}	Govind / UPRI 95-139
25.	UPR 2911-271-2-1B	L_{25}	UPRI 95-139 / IR 58025B
26.	UPR 2911-271-2-5B	L_{26}	-do-
27.	UPR 2911-271-3-3B	L_{27}	-do-
28.	UPR 2911-271-8-7B	L_{28}	-do-
29.	UPR 2911-271-12-3B	L_{29}	-do-
30.	UPRI 93-287	T_1	MRC 19340
31.	Pant Dhan 4	T_2	IR 262 / Remajda
32.	Ajaya	T_3	IET 4141 X CR 98-7216

sample, 200 μM of each dNTP (Genei, Bangalore, India), 0.2 μM of decamer primer (Operon Technologies, Alameda, USA) (Table 2), 0.2 U of Taq DNA polymerase and 1X reaction buffer containing 1.5 mM $MgCl_2$ (Genei, Bangalore, India). The thermocycler (Perkin-Elmer Gene Amp. PCR system 9600) was programmed for one cycle at 94 degrees for 5 minutes and then 40 cycles for

40 cycles at 94°C (30 seconds), 35°C (1 minute) and 72°C (2 minute). It was followed by an extended primer extension step of 5 minute at 72°C and held at 4°C until electrophoresis.

Scoring and Data Analysis

Amplified DNA samples were electrophoresed in agarose gel with 1.6% agarose. RAPDs were scored for the presence (1) and absence (0) of bands. The combined table of scored bands obtained from 32 lines using 12 decamer primers was used for clustering analysis.

Jaccard's similarity coefficient (Jaccard, 1908) values were calculated for each pair-wise comparison between genotypes and a similarity coefficient matrix was constructed. This matrix was subjected to unweighted pair-group method for arithmetic averages analysis (UPGMA) to generate a dendrogram using average linkage procedure. All the computations were carried out using NT-SYS-pc software (Rohlf, 1993).

Statistical analysis was carried out to assess the heterosis as per Fonseca and Patterson (1968). Pant Dhan 4, a widely cultivated variety was taken as check to calculate standard heterosis.

RESULTS AND DISCUSSION

RAPD analysis

Out of the 70 initially screened RAPD primers, 12 reproducible decamer primers (Table 2) were used to detect 102 RAPD markers among all the 32 genotypes (Rai, 2000). Out of these 67.64% (69 bands) were polymorphic for one or more genotypes. The total number of bands per primer ranged

Table 2 Decamer primers used and level of polymorphism detected

Primer	*Sequence (5' to 3')*	*Total number of RAPD markers*	*Polymorphic loci*	
			Frequency	*(%)*
OPA 04	AATCGGGCTG	8	5	62.50
OPA 08	GTGACGTAGG	5	4	80.00
OPA 19	CAAACGTCGG	9	7	77.77
OPB 10	CTGCTGGGAG	8	5	62.50
OPB 17	AGGGAACGAG	4	3	75.00
OPD 03	GTCGCCGTCA	10	7	70.00
OPD 08	GTGTGCCCCA	8	4	50.00
OPE 03	CCAGATGCAC	6	4	66.66
OPE 07	AGATGCAGCC	6	3	50.00
OPE 14	TGCGGCTGAG	11	8	72.72
OPE 19	ACGGCGTATG	13	10	76.92
OPE 13	GGCTGCAGAA	14	9	64.28
Total		102	69	–
Average			5.75	67.64

between 4 to 10, with an average of 8.5 per primer. No primer revealed 100% polymorphic bands. Still the primers OPA 19, OPB 17 and OPE 19 accounted for 75% or higher polymorphism. The distinct banding patterns as produced by the primer OPE 19 (Fig. 1a and 1b) are adequate enough to differentiate most of the genotypes at molecular level. In lanes 'M', DNA size markers "100 bp DNA ladder" were loaded to count the size. These results reflect that RAPD technique being technically simpler, quicker, relatively inexpensive and non-radioactive as also observed by Samec (1993). It can detect sufficient polymorphism for germplasm characterization and genetic distance studies (Naghia *et al.*, 2002).

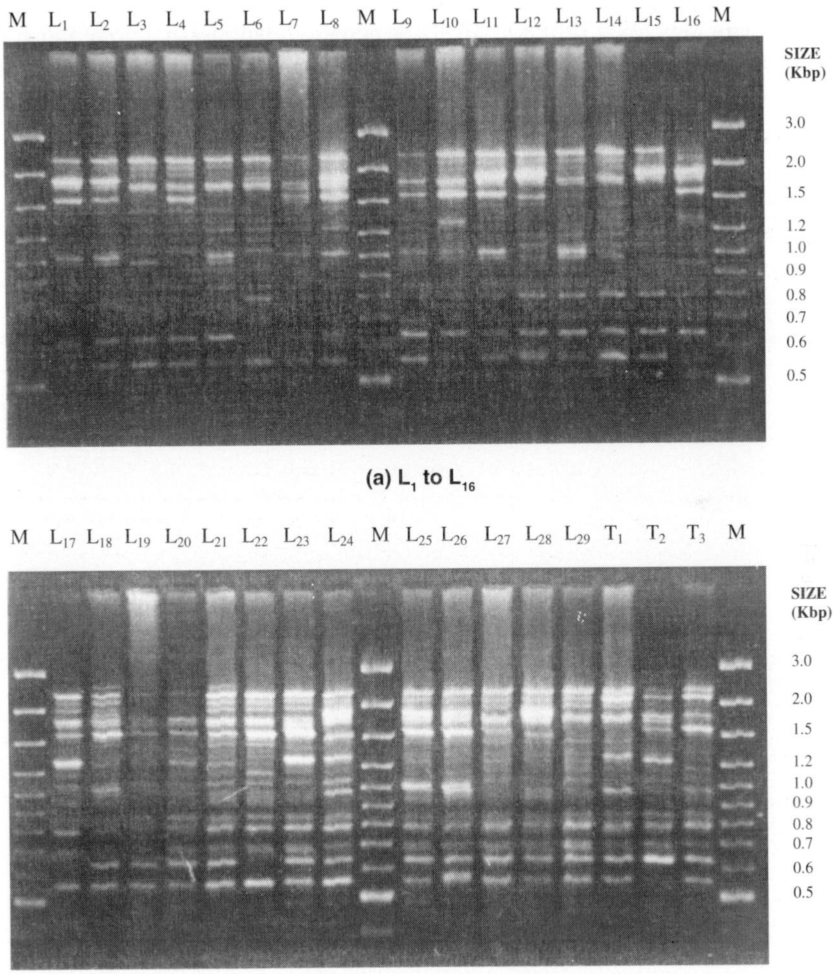

(a) L_1 to L_{16}

(b) L_{17} to L_{20} and T_1 to T_3

Fig. 1 RAPD profile of 29 maintainer lines ($L_1 - L_{29}$) and three improved cultivars with Primer OPE 19, Lane M: DNA size marker "100 bp DNA ladder".

Genetic Diversity Based on RAPD Markers

Analysis of the relationship based on 102 amplified products (RAPD markers) revealed that the genetic similarity based on Jaccard's coefficients (Table 3) ranged between 0.542 (45.8% diversity) to 0.92 (8.0% diversity). The dendrogram (Fig. 2) revealed that at about 62.7% genetic similarity, all the genotypes were grouped into 2 main clusters. The cluster I comprised lines L_{19} and L_{20} with the genetic similarity of 0.811 (18.9% diversity). Both of these lines were derived from single cross Haryana Basmati 1/UPRI 95-139. All other lines derived from this cross were placed in different subclusters of cluster II at different genetic similarity levels. The cluster II comprised 30 lines, including three improved cultivars viz., T_1, T_2 and T_3. These 30 lines were again distributed into a number of subclusters indicating different levels of genetic relationship among the genotypes. At 80% genetic similarity level, the genotypes in cluster II were redistributed into 10 subclusters, while at 85% and 90% similarity levels, the same genotypes were grouped into 20 and 28 subclusters, respectively. The lines L_{21} and L_{22} derived from Pusa Basmati 1/IR 58025 B (0.82 genetic similarity) were grouped in the same subcluster at 80% genetic relationship but in different subclusters at 90% genetic similarity. This can be attributed to the selection of typical basmati characters in the segregating population of the cross. Two different lines L_2 (UPRI 95-140/IR Basmati) and L_{11} (UPRI 95-139/IR 58025 B), although of different origin were close to each other with the genetic distance of 0.08 (92% similarity). Four of the lines L_1, L_2, L_3 and L_4 with the common parentage (UPRI 95-140/IR Basmati) grouped into two subclusters at 85% genetic relation

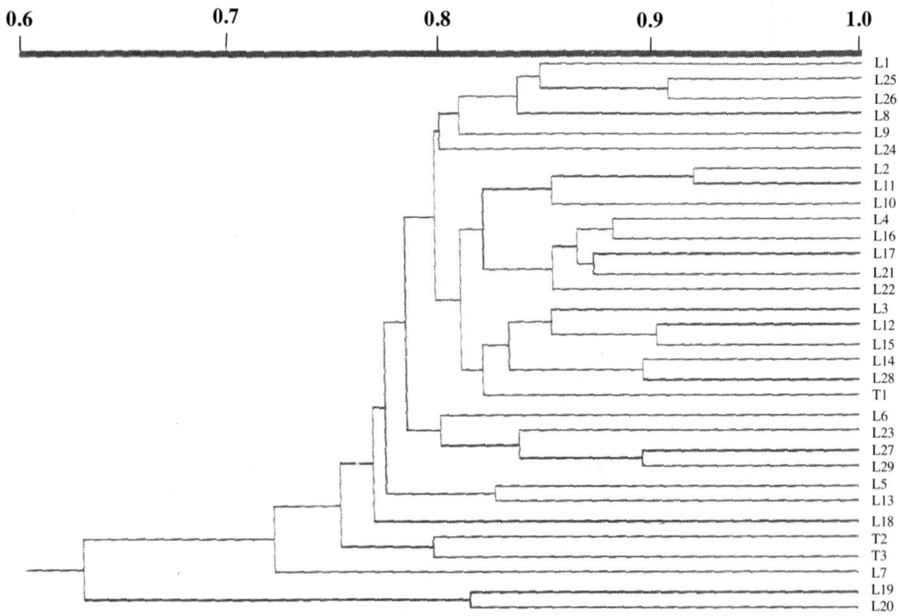

Fig. 2 Dendrogram generated using UPGMA method analysis relationship between 32 genotypes (29 elite maintainer lines and three improved cultivars) of rice using RAPD analysis.

Table 3 Jaccard's similarity coefficients among 32 rice genotypes based on RAPD markers.

	L_1	L_2	L_3	L_4	L_5	L_6	L_7	L_8	L_9	L_{10}	L_{11}	L_{12}	L_{13}	L_{14}	L_{15}	L_{16}	L_{17}	L_{18}	L_{19}	L_{20}	L_{21}	L_{22}	L_{23}	L_{24}	L_{25}	L_{26}	L_{27}	L_{28}	L_{29}	T_1	T_2	T_3
L_1	1.000																															
L_2	0.831	1.000																														
L_3	0.812	0.810	1.000																													
L_4	0.813	0.833	0.815	1.000																												
L_5	0.794	0.792	0.821	0.750	1.000																											
L_6	0.746	0.725	0.773	0.821	0.753	1.000																										
L_7	0.743	0.700	0.701	0.720	0.636	0.630	1.000																									
L_8	0.828	0.825	0.740	0.831	0.700	0.721	0.740	1.000																								
L_9	0.791	0.789	0.794	0.845	0.702	0.726	0.802	0.810	1.000																							
L_{10}	0.772	0.837	0.797	0.844	0.802	0.756	0.708	0.746	0.730	1.000																						
L_{11}	0.864	0.920	0.842	0.864	0.835	0.763	0.750	0.820	0.797	0.868	1.000																					
L_{12}	0.775	0.773	0.846	0.800	0.739	0.737	0.691	0.792	0.756	0.783	0.835	1.000																				
L_{13}	0.815	0.812	0.794	0.818	0.824	0.730	0.705	0.787	0.750	0.800	0.842	0.825	1.000																			
L_{14}	0.759	0.802	0.831	0.807	0.813	0.743	0.696	0.777	0.740	0.883	0.831	0.837	0.857	1.000																		
L_{15}	0.777	0.800	0.857	0.857	0.750	0.800	0.732	0.808	0.835	0.824	0.871	0.901	0.808	0.861	1.000																	
L_{16}	0.784	0.850	0.833	0.881	0.769	0.792	0.721	0.871	0.813	0.837	0.881	0.817	0.812	0.848	0.875	1.000																
L_{17}	0.750	0.792	0.820	0.868	0.712	0.779	0.753	0.790	0.800	0.802	0.833	0.783	0.800	0.790	0.835	0.860	1.000															
L_{18}	0.762	0.761	0.722	0.800	0.716	0.716	0.734	0.792	0.734	0.804	0.787	0.732	0.738	0.750	0.753	0.817	0.761	1.000														
L_{19}	0.555	0.581	0.569	0.614	0.542	0.597	0.630	0.611	0.630	0.625	0.567	0.586	0.554	0.567	0.611	0.602	0.581	0.608	1.000													
L_{20}	0.633	0.662	0.638	0.700	0.591	0.652	0.671	0.671	0.696	0.671	0.690	0.618	0.600	0.613	0.676	0.653	0.662	0.623	0.811	1.000												
L_{21}	0.750	0.750	0.775	0.868	0.691	0.779	0.753	0.835	0.800	0.780	0.810	0.761	0.756	0.768	0.810	0.860	0.871	0.850	0.611	0.684	1.000											
L_{22}	0.721	0.765	0.769	0.864	0.705	0.773	0.723	0.784	0.770	0.844	0.815	0.777	0.794	0.807	0.830	0.857	0.868	0.790	0.637	0.700	0.820	1.000										
L_{23}	0.753	0.740	0.756	0.813	0.701	0.794	0.733	0.759	0.756	0.818	0.759	0.722	0.746	0.815	0.780	0.772	0.784	0.762	0.605	0.720	0.794	0.813	1.000									
L_{24}	0.797	0.773	0.800	0.800	0.716	0.759	0.746	0.792	0.769	0.804	0.825	0.829	0.759	0.792	0.837	0.795	0.761	0.773	0.573	0.653	0.783	0.800	0.787	1.000								
L_{25}	0.840	0.835	0.772	0.842	0.753	0.800	0.773	0.857	0.821	0.756	0.855	0.802	0.820	0.810	0.833	0.858	0.800	0.780	0.583	0.635	0.800	0.794	0.792	0.802	1.000							
L_{26}	0.855	0.827	0.810	0.810	0.746	0.792	0.789	0.825	0.813	0.771	0.883	0.817	0.835	0.802	0.849	0.827	0.814	0.783	0.581	0.640	0.814	0.765	0.797	0.839	0.907	1.000						
L_{27}	0.776	0.820	0.802	0.851	0.783	0.808	0.733	0.772	0.780	0.831	0.756	0.614	0.685	0.807	0.826	0.849	0.765	0.853	0.820	0.700	0.812	0.700	0.767	0.800	0.900	0.800	1.000					
L_{28}	0.737	0.759	0.762	0.807	0.743	0.789	0.675	0.777	0.740	0.812	0.784	0.837	0.787	0.894	0.861	0.825	0.768	0.771	0.575	0.613	0.790	0.784	0.815	0.792	0.833	0.825	0.842	1.000				
L_{29}	0.722	0.746	0.760	0.840	0.753	0.794	0.666	0.720	0.739	0.805	0.800	0.723	0.753	0.802	0.815	0.794	0.780	0.740	0.584	0.676	0.756	0.788	0.823	0.723	0.777	0.780	0.893	0.828	1.000			
T_1	0.746	0.768	0.794	0.842	0.708	0.730	0.705	0.765	0.750	0.800	0.818	0.825	0.820	0.833	0.833	0.812	0.800	0.790	0.583	0.666	0.800	0.842	0.779	0.802	0.775	0.790	0.782	0.810	0.777	1.000		
T_2	0.697	0.753	0.756	0.819	0.689	0.760	0.662	0.810	0.757	0.763	0.746	0.766	0.760	0.786	0.794	0.837	0.786	0.721	0.621	0.671	0.786	0.830	0.753	0.753	0.783	0.740	0.756	0.773	0.746	0.773	1.000	
T_3	0.700	0.743	0.725	0.746	0.705	0.683	0.658	0.740	0.701	0.797	0.769	0.734	0.728	0.784	0.746	0.765	0.731	0.714	0.661	0.746	0.753	0.792	0.753	0.714	0.686	0.702	0.756	0.740	0.712	0.772	0.794	1.000

level with the genetic distance ranging from 0.167 (L_2 and L_4) to 0.190 (L_2 and L_3). Amongst the improved cultivars, T_1 (UPRI 93-287) was closer to most of the elite maintainer lines in comparison to Pant Dhan 4 (T_2) and Ajaya (T_3). The lines T_2 and T_3 were closer to each other with the genetic distance of 0.206 (79.4% similarity), while line T_1 was apart from T_2 and T_3 with the genetic diversity of 0.227 and 0.228, respectively. It suggests greater genetic divergence and therefore, enhanced prospects of obtaining higher standard heterosis from hybrids combining these commercial cultivars and elite maintainer converted into cytoplasmic male sterile lines. Interestingly, the study also revealed the intermingling of the lines originating from different crosses into the same cluster. This could be due to the facts that some of the lines involved one of the common parents in these crosses. For example, 12 of the lines originating from two crosses viz., L_8 to L_{16} and L_{25} to L_{29} (UPRI 95-139/IR 58025 B) and L_{21}-L_{22} (Pusa Basmati 1/IR 58025 B) involved IR 58025 B as the common parent. Similarly, the lines L_5, L_7 and L_{18} to L_{20}, all derived from the same cross Haryana Basmati 1/UPRI 95-139 and the parent, UPRI 95-139 has been common in parentage of all the lines L_8 to L_{16} and L_{25} to $L_{29.}$ Within the cluster, 11 of the lines derived from UPRI 95-139/IR 58025 B (L_8 to L_{16}) and L_{25} to L_{29} were grouped into five different subclusters, based on their relative genetic similarity. Such a variation could be attributed to the selection of diverse plant characteristics related to plant type, duration, grain and yield attributes during segregating generations (F_2-F_4) leading to divergence among these lines. The value of RAPD approach was supported by the close similarity between the dendrogram based on RAPD results. It is generally accepted that the molecular markers such as RFLP and RAPD represent genetic variation at DNA level, providing more accurate measures of relationship between the individuals without the influence of environmental variation (Miller and Tanksley, 1990). Moreover RAPD markers provide a better coverage of the genome resulting into a better estimate of relationship (Smith and Smith, 1989).

Linear relationship between heterosis of specific crosses and divergence of combining parents (data not shown) in this study could not be established as also reported by Dudley *et al.*, (1991), since the primers used were random and limited, so may not cover the complete range of genome. The results however, indicated top heterotic crosses for grain yield per plant and other characters revealing low level of genetic diversity between parents (Table 4). On the other hand, the hybrids involving parents with maximum dissimilarity coefficients ($L_{19} \times T_2$) with genetic similarity of 62.1%, display non-significant estimates of heterosis (-19.89) for grain yield per plant. It suggested the prospects of combining parents having optimum level of diversity for heterosis.

In the present study, the molecular approach using RAPD markers has reliably revealed the genetic diversity among lines and grouped them into distinct clusters and subclusters. The findings will have immediate practical implication in the hybrid breeding programme for the development of new CMS lines of varying duration, plant height and productivity traits with optimum genetic diversity, which can exhibit enhanced level of heterosis. Thus the RAPD approach maximized the opportunities to obtain superior hybrids because unrelated parents would be expected to contribute unique desirable alleles at different loci (Tatineni *et al.*, 1996).

Table 4 Best heterotic crosses and divergence among the parents of heterotic crosses for various characters under E_1 (optimum) and E_2 (high) fertility environments

Characters	*Environment*	*Heterotic cross*	*Divergence of parents in heterotic cross*
Days to 50% flowering	E_1	$L_1 \times T_1$	0.254
	E_2	$L_2 \times T_2$	0.247
Plant height	E_1	$L_{18} \times T_1$	0.210
	E_2	$L_{28} \times T_1$	0.190
Panicle number per plant	E_1	$L_{13} \times T_2$	0.240
	E_2	$L_{29} \times T_2$	0.254
Panicle length	E_1	$L_{10} \times T_2$	0.237
	E_2	$L_8 \times T_2$	0.190
Panicle exsertion	E_1	$L_2 \times T_1$	0.232
	E_2	$L_2 \times T_1$	0.232
Grain number per panicle	E_1	$L_{17} \times T_2$	0.214
	E_2	$L_{17} \times T_2$	0.214
Percent spikelet fertility	E_1	$L_{29} \times T_2$	0.254
	E_2	$L_2 \times T_2$	0.247
1000-grain weight	E_1	$L_{29} \times T_2$	0.254
	E_2	$L_{20} \times T_1$	0.334
Total dry matter	E_1	$L_{13} \times T_2$	0.240
	E_2	$L_{14} \times T_2$	0.214
Grain yield per plant	E_1	$L_{14} \times T_2$	0.214
	E_2	$L_{14} \times T_2$	0.214

REFERENCES

Cao D. & Oard J.H. (1997) Pedigree and RAPD based DNA analysis of commercial US rice cultivars. *Crop Sci*. 37: 1630-1635.

Dudley J.W., Maroof M.A.S. & Rufener G.K. (1991) Molecular markers and grouping of parents in maize breeding programme. *Crop Sci*. 31: 718-723.

Dwivedi D.K., Pandey M.P., Pandey S.K. & Rongbai Li. (1999) Combining ability over environments in rice involving *indica* and tropical *japonica* lines. *Oryza*. 36: 101-107.

Fonseca S. & Patterson F.L. (1968) Hybrid vigour in a seven parent diallel cross in common winter wheat (*Triticum aestivum* L.). *Crop Sci.* 8: 85-88.

Gopal J. & Minocha J.L. (1997) Genetic divergence for cross prediction in potato. *Euphytica*. 97: 269-275.

Jaccard P. (1908) Nouvelles recherches sur la distribution forale. *Bull. Soc. Sci. Nat*. 44: 223-270.

Miller J.C. & Tanksley S.P. (1990) RFLP analysis of polygenetic relationship and genetic variation in the genus *Lycopersicon*. *Theor. Appl. Genet.* 80: 437-448.

Naghia P.T., Malik J.P.S., Pandey M.P. & Singh N.K. (2002) Application of RAPD markers for genetic distance analysis of hybrid rice parental lines. *Indian J. Genet. Pl. Breed*. 62(1): 1-4.

Rai Maynak. (2000) Genetic analysis and divergence studies of elite maintainer lines for production of hybrid rice (*Oryza sativa* L.) Unpubl. M. Sc. Thesis, GBPUAT, Pantnagar.

Rohlf F.J. (1993) NTSYS-pc Numerical taxonomy and multivariate analysis system, version 1.80 Exeter Software, Setauket, New York.

Saghai-Maroof M.A., Soliman K.M., Jorgesen R.A. & Allard R.W. (1984) Ribosomal DNA spacer length polymorphism in barley: Mendelian inheritance, chromosomal location and population dynamics. *Proc. Natl. Acad. Sci.* (USA). 81: 8014-8018.

Samec P. (1993) DNA polymorphism and RAPD technology. Gene Slecht (Prague) 29: 291-320.

Sanghai M.A., Maroof G.P., Yang Q., Zang & Gravois K.A. (1997) Correlation between molecular marker distance and hybrid performance in US Southern long grain rice. *Crop Sci.* 37: 145-150.

Smith J.S.C. & Smith O.S. (1989) The description and assessment of distances between inbred lines of maize II. The utility of morphological, biochemical and genetic descriptors and a scheme for testing distinctiveness between inbred lines. *Maydica*. 34: 151-161.

Sun Qixin T.C., Hang Z.F., Ni M. & Procunier J.D. (1996) Study on wheat heterotic group I. Genetic diversity revealed by random amplified Polymorphic DNA (RAPD) in elite wheat cultivars. *J. Agric. Biotech.* 4(2): 103-110.

Tatineni V., Cantrell R.G. & Davis D.D. (1996) Genetic diversity in elite cotton germplasm determined by morphological characteristics and RAPDs. *Crop Sci.* 36: 186-192.

Williams J.G.K., Kubelik A.R., Livak K.J., Rafalski J.A. & Tingey S.V. (1990) DNA polymorphism amplified by arbitrary primers are useful as genetic markers. *Nucleic Acid Res.* 18: 6531-6535.

Xiao J., Li J., Yuan l., McCouch S.R. & Tunksley S.D. (1996) Genetic diversity and its relationship to hybrid performance and heterosis in rice as revealed by PCR based markers. *Theor. Appl. Genet*. 92: 637-643.

Yuan L.P., Yang Z.Y. & Young J.B. (1994) Hybrid rice in China: In: S. S. Virmani (ed.). *Hybrid Rice Technology, new developments and future prospects.* IRRI, Box 933, Manila, Philppines.

CHAPTER

11

Characterization of Candidate Clumps of *Dendrocalamus strictus* Using RAPDs

S.N. Yaradoni[1], D.P. Biradar[2], V.C. Patil[2], and M.S. Kuruninashetti[1]

ABSTRACT

India is the second largest bamboo resource in the world, next only to China, both in terms of species diversity and extent of distribution. *Dendrocalamus strictus* is predominant species of bamboo found in Western Ghats which is commercially in great demand. Eleven candidate clumps of *Dendrocalamus strictus* which were identified based on morphological parameters are used in the present investigation to characterize using random amplified polymorphic DNA-polymerase chain reaction (RAPD-PCR) technique. Protocols were standardized for isolation of DNA from field grown clones and PCR reactions to generate RAPD profiling. Out of twenty Operon primers (OPJ, OPO and OPD series), fifteen primers have exhibited good DNA polymorphism. Dendrograms constructed to mark the genetic similarity had grouped the candidate clumps into three major clusters. The first cluster comprising of elite bamboo clones i.e., Brc-1, Brc-2, Krd-1, Brc-3, Brc-4, Krd-3, Ttg-4, Ttg-1 and Ttg-2. The second and third cluster consisted of only one clone viz., Krd-2 and Ttg-3, respectively. Maximum genetic similarity of 87.80 per cent was found between elite clones of Brc-2 and Krd-1. The study suggested the usefulness of RAPD-PCR technique in characterizing the candidate clumps of bamboo.

INTRODUCTION

Bamboo, a fast growing and long-lived woody plant, is a versatile multipurpose forest product which plays a vital role in the Asian economy. Bamboos are used extensively in construction and handicrafts, for making a wide array of articles for irrigation and fuel, as food (bamboos shoots and lice), in industry mainly for paper and pulp.

Dendrocalamus strictus often referred to as "Male bamboo" in deciduous, densely tufted bamboo. It is one of the hardest and the most widely distributed bamboos in India. It is found in

*Address for correspondence: 1. Department of Biotechnology, UAS, Dharwad- 580 005
2. Department of Agronomy, UAS, Dharwad- 580 005

north-eastern region, Western Ghats and Andaman (Subramanian, 1994). It is found in deciduous dry forests of moderately dry regions all over India up to 3500 ft. withstanding drought and frost better than other bamboos (Mishra *et al.*, 2001).

Western Ghats supports about 23% of the genetic resources of bamboo (i.e. 32 species under 8 genera). Of the many genera represented in the country, only two genera, namely, *Bambusa* and *Dendrocolamus* have the overall distribution occurring in all the 5 zones in India. Considering the need for collecting information on existing genetic diversity in the Western Ghats region, selected elite clones of *D. strictus* are characterized using DNA finger printing to document polymorphism at the molecular level.

MATERIALS AND METHODS

Eleven candidate clumps of *D. strictus* identified based on morphological characteristics in the International Plant Genetic Resources Institute (IPGRI) funded project (Patil, 1999; Unpublished) are used in the present study for molecular characterization. The details of morphological parameters of the clones is furnished in Table 1.

Table 1 Morphological characters of candidate clumps of *Dendrocalamus strictus*

SL No.	*Forest range*	*Transect*	*Total No. of culms/clump*	*Clum height (m)*	*Clum diameter (cm)*
1.	Barchi	Barchi (Brc-1)	40	11.50	5.2
2.	Barchi	Barchi (Brc-2)	48	12.66	4.80
3.	Barchi	Barchi (Brc-3)	37	11.33	5.33
4.	Barchi	Barchi (Brc-4)	44	13.33	4.3
5.	Haliyal	Kerwad (Krd-1)	47	12.50	4.67
6.	Haliyal	Kerwad (Krd-2)	34	13.00	4.93
7.	Haliyal	Kerwad (Krd-3)	46	15.33	3.67
8.	Haliyal	Tatawanagi (Ttg- 1)	36	13.50	4.20
9.	Haliyal	Tatawanagi (Ttg-)	28	13.33	4.6
10.	Haliyal	Tatawanagi (Ttg- 3)	56	11.17	4.27
11.	Haliyal	Tatawanagi (Ttg- 4)	29	13.33	5.37

(Sources: Patil, 1999–Unpublished)

The leaf samples of identified clumps were collected from the forest ranges of tracts *viz.*, Barchi (Brc-1, Brc-2, Brc-3, Brc-4), Kerwad (Krd-1, Krd-2, Krd-3), Tarawanagi (Ttg-1, Ttg-2, Ttg-3, Ttg-4). Total DNA was isolated from leaves of *D. strictus* using CTAB method (Maniatis *et al.*, 1982). The polymorphism among the candidate clumps were studied with 20 primers of OPJ, OPO and OPD series of Operon primers. The procedure adopted by Lai and Hsiao (1997) with certain modification on temperature profile was employed in the RAPD study. The temperature profile for RAPD analysis was 94°C – 5 min for initial denaturation and 40 cycles of 94°C – 1 min, 36°C – 1 min, 72°C – 1 min were run before the final extension. The final extension was given at 72°C for 5 min. Amplification reactions were performed in volumes of 25 µl in 0.5 ml microfuge tube using 50 ng of template DNA, 1 µl of primer (1 µM), 1 µl of dNTP (100 µM each), 2.5 µl of *Taq* 10 X assay better, 0.5 U *Taq* DNA polymerase and 18.5 µl sterile milli Q water was added serially into each tubes to make up the volume. The tubes were placed into the thermal cyclers (Eppendorf). After completion of 40 cycles, the contents of tube were electrophoresed in 1.5 per cent agarose gel to separate DNA fragments. The separated DNA fragments were visualized on a UV-transilluminator and photographed. The bands of DNA fragments were scored as present (1) or absent (0) and the 0-1 matrix was generated for further analysis. The data for 15 of the 20 primers which produced scorable bands was used for further computation. The genetic distance was estimated on the basis of number of polymorphic bands following the procedure of Nei and Li (1979). Percentage of polymorphism was computed based on number of polymorphic bands and total number of bands produced. Distance matrices and dendrograms were constructed based on similarity co-efficients generated for individual primers and also for pooled data by using NT-sys, computer program for distance estimation.

RESULTS AND DISCUSSION

Candidate clumps of *D. strictus* have been characterized by employing PCR based RAPD technique. Of the twenty Operon primers, fifteen primers have given enough polymorphism. The photograph of one representative gel showing amplification products with OPJ-11 primer is presented in Fig. 1. These primers produced a total of 713 bands across eleven elite clones (Table 2). The number of bands produced per primer varied from 29 to 71. The average number of bands per primer was 47.53. Out of 713 bands 526 were polymorphic (74.51 per cent). The average number of polymorphic RAPD bands was 35.06 per primer. The amount of genetic similarity recognized in the studies is dependent on the primers used and the amount of genetic similarity among clones.

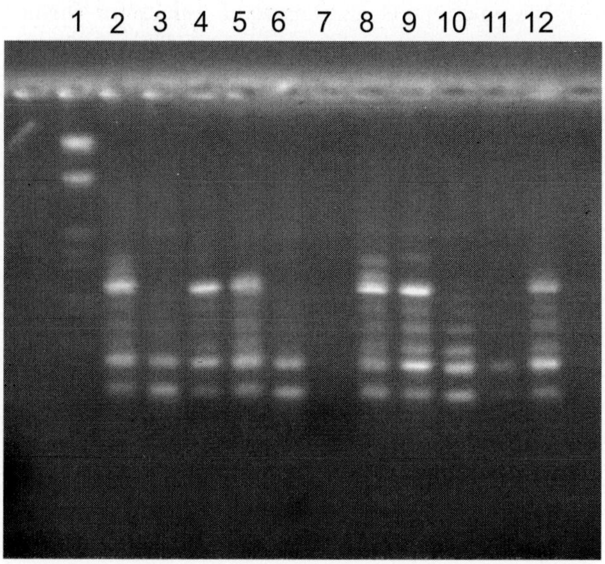

Fig. 1 Gel showing amplification profile of primer OPJ-11. Lane-1 = λ DNA digested with *Eco* RI and *Hind* III; Lane-2 = Brc-1; Lane-3 =Brc-2; Lane-4 = Brc-3; Lane-5 = Brc-4; Lane-6 = Krd-1; Lane-7 = Krd-2; Lane-8 = Krd-3; Lane-9 = Ttg-1; Lane-10 = Ttg-2; Lane-11 = Trg-3; Lane-12 = Ttg-4.

Table 2 Summary statistics of Polymorphism in RAPD analysis

Primer	*Total no. of bands*	*No. of polymorphic bands*	*Polymorphism (%)*
OPJ – 03	39	28	71.79
OPJ – 04	43	21	48.83
OPJ – 08	37	37	100.00
OPJ – 11	57	57	100.00
OPJ – 12	50	39	78.00
OPJ – 13	52	19	36.53
OPJ – 14	71	5	7.04
OPJ – 15	29	18	62.06
OPJ – 16	56	34	60.71
OPJ – 18	29	29	100.00
OPJ – 19	52	52	100.00
OPJ – 20	36	25	69.44
OPO – 13	57	57	100.00
OPO – 14	47	47	100.00
OPO – 16	58	58	100.00
Pooled	713	526	74.51
Average	47.53	35.06	74.51

Among the 15 primers that have given polymorphism, primers OPJ-08 and OPO-16 had highest (100 per cent) differentiating power. Thus, producing characteristic DNA, finger printings for different clones and they can be used for identifying the different candidate clumps of *D. strictus* (Table 3). These differentiations based on DNA markers is an important aspect for further investigation such as construction of genetic maps, gene-tagging and other manipulations.

Table 3 Differentiation power of different primers used

Primer	*Total no. of phenotypes*	*No. of unique phenotypes*	*Differentiation power (%)*
OPJ – 03	11	09	81.81
OPJ – 04	11	09	81.81
OPJ – 08	11	11	100.00
OPJ – 11	11	10	90.90
OPJ – 12	11	06	54.54
OPJ – 13	11	07	63.63
OPJ – 14	11	02	18.18
OPJ – 15	11	06	54.54
OPJ – 16	11	09	81.81

(Contd.)

(Contd.)

OPJ – 18	11	06	54.54
OPJ – 19	11	08	72.72
OPJ – 20	11	07	63.63
OPO – 13	11	07	63.63
OPO – 14	11	07	63.63
OPO – 16	11	11	100.00

The genetic similarity and the dendrograms were constructed considering all candidate clumps of *D. strictus* from the data pooled over all the fifteen primers. The maximum genetic similarity of 87.80 per cent was found between Brc-2 and Krd-1, followed by 87.00 per cent between Ttg-1 and Ttg-2 (Table 4). The lowest genetic similarity of 48.10 per cent was between Brc-3 and Ttg-3. Dendrograms were constructed for each of the primers and also for the pooled data which showed three major clusters (Fig. 2).

Conclusion

Dendrograms constructed to mark the genetic similarity had grouped the candidate clumps into three major clusters. The first cluster comprising of candidate bamboo clumps, i.e., Brc-1, Brc-2, Krd-1, Brc-3, Brc-4, Krd-3, Ttg-4, Ttg-1 and Ttg-2. The second and third cluster consisted of only one clone viz., Krd-2 and Ttg-3, respectively. Brc-2 and Krd-1 which have similar morphological characters, cluster together in the dendrogram of genetic similarity obtained from these primers and is quite evident in the pooled data. The lowest genetic similarity of 48.1 per cent was found between Brc-3 and Ttg-3. Ttg-3 which has lowest genetic similarity with the rest of the clones forms separate cluster which has highest number of culms/clump (56 in number).

Table 4 Genetic similarity (%) based RAPD profile of pooled data

	Brc-1	*Brc-2*	*Brc-3*	*Brc-4*	*Krd-1*	*Krd-2*	*Krd-3*	*Ttg-1*	*Ttg-2*	*Ttg-3*	*Ttg-4*
Brc-1	100.00										
Brc-2	79.70	100.00									
Brc-3	67.80	73.80	100.00								
Brc-4	73.30	74.40	81.90	100.00							
Krd-1	73.80	87.80	73.00	76.50	100.00						
Krd-2	61.30	72.20	64.10	67.70	75.70	100.00					
Krd-3	82.40	78.60	78.30	77.40	76.20	73.80	100.00				
Ttg-1	70.30	73.00	84.40	82.10	75.00	69.60	86.90	100.00			
Ttg-2	71.00	78.20	80.90	80.00	78.60	67.20	83.70	87.00	100.00		
Ttg-3	55.70	52.80	48.10	58.30	56.60	54.20	56.10	57.70	56.10	100.00	
Ttg-4	71.40	74.10	76.70	84.80	74.60	69.10	86.70	85.70	86.80	60.60	100.00

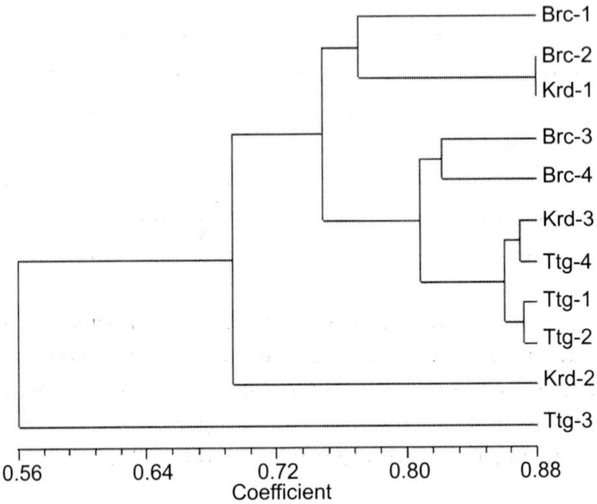

Fig. 2 Dendrogram of genetic similarity obtained from the pooled data of RAPD profiles.

Individual primers are of little value in the analysis of genetic similarity. Large number of primers have to be used to get a reasonable assessment of the status of a given clone in relation to each other and to group the clones based on DNA markers with reference to the contrasting phenotypic classes.

REFERENCES

Lai and Hsiao. (1997) Genetic variation of *Phyllostachys pubescens* (Bambusoideae, Poaceae) in Taiwan based on DNA polymorphisms. *Bot. Bull. Acad. Sin.* 38: 145-152.

Maniatis T., Fritsch E.F. & Sambrook J. (1982) *Molecular Cloning: A Laboratory Manual*. Cold Spring Harbour Laboratory. Cold Spring Harbour. USA.

Mishra Y., Rana P.K., Shirin F. & Ansari S.A. (2001) Augmenting *in vitro* shoot multiplication by vipul (tricantanol) and adventitious rhizogenesis by rice bran extract in *Dendrocalamus strictus. Ind. Jou. of Expt. Bio.* 39: 165-169.

Nei M. & Li.W. (1979) Mathematical model for studying genetic variation in terms of restriction endonucleases. *Proc. of Nat. Acad. of Sci.*, USA. 76: 5269-5273.

Subramanian K.N. (1994) Genitic improvement and conservation of bamboo in India. In: Rao IV R and Rao AN (eds) *Bamboo and Rattan Genetic Resources and Use* (pp. 23-39). Proceedings of first INBAR Biodiversity. Genetic resources and conservation working groups. Singapore: 7-9.

CHAPTER

12

Plantlet Regeneration from the Nodal Segments of Pomegranate (*Punica granatum* L.) cv. Jyoti

S.K. Singh and R.N. Khawale*

ABSTRACT

Semi-hardwood nodal segments collected just before flowering from mature tree of pomegranate (*Punica granatum* L.) cv. Jyoti was the most ideal explant for culture establishment. Among the various treatments tried to check polyphenolic exudation, half-strength Murashige and Skoog (1962) medium supplemented with 1.0 mg l^{-1} benzyl amino purine (BAP) + 1.0 mg l^{-1} kinetin along with 200 mg l^{-1} activated charcoal (AC) gave the best culture establishment. Axillary shoot buds proliferated on the original explant were excised aseptically and transferred to the proliferation medium. Shoot proliferation was noted maximum on MS medium supplemented with various concentrations of BAP, kinetin and adenine sulphate in combination. For shoot elongation and rooting, MS medium containing 2.0 mg l^{-1} indole-butyric acid (IBA), 200 mg l^{-1} AC and 40 g l^{-1} sucrose was found best. For *in vitro* hardening, rooted plantlets were transferred to glass jar filled with sterilized peat: soilrite (1:1) and covered with polypropylene cap. This treatment gave the highest plantlet survival during hardening and recorded 86.5% survival. The hardened plantlets could be transferred to greenhouse and ultimately to the field.

Key Words: *Punica granatum*, micropropagation, nodal segments.

INTRODUCTION

Pomegranate (*Punica granatum* L.) is an important horticultural crop of the tropical and sub-tropical regions grown primarily for dessert fruits. Besides, the tree is also cultivated for pharmaceutical and ornamental usages. Conventionally, pomegranate varieties are propagated by stem including both hard and softwood cuttings, however this method is not very efficient commercially as it requires about one year to raise the plants (Anon, 1982).

Micropropagation technique has been widely used for the propagation of several plant species. The clonal multiplication of pomegranate cultivars from mature trees by tissue culture has been enlightened after the work of Mascarenhas *et al.* (1981), from their work on two cvs. Ganesh and

*Address for correspondence: Central Tissue Culture Laboratory, Division of Fruits and Horticultural Technology, Indian Agricultural Research Institute, New Delhi-110 012, India

Muscat. The studies on organogenesis and micropropagation have been done by various workers such as Omura *et al.* (1987), Jaidka & Mehra (1986), Bhansali (1990), Zhang & Stoltz (1991), Yang & Ludders (1995), Naik *et al.* (1999, 2000) using different explants in various cultivars. In the present study, efforts were made to standardize the procedure for micropropagation of promising pomegranate cv. Jyoti using semi- hardwood nodal segments from the mature tree.

MATERIALS AND METHODS

Surface sterilization and culture establishment

Field-grown pomegranate cv. Jyoti trees maintained in the orchard of the Indian Agricultural Research Institute, Pusa, New Delhi were used as the explant source. The semi-hardwood shoots, just before flowering were collected and cut into two-four node cuttings. The explants were then washed under running tap water using anti-microbial soap solution containing cetrimide (0.1%) and few drops of Tween 20. This followed the agitation treatment in 500 ppm carbendazim (w/v) plus 200 ppm 8-HQC (w/v) for 30 min. The twigs were then made into small one node segments and after washing with sterile distilled water, were subsequently transferred to the laminar cabinet hood for culturing. For surface sterilization, nodal segments were agitated in 0.25% sodium hypochlorite (v/v) for 5 min. followed by rinsing in sterile double-distilled water for five times. The explants were then air-dried for few minutes and transferred to the half-strength Murashige and Skoog (MS, 1962) medium containing benzylaminopurine (BAP), 6-furfuryl aminopurine (Kin) along with activated charcoal (AC) in various treatment combinations, i.e. T1= 1.0 mg l^{-1} BAP + 200 mg l^{-1} AC, T2= 2.0 mg l^{-1} BAP + 200 mg l^{-1} AC, T3 = 4.0 mg l^{-1} BAP + 200 mg l^{-1} AC, T4= 1.0 mg l^{-1} Kin + 200 mg l^{-1} AC, T5 = 2.0 mg l^{-1} Kin + 200 mg l^{-1}AC, T6 = 4.0 mg l^{-1} Kin + 200 mg l^{-1} AC, T7 = 1.0 mg l^{-1} BAP + 1.0 mg l^{-1} Kin + 200 mg l^{-1}AC, T8 = 2.0 mg l^{-1}BAP + 2.0 mg l^{-1} Kin + 200 mg l^{-1} AC, T9 = 4.0 mg l^{-1} BAP + 4.0 mg l^{-1} Kin + 200 mg l^{-1}AC. All the growth regulators, sucrose and agar-agar use for the study were from Qualigens Chemicals, Mumbai. The pH of the media was adjusted to 5.8 prior to autoclaving (1.06 Kg cm^{-2}) for 15 min. The cultures were then maintained under 16/8 h photoperiod and temperature regime of 25 ± 2°C.

Shoot proliferation

From the established cultures, 1-2 cm long axillary proliferated shoots were aseptically isolated and transferred to proliferation media comprising MS medium supplemented with various concentrations of BAP (0.5, 1.0, 2.0 mg l^{-1}), kinetin (0.5, 1.0, 2.0 mg l^{-1}) and 40 mg l^{-1} adenine sulphate (Ads) (Table 1). For shoot multiplication subculture was made on similar medium after 3 weeks of culture.

Rooting

After the 30 days of subculture, individual shoots were transferred on to rooting-cum-shoot elongation medium. Different media were employed for this purpose, i.e. MS medium containing 40 g l^{-1} sucrose and supplemented with different concentrations of auxin (IBA and NAA) at 0.5, 1.0, and 2.0 mg l^{-1} with or without 200 mg l^{-1} activated charcoal (Table 2).

In vitro hardening and field transfer

After rooting and shoot elongation, the plantlets were transferred to different *in vitro* hardening treatments for improving their *ex vitro* survival. A mixture of soilrite: peat (1:1) was used after moistening with 1/4th MS salts minus organics. All the hardening treatments were maintained at 27 ± 2°C up to complete hardening under 16/8 h photoperiodic regime. After two weeks the plants under *in vitro* hardening were regularly watered at two-three days interval with gradual removal of cover. Hardened plants were successfully transferred to greenhouse and subsequently to the field.

The experiment was laid out in complete block design with minimum thirty explants per treatment replicated thrice. The percentage data was subjected to arc sin% transformation before analysis.

Observations

Culture establishment and proliferation

When attempts were made to establish the *in vitro* culture from woody species lead to heavy phenol exudation, which resulted in explant necrosis. The effect of activated charcoal supplementation and growth regulators on culture establishment from nodal segment is presented in Fig. 1. It is clear that the treatment T7 recorded the best response for culture establishment and also axillary bud induction. Whereas, the treatment T6 showed poor response. The phenol exudation was not observed as a problem as all the treatments had the supplementation of activated charcoal and fast subculturing on to the fresh medium.

After 40-45 days, the axillary shoots aseptically transferred to the shoot proliferation medium having treatments of cytokinin(s) and adenine sulphate (Table 1). Treatment of MS medium supplemented with 1.0 mg l^{-1} BAP + 1.0 mg l^{-1} Kin + 40 mg l^{-1} Ads showed the highest proliferation (89.8%). Minimum (21.3) days were required for shoot proliferation in this treatment. The number (1.3) of shoots and mean shoot length (0.75cm) were also registered highest for the above treatment.

Table 1 Effect of cytokinin(s) and adenine sulphate on shoot proliferation in pomegranate

Treatment (mg l^{-1})	*Proliferation (%)*	*Days to shoot proliferation*	*No. of shoots per explant*	*Mean length of shoots (cm)*
BAP-0.5	60.4 (51.0)*	22.4	1.7	1.2
BAP-1.0	53.5 (59.0)	20.8	2.1	1.8
BAP-2.0	56.4 (37.1)	28.4	1.9	0.9
Kin-0.5	45.8 (42.6)	26.7	1.6	1.0
Kin-1.0	59.6 (50.5)	29.5	1.3	0.9
Kin-2.0	28.7 (32.3)	31.4	0.8	0.5
BAP-0.5 + Kin- 0.5 + Ads-40	75.6 (60.4)	19.6	2.9	1.5
BAP-1.0 + Kin-1.0 + Ads-40	89.8 (71.3)	21.3	3.4	2.2
BAP-2.0 + Kin 2.0 + Ads- 40	83.4 (69.8)	33.3	1.3	0.75
C.D. at 5%	4.52	2.85	0.43	0.21

*Transformed values

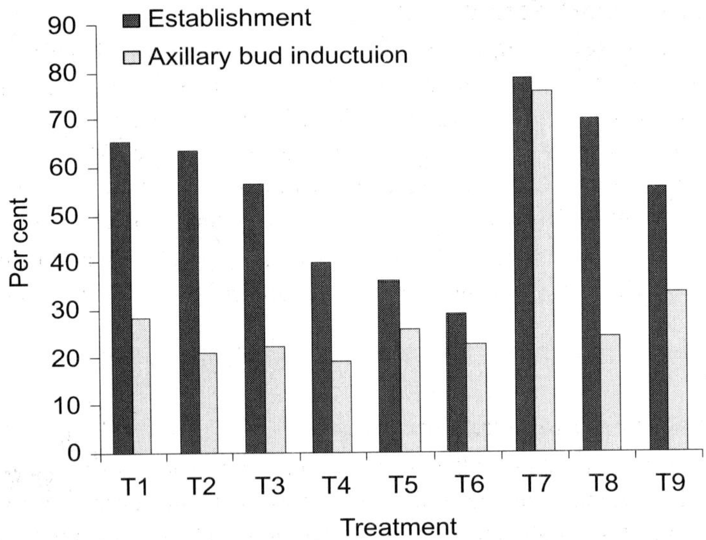

Fig. 1 Effect of activated charcoal and growth regulators on culture establishment from nodal segment in pomegranate.

Rooting and hardening

The effect of IBA and NAA with or without activated charcoal on rootability of microshoots is shown in Table 2. The MS medium containing 2.0 mg l^{-1} IBA + 200 mg l^{-1} AC recorded highest rooting percentage (88.7) having 4.2 roots per shoot and mean shoot length as 3.4 cm. Whereas, MS

Table 2 Effect of auxin with or without activated charcoal on root and shoot elongation of microshoots

Treatment (mg l^{-1})	*Rooting (%)*	*No. of roots per microshoot*	*Mean length of roots (cm)*
IBA-0.5	47.8 (43.7)*	2.3	2.3
IBA- 1.0	60.0 (50.8)	2.1	2.7
IBA-2.0	40.5 (39.5)	2.6	2.1
NAA-0.5	19.5 (26.2)	2.8	1.4
NAA-1.0	21.4 (27.5)	3.1	1.1
NAA-2.0	31.3 (34.2)	3.4	0.8
IBA-0.5 + AC- 200	67.5 (55.2)	3.8	3.2
IBA-1.0 + AC- 200	78.5 (62.3)	4.2	3.9
IBA-2.0 + AC- 200	88.7 (69.4)	4.2	3.4
NAA-0.5 + AC-200	50.4 (45.2)	4.4	1.5
NAA-1.0 + AC-200	40.8 (39.7)	4.6	1.0
NAA-2.0 + AC-200	34.5 (35.9)	5.2	1.2
C.D. at 5%	4.27	1.12	1.02

*Transformed values

medium supplemented with 2.0 mg l^{-1} NAA +200 mg l^{-1} AC resulted in only 34.5% rooting but highest number of short roots per shoot. MS medium supplemented singly with IBA or NAA showed poor rooting response. When both the auxins were supplied in combination with 200 mg l^{-1} activated charcoal higher number of roots and over all improvement in rooting was noted.

The rooted plantlets upon transfer to the glass jar with polypropylene cap containing sterile mixture of peat: soilrite (1:1), moistened with ¼ MS (1962) macro and micro-salts without organics recorded the highest (86.5 and 78.7 %) success after 30 and 45 days of transfer (Table 3). Plastic pots containing the sterile potting mixture covered with polythene showed very high mortality of plants, whereas plastic pot containing the same sterile potting mixture covered with inverted beaker showed moderate mortality rate.

Table 3 Effect of different hardening methods on acclimatization of *in vitro* rooted plantlets

Method	*Survival (%)*	
	30 days	*45 days*
I. Plastic pots with polythene cover (Peat; soilrite: 1:1)	45.8 (42.6)*	36.8 (37.3)*
II. Plastic pots with inverted beaker (Peat: soilrite; 1:1)	68.9 (56.1)	50.4 (45.2)
III. Glass jar with Polypropylene cap(Peat: soilrite; 1:1)	86.5 (67.6)	78.7 (62.5)
C.D. at 5%	11.2	8.9

*Transformed values

Discussion

Culture establishment from field-grown plants is a very critical process as it is met with an array of problems viz. phenol exudation, microbial contamination etc. These contaminations can be avoided by using the surface sterilization treatments. Phenol secreted from the cut ends of explant lead to medium browning and other problems related to culture and explant growth and development. To avoid/ minimize phenol exudation problem, various workers tried different methods. The treatment comprising antioxidants as pretreatment or addition to culture medium or transferring the explant to fresh medium after short interval or sealing of the cut ends with paraffin wax and/or addition of activated charcoal in the culture medium as an adsorbent. In the present study, culture initiation was improved when ½ strength MS medium with the addition of 200 mg l^{-1} activated charcoal. Use of such compounds has been suggested to control phenol exudation during culture establishment by different workers (Naik *et al.* 1999; Murkute *et al.*, 2002).

Wickson and Thiman (1958) discovered the cytokinins which could be used to release apical dominance in the lateral buds when they are made into small form in an intact stem tip. The dormant buds of vegetative apex are stimulated to grow and elongate into presence of cytokinins and also produce new axes. This effect was effectively utilized to initiate cultures from dormant axillary buds.

For the shoot proliferation BAP is effective when used in combination with an auxin. Naik *et al.* (1999) reported the efficacy of BA and zeatin for effective shoot proliferation and found zeatin to be more effective for shoot proliferation. In the present study, the highest number of shoots (3.4)

per explant was observed in the BAP 1.0 mg l^{-1} + Kin 1.0 mg l^{-1} + 40 mg l^{-1} Ads + 40 g l^{-1} sucrose. Both BAP and kinetin levels when tried alone in MS medium showed poor shoot proliferation response. Earlier, Naik *et al.* (2000) reported the proliferation from the cotyledonary nodes of *in vitro* raised plantlets. While, Nataraja & Neelambika (1996) reported somatic embryogenesis and plantlet regeneration from petal culture of pomegranate cv. Jyoti on to the half-strength MS medium and sucrose (4%). Concurrent results were also recorded in the present experiment when proliferated shoots were subjected to the *in vitro* rooting and shoot elongation. MS medium containing 2 mg l^{-1} IBA + 200 mg l^{-1} AC + 4% sucrose.

Fig. 2 Micropropagation stages in pomegranate cv. Jyoti (a) culture initiation and axillary bud induction from nodal segment (b) rooting of the microshoot (c) hardened plantlet in glass jar filled with peat: soilrite (1:1) (d) plantlets growing in the glasshouse.

The ultimate goal of plant propagation is the large-scale transfer of the *in vitro* raised plantlets the natural environments at low cost along with high survival rate. To achieve this, it has been und that when *in vitro* raised plantlets were subjected to *in vitro* hardening treatments they tablish better as earlier reported for different grape genotypes (Singh *et al*, 2000). The peat and ilrite used in hardening mixture can be reused, which may thereby minimize the cost of plantlet oduction.

The results of the present investigation reveal that though the multiplication rate in cv. Jyoti moderate it could be enhanced with careful manipulation of culture conditions and the right oice of growth hormones in combinations which could provide a reliable protocol for large scale ant production in pomegranate. The different stages of the protocol developed have been given in g. 2.

EFERENCES

on. (1982) The Wealth of India: A dictionary of Indian Raw Materials and Industrial Products. Vol. VIII (Ph-Re), CSIR, New Delhi, pp. 317-324.

ansali R.R. (1990) Somatic embryogenesis and regeneration of plantlets in pomegranate, *Annals of Bot.* 66: 249-253.

dka K. & Mehra P.M. (1986) Morphogenesis in *Punica granatum* (pomegranate); *Canadian J. Bot.* 64: 1644-1653.

ırkute A.A., Patil S., Patil B.N. & Mayakumari. (2002) Micropropagation in pomegranate callus induction and differentiation; *South Indian Hort.* 50(1-3): 49-55.

ırashige T. & Skoog F. (1962) A revised medium for rapid growth and bioassays with tobacco tissue culture. *Physiol. Plant.* 15: 473-497.

ıscarenhas A.F., Gupta P.K., Kulkarni V.M., Mehta U.J., Iyer R.S., Khupse S.S. & Jagannathan V. (1981) Propagation of trees by tissue culture; In: *Proc. Costed Symp. Tissue Culture of Econ. Imp. Plants.* Singapore pp. 175-179.

ik S.K., Pattanaik S & Chand P.K. (1999) *In vitro* propagation of pomegranate (*Punica granatum* L. cv. Ganesh) through axillary shoot proliferation from nodal segment of mature tree; *Scientia Hort.* 79: 175-183.

ik S.K., Pattanaik S. & Chand P.K. (2000) High frequency shoot proliferation and plant regeneration from cotyledonary nodes of pomegranate (*Punica granatum* L); *Scientia Hort.* 85: 261-270.

taraja K. & Neelambika G.K. (1996) Somatic embryogenesis and plantlet regeneration from petal culture of pomegranate, *Punica granatum* L; *Indian J. of Exp. Biol.* 34: 719-134.

ıura M., Matsuta N., Moraguchi T. & Kozaki I. (1987) Adventitious shoot and plantlet formation from cultured pomegranate leaf explants; *Hort. Sci.* 22(1): 133-134.

gh S.K., Sharma H.C., Singh S.P. & Sharma R.R. (2000) Propagating grapes through repetitive microcutting technique. *Indian Horticulture.* 3 pp. 14-15.

ckson M.E. & Thimann. (1958) The antagonism of auxin and kinetin in apical dominance. *Physiol. Plant.* 11: 62-74.

ıg Z.H. & Ludders P. (1995) Organogenesis of *Punica granatum* L var. Nana; *Angewandte Botanike* 67(5-6): 151-156.

ang B. & Stoltz L.P. (1991) *In vitro* shoot formation and elongation of dwarf pomegranate; *Hort. Sci.* 26(8): 10-84.

CHAPTER

13

Hardening of *Garcinia indica*

Meera Chabukswar, Manjushri Deodhar and Mohan Heble*

ABSTRACT

Fruits of *Garcinia indica* chois. (Family: Clusiaccac) are a rich source of antiobesity drug hydroxycitric acid (HCA). The fruits are also used to prepare pleasant, attractive red beverage which has bilious action. The fats extracted from seeds are used in the several pharmaceutical and cosmetic products. The plant is in great demand. In the present paper, attempts have been made for micropropagation of the tree species.

Shoot initiation was obtained from immature seed explants when cultured on MS medium containing NAA (0.5 mg/l) + BAP (2 mg/l) + kin (0.2 mg/l). Individual shoots were then transferred to ½ MS medium with BAP (0.2 mg/l) + kin (0.2 mg/) for elongation. For root initiation the shoots (about 2 cm height) were subjected to 2000 ppm of IBA dip for 30 sec. 100% plants showed rooting and mean number of roots were 2-3 per plant.

The rooted plants of about 2-4 cm height (2-3 nodes) were subsequently transferred to different media composition for acclimatization. (A) 100% cocopeat (B) 50% cocopeat +50% sand (C) 25% cocopeat +50% sand +25% soil. The plants grown in 100% cocopeat showed 75% of survival. The plants in 50% cocopeat and 50% sand showed 74% of survival. The hardened plantlets were subsequently acclimatized in the greenhouse and field conditions.

Key Words: *Garcinia indica*, Micropropagation, Hardening:

INTRODUCTION

Garcinia indica Chois. (Family–Clusiaceae) occurs in tropical rain forests of Western Ghats of India. The natural flora shows wide variation in the fruit yield and quality. Fruits of *G. indica* obtained from coastal regions of Maharashtra, are a rich source of hydroxycitric acid (HCA), an important biologically active plant metabolite used as an antiobesity drug (Watson and Lowenstein,

*Address for correspondence: Dept. of Botany, KET's V.G. Vaze College, Mulund (E.), Mumbai-400 081.
*KET' s Scientific Research Centre, Mithagar Rd., Mulund (E.), Mumbai 400081,India, Tel: +91-22-25900392/93, +91-22-25671422; Fax: +91-22-25647334; E-mail: meera_cha@yahoo.com (Meera Chabukswar), kelkar_srcmulund@hotmail.com (M.A. Deodhar).

1970). The plants are polygamodioeious (Anonymous, 1956) and differentiation between male a female plants can only be done at the flowering stage. Tissue culture methods therefore not on offer a source for cloning the elite explants of *Garcinia indica* but also multiply the female proge through grafting procedures. In the present paper, attempts have been made for micropropagation the tree species.

EXPERIMENTAL

In Vitro Regeneration of the Plant

In vitro regeneration of the plantlets was achieved by the method described by (Kulkarni ar Deodhar, 2002).

The immature fruits of about 12 weeks old were collected from Konkan Krishi Vidyapeet Dapoli, Maharashtra, India. The immature fruits about 1.5 to 2 cm in diameter were washed running tap water, immersed in 70% alcohol and flamed for surface sterilization. The fruits were c open aseptically. The seeds from the fruits were cut into two segments and inoculated in M medium (Murashige and Skoog, 1962) supplemented with 3% (w/v) sucrose, 0.8% (w/v) agar ar containing NAA (0.5 mg/l) BAP (2 mg/l) and Kin (0.2 mg/l). The cultures were incubated at 25 20°C under illumination 40 μMol m^{-2} S^{-1} (2000lux) with 16 hrs day light photoperiod. After subcultures in the same medium the explants produced multiple shoots at the rate of 16.7 p explant. After 4/5 subcultures the shoot initials about 0.5 cm in height were separated a inoculated on elongation medium containing ½ MS + BAP (0.2 mg/l) + Kin (0.2 mg/l). For ro initiation the *in vitro* shoots (about 2 cm in height) were immersed in 2000 ppm IBA solution for 3 second, and then transferred to half MS medium containing 2% sucrose and 0.6% Agar. Near 100% plants rooted in about 15 days. The average number of roots was 2 to 3.

Hardening

Hardening of rooted plantlets

Rooted plantlets with 2/3 nodes and having at least 2 roots were washed carefully to remove trac of agar then transferred to the pots containing different autoclaved potting mixture thorough drenched with bavistin.

(A) 100% cocopeat

(B) 50% cocopeat + 50% sand

(C) 25% cocopeat + 50% sand + 25% soil

The potted plantlets were acclimatized in the mist house at 20 to 30°C temperature and 14 hr illumination. The pots were covered with tight plastic cap to prevent desiccation to avoid rap changes in environment. Plastic cover was gradually perforated after 15 days and after 1 mon plastic cover was entirely removed.

After 2 months from weaning in the mist house the plants were repotted in pots containin mixture of soil, sand and compost (2: 1: 1) and keep in net shade house for 1 month. During first 1 days of acclimatization no watering or any fertigation was provided to plants. After removing th plastic bags watering was given to the plant every morning and evening by only sprayer. The plants were then transferred to nursery garden. The plants are ready for plantation.

Hardening of unrooted plantlets

In vitro shoots (about 2/3 cm) were first subjected to:

(1) Keradix, (2) Rootex-1, (3) Rootex-3 dip and then transferred to pots containing sterile 100% cocopeat. The potted plantlets were acclimatized in the mist house at 20 to 30°C temperature and 14 hrs. illumination. The pots were covered with tight plastic cap to prevent desiccation to avoid rapid changes in environment. Plastic cover was gradually perforated after 30 days and after 2 month plastic cover was entirely removed.

After 3 months from weaning in the mist house the plants were repotted in pots containing mixture of soil, sand and compost (2: 1: 1) and kept in net shade house for 1 month. These plantlets were kept outside from net shade house after 1 month. The plants are ready for plantation.

RESULTS

Shoot Establishment from *In Vitro* Explants

The maximum shoots were obtained in MS medium containing NAA (0.2 mg/l) + BAP (2 mg/l) + Kin (0.2 mg/l). After 3 subcultures in the same medium the explants produced multiple shoots at the rate of 16.7 per explant. After 4/5 subcultures the shoot initials about 0.5 cm in height were separated and inoculated on elongation medium containing ½ MS + BAP (0.2 mg/l) + Kin (0.2 mg/l) (Fig. 1).

For root initiation the *in vitro* shoots (about 2 cm in height) were immersed in 2000 ppm IBA solution for 30 seconds and then transferred to half MS medium containing 2% sucrose and 0.6% agar. Nearly 100% plants rooted in about 15 days (Fig. 1). The average number of roots was 2/3. After 1 month from rooting these rooted plantlets were subsequently transferred to different media combinations for hardening.

Hardening of the Plants

A) 100% cocopeat.

B) 50% cocopeat + 50% sand.

C) 25% cocopeat + 50% sand + 25% soil.

Table 1 Observation for rooted plantlets

Combination of different types of medium	*Survival % after 20 days*	*Survival % after removal of plastic bags i.e. 30 days*	*Survival % in net shade after 2 months*	*% of new leaf formation*	*Days required new leaf*
A	96.25%	75.3%	75.3%	99%	10
B	82.5%	74.2%	74.2%	75%	17
C	81.2%	67.6%	67.6%	75%	17

*240 plants were taken for this experiment.

Fig. 1 **(a)** Shoot initiation from seed explant (MS+2 mg/l BAP+0.2 mg/l NAA+0.2 mg/l Kin) **(b)** Multiple shoots developed from the seed explant (MS+2 mg/l BAP+0.2mg/l NAA+0.2 mg/l Kin) **(c)** Rooted plantlet (1/2 MS+20% sucrose+0.6% agar+2000 ppm IBA dip 30 sec.) **(d)** Hardened plant with secondary and tertiary roots.

Out of 3 potting mixtures utilized for hardening procedure the better survival percentage was found in combination 'A' and 'B' (75.3%, 74.2% respectively) as compared to 'C' (67.6%) combination. But in combination 'A' the new leaf pair was formed after 10 days contrary to this in 'B' and 'C' combinations the new pair appeared after about 17 days. In combination 'A' almost 99% plants showed new leaf formation (Fig. 2).

Fig. 2 **(a)** Hardened (rooted) plantlets in different potting mixtures (A) 100% cocopeat (B) 50% cocopeat + 50% sand (C) 25% cocopeat + 50% sand + 25% soil **(b)** Hardened plants in the potting mixture of soil, sand and compost (2:1:1). **(c)** Hardened (unrooted) plantlets showed rooting after 2 months (1) Keradix (2) Rootex-1 (3) Rootex-3

Table 2 Observation for unrooted plantlets.

Different rooting hormones	*Survival % after 30 days*	*Survival % after removal of plastic bags i.e. 60 days*	*Rooting % in particular rooting hormone powder*
Keradix	67.5%	20%	20%
Rootex-1	92.5%	45%	45%
Rootex-3	90.0%	55%	55%

*120 plants were taken for this experiment.

In vitro shooting when subjected to synthetic rooting mixtures like Keradix, Rootex-1 and Rootex-3 and then hardening also showed 20%, 45%, 55% rooting respectively.

DISCUSSION

A mass propagation of *Garcinia indica* from seed explants can be achieved by following different cultured steps—shoot initiation, elongation and rooting. Using single seed explants as high as 16.7 shoot were produced and all these shoots were successfully rooted.

Hardening is a crucial step prior to transplantation of plants to the soil. The *in vitro*-regenerated plantlets are very delicate and show wilting when exposed to environment in the mist house. Plants are therefore allowed to grow on rooting media for about 1 month after root initiation. During this phase the nutrients in the culture go on gradually depleting and plants become sturdy and easy to acclimatize in mist house (Purohit *et al.*, 1998).

Out of 3 potting mixtures utilized for hardening procedure the better survival percentage was found in combination 'A' and 'B' (75.3%, 74.2% respectively) as compared to 'C' (67.6%) combination. But in combination 'A' the new leaf pair was formed after 10 days contrary to this in 'B' and 'C' combinations the new pair appeared after about 17 days. In combination 'A' almost 99% plant showed new leaf formation (Fig. 2).

In vitro rooted plantlets generated through IBA dip showed 2/3 adventitious roots from the lower nodes. During weaning procedure when the plants were transferred from pot to soil the hardened plantlets in all combinations 'A', 'B' and 'C' showed well-developed roots with large number of secondary and tertiary branches (Fig. 1).

In vitro shoots when subjected to synthetic rooting mixtures like Keradix, Rootex-1 and Rootex-3 showed variable degree of rooting, the plants were hardened in pots containing soil, sand and compost (2:1:1) showed 100% survival. Root hormone powder promotes rooting in *Garcinia indica* but mortality of plantlets was more as compared to rooted plantlets and duration of hardening was comparatively more. Efforts are still going on in improvement of the procedures. The tree species of *Garcinia indica* can be successfully grown by tissue culture method.

Acknowledgements

This research is supported by research grant SAN No. 102/ IFD/ SAN/ 1293/ 2002-2003 from Govt. of India, Department of Biotechnology. Authors thank Mr. G.D. Kelkar, Chairman, Kelkar

Education Trust's Scientific Research Centre and Dr. M. R. Kurup, Principal of KET's V. G. Vaze College for encouragement. They also thank Dr. D. G.Joshi, Konkan Krushi Vidyapeeth, Dapoli, for providing plant material.

REFERENCES

Anonymous. (1956) *The Wealth Of India—Raw Materials.* Council of Scientific and Industrial Research, New Delhi, India, 4: 102.

Kulkarni M.D. & Deodhar M.A. (2002) *In vitro* regeneration and hydroxycitric acid production in tissue cultures of *Garcinia indica. Indian Journal of Biotechnology*, Vol. 1, pp 301-304.

Purohit S.D., Tak K., Joshi P. & Samar D. (1998) Micropropagation of some woody tree species of Aravallis in Rajastan. In: *Plant Tissue Culture and Molecular Biology,* Srivastava PS (Ed.). 84-125.

Watson J.A. & Lowenstein J.M. (1970) Citrate and the conversion of carbohydrate in to fat. *Jour. Biol. Chem.* 245(22): 5993-6002.

CHAPTER

14

Production of Plumbagin—A Multifarious Medicinal Compound from Cultured Cells of *Plumbago rosea* L.

P. Komaraiah, R. Naga Amrutha, G. Jogeswar, N. Sita Lakshmi, S.V. Rama Krishna[1] and P.B. Kavi Kishor

ABSTRACT

Callus and suspension cultures derived from leaf explants of *Plumbago rosea* were established and plumbagin, a naphthoquinone, was isolated from them and confirmed by ^{1}H NMR and electron-ionization mass spectroscopy. Maximum content of plumbagin was obtained in the stationary phase of growth (4.3 mg/g dry cell weight). Media pH, phytohormones and carbon source were optimized for biomass and plumbagin accumulation. Cell aggregates measuring 500 mm in diameter produced 8.2 g dry cell weight/l, but larger aggregates (above 500 mm) favoured plumbagin accumulation with an yield of 4.5 mg/g dry cell weight. Cell cultures were immobilized in calcium alginate and cultured in Murashige and Skoog's basal medium containing 10 mM $CaCl_2$ for the production of plumbagin. Studies were carried to find out the impact of immobilization on the increased accumulation of this secondary metabolite. Immobilization in calcium alginate enhanced the production of plumbagin by 3, 2 and 1-folds compared to that of control, uncrosslinked alginate and $CaCl_2$ treated cells respectively. Cell loading at a level of 20% to the polymer volume (sodium alginate) was optimal and maximum plumbagin was obtained. At higher cell loading (40-50%), lower plumbagin accumulation was noticed. Addition of 200 mg/l chitosan as an elicitor to the immobilized cells resulted in 8 and 2-folds higher accumulation of plumbagin over control and immobilized cells. Also, more than 70% of the plumbagin was released into the medium, which is highly desirable for easy recovery of the product. Sucrose utilization rate of the cells was higher when cells were subjected to *in situ* product removal using the resin Amberlite XAD-7. This may indicate that the toxicity of plumbagin was reduced on cells when it was removed from the medium. Cells subjected to combined treatments of chitosan, immobilization and *in situ* extraction showed a synergistic effect and yielded 92.13 mg/g dry cell weight of plumbagin which is 21, 5.7, 2.5-times higher than control, immobilized, immobilized and elicited cells respectively.

*Address for correspondence: Department of Genetics, Osmania University, Hyderabad 500 007.
[1]Biochemical and Environmental Engineering Lab, Indian Institute of Chemical Technology, Hyderabad 500 007.

INTRODUCTION

Plumbago rosea Linn (synonym *P. indica*) belongs to the family Plumbaginaceae. Bright red flowers in axilary or terminal racemes characterize this plant. The genus *Plumbago* includes about 10 species (Deb, 1983), namely, *P. auriculata*, *P. coerulea*, *P. europea*, *P. pearsonii*, *P. pulchella*, *P. rosea*, *P. scandans*, *P. tristis*, *P. wessi*, and *P. zeylanica*. *P. auriculata*, *P. rosea* and *P. zeylanica* are found growing in India. Among them, *P. zeylanica* is grown widely in tropical and subtropical regions of Asia, Australia and Africa and hence, more extensive studies on this species have been carried out. Morphologically, *P. rosea* is distinctive with red petals and *P. auriculata* with *blue petals*, while all others have white petals. Several commercially important compounds have been isolated from these plants including plumbagin. Plumbagin (5-hydroxy, 2-methyl, 1,4-naphthoquinone) is an yellow coloured naphthoquinone and has anti-cancer (Parimala and Sachdanandam, 1993), anti-bacterial and anti-fungal activities (Krishnaswamy and Purushothaman, 1980, Didry *et al.*, 1994). Plumbagin is also found in some species of the families Droseraceae (Crouch *et al.*, 1990), Ebenaceae (Robinson, 1991) and Euphorbiaceae (Thomson, 1971). However, *P. rosea* grows slowly and it may take years to get plumbagin from the mature roots of intact plants (Kitanov and Poshankov, 1994). The endemic nature of *P. rosea*, long duration of cultivation and low yields of plumbagin limits its utility in the pharmaceutical industry (Sateeshkumar and Bhavanandan, 1988). These problems could be overcome by adopting cell culture techniques which offer an alternative method to produce pharmaceutically important compounds (Nahalka *et al.*, 1996). Earlier studies on *P. rosea* revealed the isolation and identification of several naphthoquinones, flavonoids, anthocyanidins, tannins and β-sitosterol from various plant parts such as leaf, root etc. In addition to plumbagin, several important naphthoquinones such as drocerone, isoshinanolone, 2-methyl naphthazarin, elliptinone, chitanone, chitranone, 3,3-biplumbagin, plumbazeylanone etc. were isolated from these plants (Dinda *et al.*, 1997).

Uses of Plumbagin

The roots of *P. rosea* are used as an important indigenous drug in the traditional medicinal systems. Plumbagin is a natural naphthoquinone showing a broad range of medicinal properties. Plumbagin is an effective chitin-synthase inhibitor and recently shown to have letshmanicidal activity. It can also be utilized for insect killing in agriculture (Kubo *et al.*, 1983) and also as an antileishmanial drug (Kayser *et al.*, 2000). Other important medicinal properties of plumbagin are given in Table 1.

Biogenesis of Plumbagin

In higher plants, four separate pathways, shikimate, homogentisate, the polyacetate-malonate and mevalonate might exist for the biogenesis of naphthoquinones. But, feeding experiments with acetate-1^{14}C-2-^{14}C and malonate-2-^{14}C in young shoots of *P. europea* suggested that plumbagin and its analogs are produced in higher plants via the polyacetate-malonate pathway (Durand and Zenk 1971, 1974). Plumbagin synthesis via the acetate-polymalonate (polyketide) pathway converts one molecule of acetyl-CoA and five molecules of malonyl-CoA to one molecule of plumbagin and is catalyzed by yet an unknown polyketide synthase in plants. But, complete biosynthetic pathway of plumbagin is not yet known.

Table 1 Medicinal and other properties of plumbagin

Use	*Reference*
Antimicrobial	Lima *et al.*, 1968; Durga *et al.*, 1990; Fuji *et al.*, 1992 Lakshmi *et al.*, 1987; Ray and Majumdar, 1976
Antileishmanial	Craft *et al.*, 1985
Anticancer	Krishnaswamy and Purushothaman, 1980 Chandrasekharan *et al.*, 1982; Krehar *et al.*, 1990
Antifertility, Abortifacient	Bhargava 1984; Bhargava and Dixit, 1985
Anticoagulant	Santhakumari *et al.*, 1978
Bronchial infection	Denoel, 1949
Antifeedant	Antarkar *et al.*, 1980
Insecticidal	Kubo *et al.*, 1983
Molluscicidal	Marston *et al.*, 1984
Marston and Hostettmann 1985	
Adult sterilant effect	Joshi *et al.*, 1988
Larval mortality effect	Ghosh *et al.*, 1994
Cardiotonin	Itoigawa *et al.*, 1991
Antinematodes	Fetterer and Fleming, 1991
Antimutagenic	Farr *et al.*, 1985; Durga *et al.*, 1992

In Vitro Studies on *P. rosea*

P. rosea is highly endemic in nature and grows in parts of Kerala, Assam, Tripura and West Bengal. Added to this, the most exploited source of plumbagin is the tuberous roots of *Plumbago* species (*P. europea*, *P. rosea* and *P. zeylanica*). However, these plants grow quite slowly and the roots suitable for extraction may take years to grow (Kitanow and Pashankov, 1994). Frequent harvesting of the natural populations has resulted in the extinction of many of the important plants. Therefore, the search for alternative and more effective source of plumbagin is necessary. One of the possibilities was a synthetic approach (Wurm and Gurka, 1986), but each method published so far has been ineffective from the commercial point of view. Due to low productivity or complicated cultivation, extraction of plumbagin from other sources also proved to be uneconomical. Tissue culture technique has widely been accepted as a tool for the production of medicinal products (Bajaj *et al.*, 1988; Bajaj, 1989). The roots of *P. rosea* are the accepted source of drug in Kerala and are locally known by the vernacular name 'chettikotuveli'. The plant is propagated vegetatively by offsets or small cuttings and has never been known to bear fruits (Anonymous, 1969). Apart from pharmacological and photochemical aspects, studies on cell cultures of *P. rosea* were very few. Tissue culture studies have been attempted earlier for micropropagation of *P. rosea* cultures and were raised from young leaf and stem on MS medium supplemented with various concentrations of growth hormones (Sateeshkumar and Bhavanandan 1989; Harikrishnan and Hariharan, 1996). Plant regeneration directly from leaves or internodes (Harikrishnan and Hariharan, 1996) or from callus cultures has also been reported (Komaraiah *et al.*, 2006).

Production of Plumbagin from Cell Cultures of *Plumbago rosea*

The production of low amounts of pharmaceutically important compounds in intact plants can be overcome by adopting cell culture techniques which offer an alternative method to produce secondary products (Nahalka *et al.*, 1996). Heble *et al.* (1974) reported production of plumbagin from callus cultures of *Plumbago zeylanica* and from cell and hairy root cultures of *P. rosea* by Sateeshkumar and Seeni (2002) and by Komaraiah *et al.* (2001; 2002a; 2002b; 2003). In the present investigations, an attempt was made to increase the production of plumbagin using strategies viz. optimization of media components, cell line development by single cell cloning, elicitation, permeabilization, immobilization, *in situ* extraction of product and cultivation in different modified bioreactors in different modes of operation etc. This review focuses on recent progress made in the production of plumbagin from cultured cells.

Callus and suspension cultures were established from the leaves of *P. rosea* in our laboratory earlier. The quantity of plumbagin content varied with the culture conditions. Plumbagin production was growth-associated and the content gradually increased up to 4.3 mg/g dry cell wt (DCW)) along with an increase in biomass up to 16th day (Fig. 1). Both biomass and plumbagin were optimal when the pH of the medium was between 5.2 to 6.0 (Fig. 2). Phytohormonal concentrations and combinations considerably influenced the growth and metabolite production in *P. rosea*. Optimum hormonal combination for cell growth (8.3 g DCW/l) was 1.5 mg/l IAA, 0.5 mg/l NAA and 0.3 mg/l BAP, whereas for plumbagin production (4.9 mg/g DCW) it was 1 mg/l IAA plus 0.3 mg/l BAP. Nahalka *et al.* (1996) reported enhanced plumbagin accumulation in cell suspensions of *Drosophyllum* when a combination of NAA, IBA and BAP were used in the medium. Glucose and fructose showed a good effect on biomass and plumbagin production and

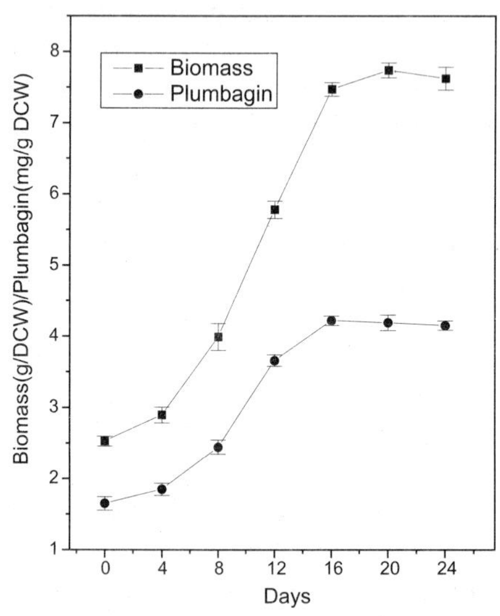

Fig. 1 Time course study of *P. rosea* cell cultures for the production of biomass (g dry cell wt/l) and plumbagin (mg g-1 dry cell wt).

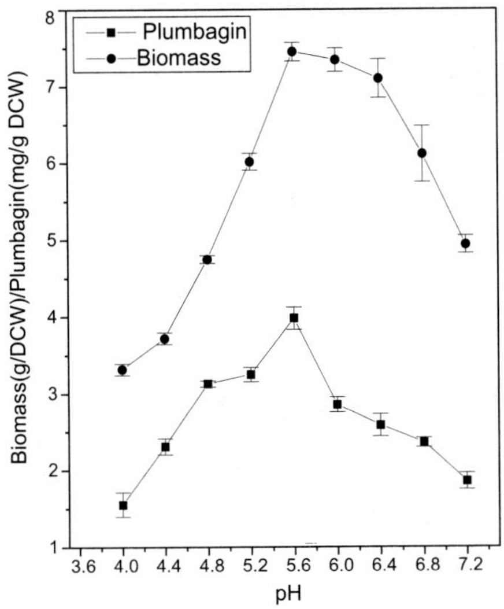

Fig. 2 Influence of pH on the production of biomass (g dry cell wt/l) and plumbagin (mg g-1 dry cell wt).

accumulation, which is comparable to that of sucrose. Low biomass and metabolite accumulation was recorded in cultures grown on lactose or glycerol as carbon sources. Carbon sources are known to influence growth of callus (Kavi Kishor and Mehta, 1987) as well as secondary metabolites like lignins in *Ipomoea* (Paska *et al.*, 1999).

The role of cell aggregate size on culture growth and metabolite synthesis was investigated by inoculating cell aggregates of different sizes (Table 2). Cell aggregates at 500μm diameter favoured biomass accumulation (8.3 g DCW/l) and as cell aggregate size increased to 2000μm, biomass accumulation decreased (6.5 g DCW/l). Conversely, plumbagin accumulation increased (4.5 mg/g DCW) with an increase in cell aggregate size (1500 μm). Low plumbagin concentration (3.1 mg/g

Table 2 Effect of cell aggregate sizes on biomass and plumbagin production

Days	*Cell aggregate sizes in diameter*							
	500 μm		*1000 μm*		*1500 μm*		*2000 μm*	
	a	b	a	b	a	b	a	b
0	2.49	0.36	2.34	1.34	2.40	1.60	2.44	1.32
4	3.32	0.43	2.74	1.71	2.77	1.80	2.55	1.51
8	4.51	0.58	3.60	1.95	3.57	2.26	3.17	1.67
12	5.69	0.76	4.90	2.46	5.40	3.42	4.07	2.47
16	7.40	0.94	6.84	2.74	6.50	3.97	5.58	3.04
20	7.91	1.43	7.50	3.28	7.53	4.39	6.35	3.17
24	8.29	1.83	7.78	3.56	7.64	4.57	6.53	3.09

(a) Biomass, gram dry cell wt 1^{-1} (b) plumbagin, mg g^{-1} dry cell wt.

DCW) at 2000μm cell aggregate size indicates that the mass transfer effects might govern the metabolite synthesis. Therefore, a certain degree of cell aggregation may be required for secondary metabolite synthesis as also pointed out by Madhusudhan and Ravishankar (1996).

Detection of Plumbagin

Lyophilized root tissues or callus or suspension cultures were extracted with methyl alcohol in a Soxhlet extraction apparatus for 24 h and the extracts were concentrated under vacuum. The crude extract was loaded into a silica column and eluted with hexane and ethyl acetate in the ratio of 10:1. Purity of the fractions was checked by both TLC and HPLC. Plumbagin was identified by ^{1}H NMR and electron-ionization mass spectra and the data were consistent with the authentic compound obtained from the Sigma Chemical Company, USA. After every experiment, the medium and roots were separated and the medium was extracted three times with equal volumes of ethyl acetate. The pooled fractions were treated with anhydrous sodium sulfate and reduced under vacuum. The residues were re-dissolved in 1 ml of HPLC mobile phase and stored at 4°C. The cells were extracted thrice with methanol in a porcelain mortar using neutralized sand. Methanol was reduced under vacuum and re-dissolved in 1 ml of HPLC mobile phase and stored at 4°C till they were analyzed. Plumbagin content from different samples was determined by HPLC using two Shimadzu LC 10 AD pumps; SPD 10A UV-VIS detector, the column m Bondapak C18 (3.9 × 300 mm) with the detection wavelength of 254 nm. The mobile phase was methanol, water (80:20) with 0.1% trifluoroacetic acid and the flow rate was set to 1 ml/min. Quantitative analysis was carried out by comparing the peak area with the authentic sample of plumbagin from the Sigma Chemical Company, USA. Erlenmeyer flasks (250 ml capacity) containing 50 ml of growth medium were used for growing isolated roots. Fresh roots (500 mg) were used as inoculum and cultures were harvested after 20 days of incubation. Plumbagin is expressed in mg/g DW. The data presented are mean of 6 replicates and standard error is presented as error bars.

Screening and Selection of *P. rosea* Cell Cultures for Stable Cell Lines

In plant cell cultures genetic variation occurs spontaneously. In order to find the best possible variant cell line, large number of cultures needs to be screened for higher accumulation of secondary metabolite or pharmacologically important compounds. Subsequently, the variants need to be isolated from the established lines, maintained and tested for their stability of production. In *P. rosea* cell cultures, initially a large degree of variation in growth and plumbagin production pattern was observed and the results were not comparable with one culture passage to the other. To avoid this instability in growth and metabolite production, several cell lines were screened using cell aggregate cloning by visual and analytical methods. The biomass and plumbagin accumulation pattern of several cell lines from single leaf derived callus were studied for a period of six months. The cell lines PR8 and PR9 accumulated the highest biomass ranging from 12 to 15 g DCW/l but plumbagin synthesis in these cell lines was very less (1 to 2.5 mg/g DCW). In case of cell lines PR3 and PR10, plumbagin biosynthesis was reasonably good but the biomass accumulation was limited to 5 to 7 g DCW/l. Of all the cell lines studied, PR1 produced good amount of both biomass and plumbagin (Fig. 3).

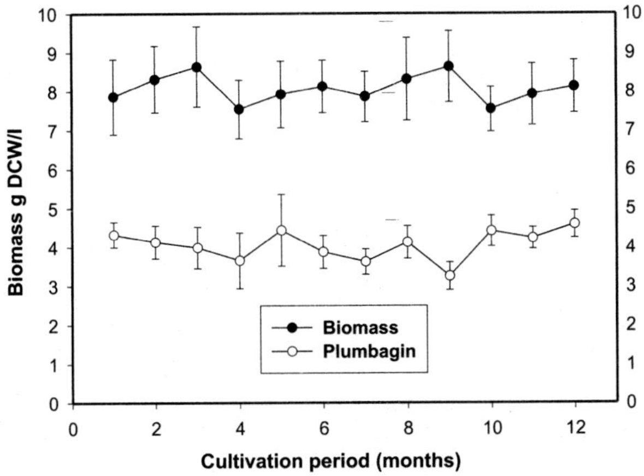

Fig. 3 Stable production of biomass and plumbagin by *P. rosea* cell line PR No. 1

Moreover, this cell line was stable and the stability in both biomass and plumbagin accumulation was maintained for a period of 12 months with regular subcultures. This cell line was therefore, selected and used for further studies by testing its stability for growth and plumbagin accumulation now and then. In all other cultures, the degree of variation in biomass and plumbagin accumulation was very high and hence could not be used further.

Elicitor Enhanced Production of Plumbagin from Cell Cultures

In recent years, various plant cell culture systems were exploited for the enhancement of high value metabolites. Elicitation was most successful in cell cultures of various plant species (Eliert, 1987; Asada and Shuler, 1989). For example, the elicitors prepared from *Rhizoctonia solani* enhanced the solavetivone production in *Hyoscyamus muticus* (Ramakrishna *et al.*, 1993) and *Asperigillus flavus* mycelial extract elicited anthocyanin content in *D. carota* cell cultures (Rajendran *et al.*, 1994). Increased levels of benzylacetone and raspberry ketone in suspension cultures of *Rubus idaeus* by methyl jasmonate (Pedapudi *et al.*, 2000) and chitosan induced indurubin production in *Polygonum tinctorum* suspension cultures (Kim *et al.*, 1997) was also reported. Most of the elicitors used were obtained from fungal, bacterial and yeast cell wall extracts or their purified fractions. In addition, some abiotic stress agents like heavy metals, UV light, osmotic stress and pH are also known to induce secondary metabolite production in cultured cells (Eliert, 1987). However, no specific elicitor showed a general effect on different plant species. Also, there is no way to predict that an elicitor will be effective in a specific cell system on metabolite accumulation. The effect of different elicitors prepared from fungi (*Aspergillus niger* and *Rhizopus oryzae*), bacteria (*Bacillus subtilis* and *Pseudomonas aeruginosa*), yeast and chitosan were tested on plumbagin accumulation. As an extension to this dose response of elicitors, effect of culture age on elicitor treatment (Fig. 4) time course of the elicitors (Fig. 5) and their influence on extracellular metabolite production was studied. In preliminary experiments, all the elicitors (biomass and spent medium) were screened for their influence on plumbagin production in *P. rosea* cell cultures. In all the cases elicitors prepared

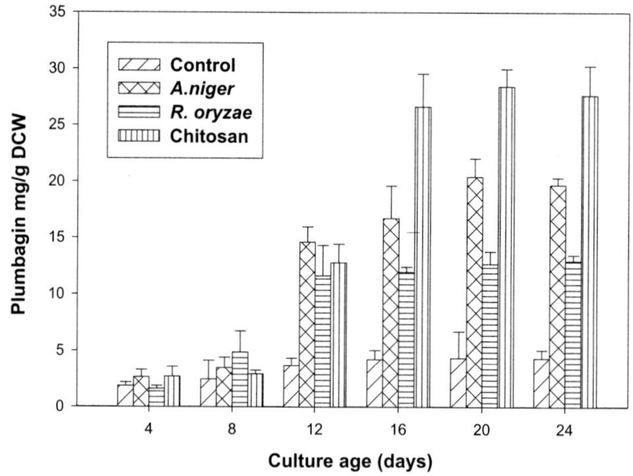

Fig. 4 Elicitors added at their optimum dose levels. Control cells received same amount of distilled water and harvested after 48 h of incubation. Data shown are mean of six replicates and ± S.D. values presented as error bars.

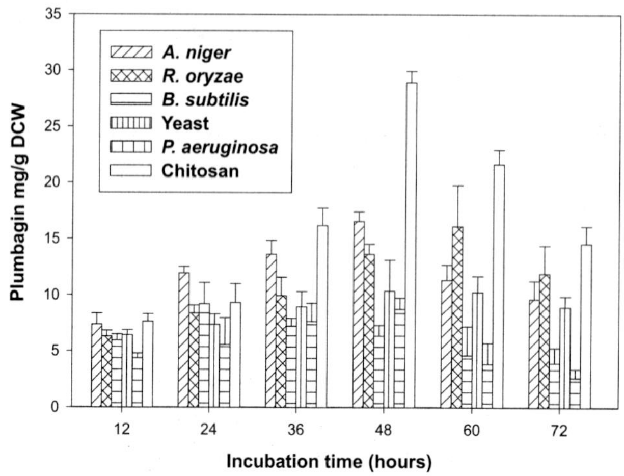

Fig. 5 Optimum concentrations of the individual elicitors to 16-day-old cultures. Control cultures were added same amounts of distilled water. Control cells produced 4.31 (± 0.24) mg/g DCW. The data shown are mean of six replicates and ±S.D. values presented as error bars.

from mycelia or cell extracts showed better effect over their spent media (results not shown). The elicitors prepared from microorganisms were diluted appropriately to give the test range with sterile distilled water. The best doses of the elicitors were selected based on their maximum effect and used for further studies. All the six elicitors used in this study showed positive effect on plumbagin accumulation, at their optimum doses.

Effect of Culture Age

The response to elicitation depends on growth phase of the culture, which not only affects the quantitative response but also the product pattern in general (Hahn and Albersheim, 1978). Figure 4 represents the relation between culture age and the effect of elicitation on plumbagin production. When the cultures were at an early stationary phase, the effect of elicitation was maximum and more or less the same pattern continued till the end of culture period. Several workers studied the optimum age of the culture for elicitation in different plant cell systems. In majority of the cases, response of the cells was highest if they were challenged at the end of growth phase or at the onset of stationary phase (Eliert, 1987). On the other hand, thiophenes were produced during the exponential phase of growth in *T. patula* hairy root cultures (Buitelaar and Trampar, 1992). In the present study, it was observed that the optimum age of the culture for elicitation was 16-24 days, by which time the growth phase sieged and the cells entered into the stationary phase. Elicitors that resulted in more than threefold increase in production of plumbagin over control were used for further studies.

Incubation Time

The effect of treatment time with elicitors is presented in Figure 5. Effective induction of secondary metabolite in *P. rosea* cultures varied from 12 to 72 h. Optimum incubation time was 48 h for obtaining the highest levels of metabolite with chitosan and *A. niger*. In case of *B. subtilis,* a maximum yield of 9.21 mg/g DCW was obtained after 24 h, whereas in other elicitors the optimum incubation time was extended up to 60 h. In all the cases, plumbagin content declined during an extended time of incubation. The results show that the amount of plumbagin produced varied with duration of incubation in the presence of elicitors. This reveals that the duration of cultivation in the presence of elicitor was rather important with respect to plumbagin production.

Elicitor Doses

In the preliminary experiments, all the elicitors were screened at various doses, 0.5–2.0% (v/v) to optimize their concentrations for maximum plumbagin accumulation (Fig. 6). Addition of *A. niger* elicitor to the *P. rosea* cells enhanced the metabolite content with an increase in the dose up to 1.5%. Above this level, decreased plumbagin accumulation was observed. Maximum accumulation of plumbagin (16.56 mg/g DCW) was observed with this elicitor at a level of 1.5%, almost fourfold increase over control cells (4.31 mg/g DCW). The elicitor prepared from yeast extract also displayed a maximum yield of 10.38 mg/g DCW at a concentration of 1.5%. Treatment with *R. oryzae* derived elicitor at the dose level of 1.0%, exhibited more than threefold increase in the metabolite production (15.18 mg/g DCW). In case of bacterial elicitors, *B. subtilis* showed a maximum effect at 1.0% and the yield was 6.34 mg/g DCW and *P. aeruginosa* at 0.5% (v/v) gave 8.76 mg/g DCW of plumbagin. When cells were treated with different concentrations of chitosan (50 to 250 mg/l), more than sixfold increase of plumbagin over control cells was observed (28.94 mg/g DCW) at 150 mg/l level (Fig. 4). This is the highest amount of plumbagin accumulated when compared to the other elicitors used in this study. Though, bacterial elicitors showed positive influence on plumbagin production, but the yields were not significant in comparison with other

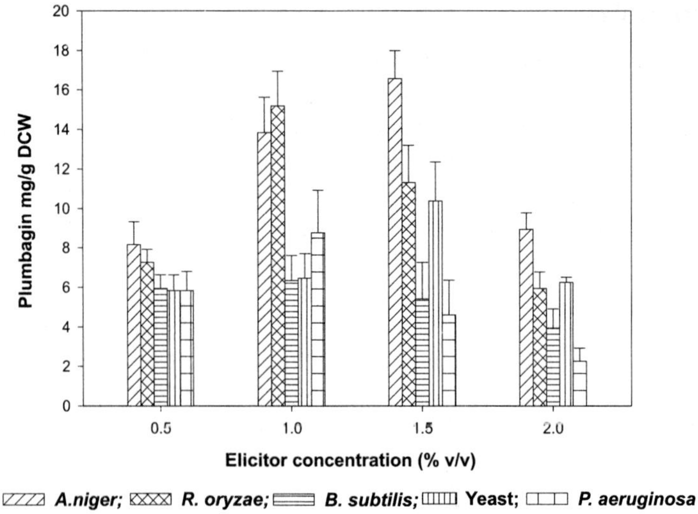

Fig. 6 Elicitors were added to 16-day-old cultures and harvested after 48 h of incubation. Elicitors were added to 16-day-old cultures and control received same amounts of distilled water. Cultures were harvested after 48 h of incubation. Control cultures produced 4.31 (± 0.24) mg/g DCW. The data shown are mean of six replicates and ± S.D. values presented as error bars.

elicitors. Elicitor induced accumulation of secondary metabolites received wide acceptance because of its ability to improve the productivity of the plant cell systems significantly. Our results showed dose-dependent response of *P. rosea* cell cultures for elicitor treatment. This tendency is similar to that reported for induction of enzymes involved in phenylpropanoid metabolism by chitosan in *Vanilla planifolia* (Funk and Brodelius 1990).

Effect of Elicitors on Extracellular Product Formation

It is always desirable to harvest secondary metabolites without destroying the cells. Therefore, the effect of elicitors on extracellular product formation was attempted and the results are represented in Table 3. In addition to its elicitor's effect, chitosan can also be used as a permeabilizing agent (Knorr and Teutonicw, 1986) and when used more than 70% of the plumbagin was produced extracellularly. But the other two fungal elicitors were less effective in mimicking a similar response. Chitosan has been proved to enhance the production of secondary metabolites in suspension cultures of *Mentha piperita* (Chang and Lee, 1992), *Vanilla planifolia* (Funk *and* Brodelius, 1990) and *Polygonum tinctorum* (Kim *et al.*, 1997). In the cultures of *P. tinctorum*, 72% higher indirubin was accumulated when subjected to elicitation with chitosan for a period of fivedays (Kim *et al.*, 1997). In our study, chitosan was found the most effective and highest amount of plumbagin accumulation (671%) was obtained in chitosan as compared to the other elicitors. However, in case of other elicitors, extracellular accumulation of metabolite was minimum. It was also observed that the dosage levels of chitosan used in the present study did not much affect the cell viability (Table 3). This indicates that chitosan can be used as a permeabilizing agent to obtain plumbagin on a large scale without sacrificing the cells.

Table 3 Effect of elicitors on extracellular product formation and cell viability

Elicitor	*% Release of plumbagin*	*% Viability*	*% Production of plumbagin*
Control	18.12	97.34	100*
A. niger	27.09	90.39	384
R. oryzae	35.34	87.61	352
Chitosan	73.56	82.75	671

*The control value (100%) was about 4.31 mg/g DCW.

Effect of Cell Immobilization in Calcium Alginate

P. rosea cells were immobilized in calcium alginate and cultured in MS medium containing 10 mM $CaCl_2$ to give strength to the beads. To study the impact of immobilization on *P. rosea* cells, the non-gelled alginate and $CaCl_2$ on cell growth and plumbagin production were examined (Figs. 7 and 8). Cells grown in 10 mM $CaCl_2$ exhibited normal growth, but the biomass was not as high as that of control cells. It was reported that uncrosslinked alginate can act as carbohydrate elicitor of secondary metabolism (Johnson *et al.*, 1991). Though in the present study uncrosslinked alginate stimulated plumbagin synthesis almost 2-folds than the control cells, cell growth was inhibited and over 40% of the cells lost their viability (data not shown). This may be due to the effect of alginate, which modifies the medium viscosity leading to poor oxygen transfer (Curtis *et al.*, 1995). Immobilization of *P. rosea* in calcium alginate substantially enhanced the production of plumbagin. Maximum amount of plumbagin was recorded on 8th day of culture but the concentration decreased

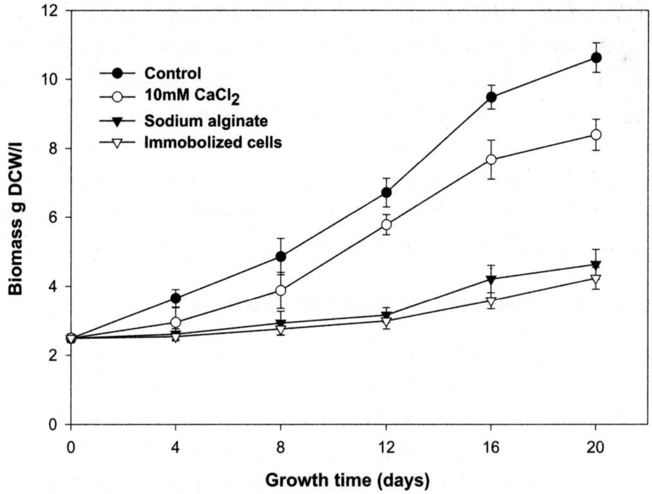

Control: Standard Medium (SM) with MS basal salts, 3% sucrose, 1 mg/l IAA, 0.5 mg/l NAA, 0.3 mg/l BAP; **10 mM $CaCl_2$:** SM containing 10 mM $CaCl_2$; **Sodium alginate:** SM + 2% uncross linked sodium alginate; **Immobilization:** 20% cells immobilized in calcium alginate and cultured in SM containing 10 mM $CaCl_2$: each data is mean of six independent replicates standard error presented as error bars.

Fig. 7 Effect of immobilization and components on growth of *P. rosea* cell cultures.

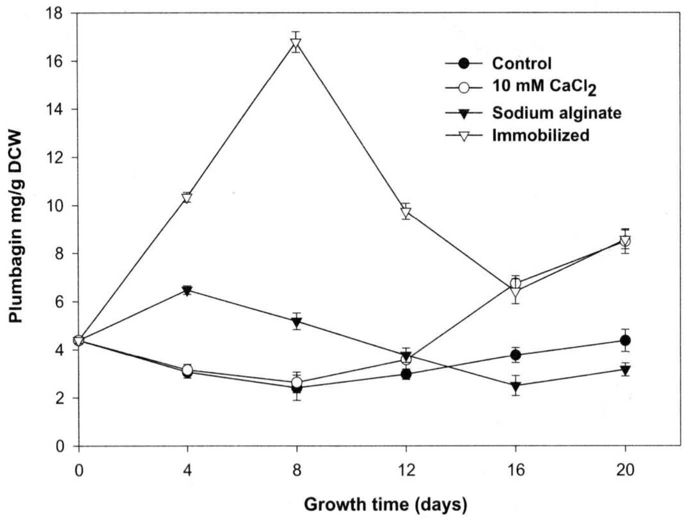

Control: Standard Medium (SM) with MS basal salts, 3% sucrose, 1 mg/l IAA, 0.5 mg/l NAA, 0.3 mg/l BAP; **$CaCl_2$:** SM containing 10 mM $CaCl_2$; **Sodium alginate:** SM + 2% uncross linked sodium alginate; **Immobilization:** 20% of cells immobilized in calcium alginate and cultured in SM containing 10 mM CaCl: each data is mean of six independent replicates and standard error presented as error bars.

Fig. 8 Effect of immobilization and components on plumbagin accumulation of *P. rosea* cell cultures.

thereafter rapidly. This could be due to degeneration of secondary metabolite when it is present in higher concentrations (Wu *et al.*, 1999, Yuan *et al.*, 2001). Immobilized cells produce three-, two- and onefold higher plumbagin to that of the control, uncrosslinked alginate and 10 mM $CaCl_2$ treated cells respectively (Fig. 8). This could be due to the effect of immobilization, which provides better cell-to-cell contact and also helps in cell differentiation to produce secondary metabolites (Brodelius, 1983; Curtis *et al.*, 1995). A similar pattern was observed in *Datura innoxia* cell cultures (Gontier *et al.*, 1994). Besides, immobilization also influences the extra cellular product formation and more than 40% of the product was released into the culture medium. This shows that immobilization can effectively influence the physiology of the cells (Curtis *et al.*, 1995).

Cell Concentration in Calcium Alginate Beads

In order to optimize cell loading or concentration in alginate beads experiments were carried out using various cell densities to the polymer concentration. Cell loading at 20% level was optimal and maximum plumbagin was noticed, but in the case of higher concentration (40-50%), lower metabolite accumulation was recorded. At this stage, cells started coming out from the beads and the bead strength were also affected. The lower metabolite accumulation at higher concentration of cell loading may be due to the limitation of nutrients and availability of dissolved oxygen to cells.

Elicitation of Immobilized Cells

Optimum concentration of chitosan for immobilized cells was tested (data not shown). In the case of free cells it was effective at 150 mg/l and this may be due to the diffusion barrier of immobilized beads to reach the cells. Addition of 200 mg/l chitosan to the immobilized cells produced eight and twofolds higher plumbagin over control and immobilized cells when considered the most suitable time for plumbagin production in each case. As a natural, incompatible, cationic biopolymer, chitosan is safe to use and advantageous because of its effect on cell permeabilization (Johnson *et al.*, 1991). This is highly desirable for the extraction of the metabolite more economically.

Sucrose Utilization

The effect of sucrose utilization rate of *P. rosea* cell cultures is presented in Fig. 9. The cells subjected to *in situ* extraction consumed more sucrose than the cells that have not been added XAD-7. *In situ* adsorption did not show any adverse effects on cell metabolism. In fact, the cells utilized sucrose more rapidly and this indicates the reduced toxicity of secondary metabolite on cells due to adsorption by XAD-7 as pointed out by Pavao *et al.* (1996) and Kown *et al.* (1998).

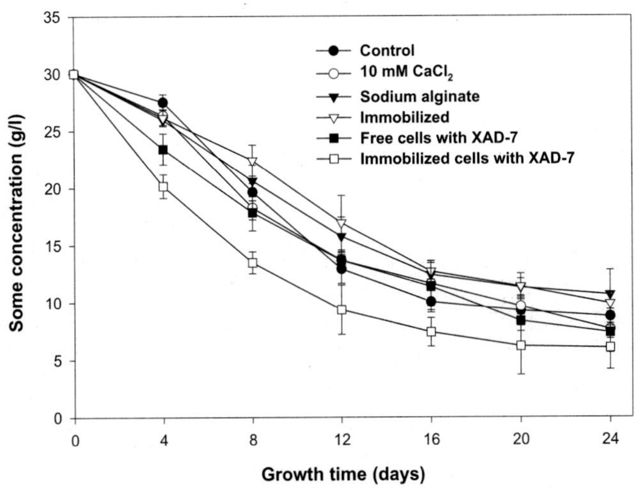

Control: Standard Medium (SM) with MS basal salts, 3% sucrose, 1mg/l IAA, 0.5 mg/l NAA, 0.3 mg/l BAP; **10 mM $CaCl_2$:** SM containing 10 mM$CaCl_2$; **Sodium alginate:** SM + 2% sodium alginate;
Immobilization: Cells immobilized in calcium alginate and cultured in SM containing 10 mM$CaCl_2$.
Free cells with XAD-7: SM + 1g of XAD-7 beads were added to the medium;
Immobilized cells with XAD-7: Immobilized cells (20% cell load) cultured in SM containing 10 mM $CaCl_2$ and 1g of XAD-7 beads added to medium each data is mean of six independent replicates and standard error presented as error bars.

Fig. 9 Effect of immobilization and *in situ* extraction of plumbagin on sucrose consumption in *P. rosea* cell cultures.

In Situ Plumbagin Adsorption

It is important to remove the water insoluble compounds/metabolites from the medium, which are not only toxic to cells at higher concentrations, but also their presence for longer duration may eventually lead to product degeneration. This demonstrates that *in situ* extraction of metabolites with organic solvents or adsorption onto nonionic polycarbonated beads is more important (Kov *et al.*, 1998; Fu and Lu 1999). Synergistic effect of immobilization, elicitation and *in situ* produ adsorption yielded a maximum amount of 92.13 mg of plumbagin/g DCW. The total metaboli production was 21, 5.7, 3.1 and 2.5-times higher than free cells, immobilized cells, free cells wi chitosan, elicitation and immobilized cells with chitosan respectively. These results suggested th *in situ* removal of plumbagin using the resin XAD-7 as second phase reduced feedback inhibition secondary metabolite synthesis. Similar pattern of results were reported in *Rubia akane* and *Tax* suspension cultures when they were subjected to *in situ* product recovery (Kown *et al.*, 1998; Shi *et al.*, 1999).

Hairy Root Cultures from *P. rosea* and Detection of Plumbagin

The attractiveness of hairy root cultures in bioprocessing can be attributed to their genetic a biochemical stability, rapid growth in phytohormone-free medium and morphologic differentiation (Flores and Curtis, 1992). Many important plant secondary metabolites a synthesized in root tissues and are stored *in situ* or transported to other organs (Waller and Nowack 1978). A vast amount of literature is available on production of high volume phytochemicals su as, alkaloids and terpenoids (Weathers *et al.*, 1997) in *Agrobacterium* transformed hairy ro cultures. From *in vitro* grown shoots, hairy roots were induced on MS basal medium and cultured four different nutrient media (MS, B_5, SH and LS) to assess their suitability for growth a plumbagin production. The cultures started growing exponentially after four days of initial l period and reached the stationary phase within 20 days of incubation with a yield of 4.16 g/l DW root biomass in MS medium. The specific growth rate of the cultures was calculated as 0.15 per d with a doubling time of 4.5 days. The roots grew in all the media with different growth rat showing a large degree of lateral branching. The growth and plumbagin accumulation varied different culture media. Of the four media used, only roots grown in MS medium produced 4.16

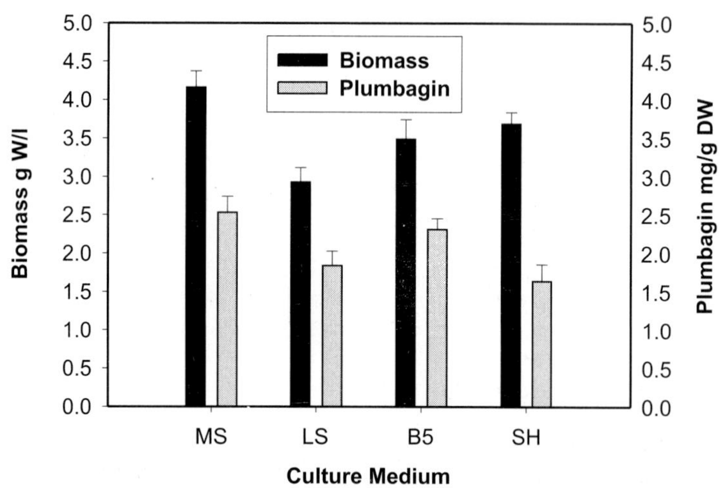

Fig. 10 Effect of different culture media on biomass and plumbagin accumulation in hairy root cultures.

DW/l biomass and 2.53 mg/g DW plumbagin. Other media showed the same growth pattern but the yields of plumbagin were not as high as that of MS medium (Fig. 10). Plumbagin content in hairy root cultures was higher than those of roots from intact plants (0.09% DW), indicating an interesting biosynthetic potential of this culture system. Plumbagin biosynthesis also followed the pattern of root growth and the highest metabolite accumulation was observed during stationary phase of the root cultures.

Effect of media pH on growth of hairy root cultures and plumbagin biosynthesis is also worked out. Cultures showed better growth and plumbagin in the pH range of 5.2 to 6.0. At lower media pH (4-5.2), both growth and plumbagin accumulation was suppressed, and the roots turned brownish yellow and lost the viability. This could be due to the lower pH and reduced sucrose utilization rate of the cultures (Paska *et al.*, 1999). When the media pH was above 6.0, a decrease in growth and plumbagin accumulation was recorded. Lower ion uptake at higher pHs may be one of the factors for this. Hairy root cultures showed an enhancement in biomass and metabolite production with an increase in the concentration of sucrose till 3% (w/v) level. Plumbagin content increased only slightly in presence of 5% (w/v) sucrose. Higher concentration of sucrose was also reported to produce optimum biomass in *C. roseus* hairy root cultures (Toivonen *et al.*, 1991). On the other hand, Nahalka *et al* (1996) reported higher plumbagin content at lower sucrose levels (1%) in suspension cultures of *Drosophyllum*. The results presented here showed that *P. rosea* hairy roots can biosynthesize and accumulate plumbagin, which can be exploited as an alternative source for the production of this important naphthoquinone.

Conclusions

Increase in metabolite yield was found to be 20-25 times higher than control cultures when the cells were subjected to different treatments such as elicitation, permeabilization, immobilization and *in situ* extraction of the product. Suspension and hairy root cultures of *Plumbago rosea* represent an excellent method to produce plumbagin of commercial quality. However, further work to large scale culturing is needed before evaluating this methodology for commercial and economic feasibility.

Acknowledgements

We thank the UGC, New Delhi for financial assistance in the form of SAP (DSA) to the Department of Genetics.

REFERENCES

Anonymous. (1969) The Wealth of India, Vol. VI, New Delhi: CSIR.

Asada M. & Shuler M.L. (1989) *Appl. Microbiol. Biotechnol.* 30: 475-481.

Bajaj Y.P.S., Furmanowa M. & Olszowska O. (1988) In: *Biotechnology in Agriculture and Forestry, 4, Medicinal and Aromatic Plants I,* Bajaj Y.P.S. (Ed), Springer-Verlag, Berlin, pp. 60-103.

Bajaj Y.P.S. (1989) *Biotechnology in Agriculture and Forestry 1, Trees-1*, Springer-Verlag, Berlin, pp. 1-23.

Brodelius P. (1983) In: *Biochemical Engineering III* (Venkatasubramanian K., Constantinides A. and Veith W.R. eds) *Ann. N.Y. Acad. Sci.* 413: 383-393.

Buitelaar R.M. & Trampar J. (1992) *Enz. Microb. Technol.* 14: 2-7.

Chang J.H. & Lee H.J. (1992) *Ann. Res. Center for New Bio-Mater. Agric.* 111-117.

Crouch I.L., Finnie J.F. & Staden J.V. (1990) *Plant Cell, Tissue Org. Cult.* 21: 79-82.

Curtis W.R., Wang P. & Humphrey A.E. (1995) *Enzyme Microb. Technol.* 17: 554-557.

Deb D.B. (1983) The flora of Tripura State, Today and Tomorrow, New Delhi, Vol. 2, pp. 184.

Didry N., Dubrevil L. & Pinkas M. (1994) *Pharmazie.* 49: 681-683.

Dinda B., Hajra A.K. & Chel G. (1997) *J. Indian Chem. Soc.* 74: 974-979.

Durand R. & Zenk M.H. (1971) *Tetrahedron Lett.* 32: 3009-3012.

Durand R. & Zenk M.H. (1974) *Phytochemistry* 13: 1483-1492.

Eliert U. (1987) In: Cell cultures and somatic cell genetics of plants. Vol. 4 (Vasil I.K. and Constable F., Eds.), Academic Press, San Diego. 153-196.

Flores H.E. & Curtis W.R. (1992) In*: Biochemical Engineering VII: Cellular and Reaction Engineering*, (Pederson H., Mutharsan R. and Di Biasio D., Eds.) Proceedings of the New York Academy of Sciences, New York, 665: 188-209.

Fu X.Q. & Lu D.W. (1999) *Enzy. Microb. Technol.* 24: 243-246.

Funk C. & Brodelius P. (1990) *Phytochemistry.* 29: 845-848.

Gontier E., Sangwan B.S. & Barbotin J.N. (1994) *Plant Cell Rep.* 13: 533-536.

Hahn M.G. & Albersheim P. (1978) *Plant Physiol.* 62: 107-111.

Harikrishnan K.N. & Hariharan M. (1996) *Phytomorphology.* 46: 53-58.

Heble M.R., Narayanaswamy S. & Chadha M.S. (1974) *Plant Sci. Lett.* 2: 405-409.

Johnson T.S., Ravishankar G.A. and Venkataraman L.V. (1991) *Food Biotechnol.* 5: 197-205.

Kavi Kishor P. B. & Mehta A.R. (1987) *J. Rec. Adv. Appl. Sci.* 2: 244-253.

Kayser O., Kiderlen A.F., Laatsch H. & Croft L.S. (2000) *Acta. Trop.* 77: 307-314.

Kim J.H., Shin J.H., Lee H.J. & Chung I.S. (1997) *J. Fermentation Bioengg.* 83: 206-208.

Kitanov G.M. & Pashankov P.P. (1994) *Pharmazie.* 49: 462.

Knorr D. & Teutonico. (1986) *J. Food. Sci.* 52: 1397.

Komaraiah P., Kavi Kishor P.B. & Ramakrishna S.V. (2001) *Biotechnol. Lett.* 23: 1269-1272.

Komaraiah P., Naga Amrutha R., Kavi Kishor P.B. & Ramakrishna S.V. (2002a) *Enzyme and Microb. Tech.* 31: 634-639.

Komaraiah P., Naga Amrutha R., Jogeswar G., Ramakrishna S.V. & Kavi Kishor P.B. (2002b) *Plant Cell Biotechnology and Mol. Biol.* 3: 65-68.

Komaraiah P., Ramakrishna S.V., Reddanna P. & Kavi Kishor P.B. (2003) *J. Biotech.* 101: 181-187.

Kown L.C., Yoo Y.J., Lee J.H. & Hyun J.O. (1998) *Process Biochem.* 33: 701-707.

Kubo I., Uchida M. & Klocke J.A. (1983) *Agri. Biol. Chem.* 47: 911-913.

Krishnaswamy M. & Purushothaman K.K. (1980) *Ind. J. Exp. Biol.* 18: 876-877.

Madhusudhan R. & Ravishankar G.A. (1996) *Biotechnol. Lett.* 18: 1253-1256.

Nahalka J., Blanarik P., Gemeiner P., Matusova E. & Partlova I. (1996) *Biotechnol. Lett.* 18: 1453-1458.

Parimala R. & Sachdanandam P. (1993) *Mol. Cell. Biochem.* 125: 59-63.

Paska C., Innocenti G., Kunavari M., Lasizo M. & Szilagyi L. (1999) *Phytochemistry.* 52: 59-63.

Pavao M.C., Chin C.K. & Pedersen H. (1996) *J. Biotechnol.* 49: 95-100.

Pedapudi S., Chin C.K. & Pedersen H. (2000) *Biotechnol. Prog.* 16: 346-349.

Rajendran L., Suvarnalatha G., Ravishankar G.A. & Venkataraman L.V. (1994) *Appl. Microbiol. Biotechnol.* 42: 227-231.

Ramakrishna S.V., Reddy R.G., Curtis W.R. & Humphry A.E. (1993) *Biotechnol. Lett.* 15: 307-310.

Robinson T. (1991) In: Organic constituents of higher plants, (6th ed.), North Amherst: Cordus Press.

Sateeshkumar K. & Bhavanandan K.V. (1988) *Plant Cell Tissue Organ. Cult.* 15: 275-278.

Sateeshkumar K. & Bhavanandan K.V. (1989) *Ind. J. Exp. Biol.* 27: 368-369.

Sateeshkumar K. & Seeni S. (2002) *Ind. J. Biotech.* 1: 305-308.

Shim J.J., Shin J.H., Pai T., Chung I.S. & Lee H.J. (1999) *Biotechnol. Lett.* 13: 249-252.

Thomson R.H. (1971) Naturally occurring napthoquinones. 2nd Edn., Academic Press, London.

Toivonen L., Ojala M. & Kauppinen V. (1991) *Biotech. Bioengg.* 37: 673-680.

Waller G.R. & Nowacki E.K. (1978) In: Alkaloid biology and metabolism in plants. Plenum Press, New York.

Weathers P.J., Hemavanh D.D., Walcerz D.B. & Cheetham R.D. (1997) *In Vitro Cell Dev. Biol. Plant.* 33: 306-312.

Wu Z.L., Yuan Y.J., Liu J.X., Xuan H.Y., Hu Z.D., Sun A.C. & Hu C.X. (1999) *Acta Bot. Sin.* 41: 1108-1113.

Wurm G. and Gurka H.J. (1986) *Arch. Pharma.* 319: 190-191.

Yuan Y.J., Wei Z.J., Wu Z.L. & Wu J.C. (2001) *Biotech. Lett.* 23: 1659-1662.

CHAPTER

15

econdary Metabolites from Tissue Cultures of ome Arid Zone Plants

N. Nag, Sapna Tyagi, Navdeep Singh Baina and Neelam Chouhan

BSTRACT

rid zone plants are good candidates for the production of various types of secondary metabolites which makes iem resistant against drought, salinity and pathogens. Since plant cells can be grown *in vitro*, hence such 'plant ell cultures' have been used for production of secondary metabolites. Tissue cultures of *Capparis decidua*, four *'igna aconitifolia* cultivars- –RMO-257, RMO-40, IPCMO-880, Jwala, *Zizyphus mauritiana* cultivars–Gola, eb and *Withania somnifera* were established and maintained on Murashige and Skoog's (1962) medium upplemented with various concentrations and combinations of PGR's for 12-18 months by frequent subculturing fter 4-6 weeks. The tissues were harvested at 2, 4, 6, 8 and 10 weeks and growth indices were calculated. The ssues were dried, weighed, extracted in different solvents and analyzed for various secondary metabolites viz. avonoids, steroids, alkaloids and withanolides.

NTRODUCTION

id zone plants are good candidates for the production of various types of secondary metabolites iich makes them resistant against drought, salinity and pathogens hence screening of plants of d areas *in vivo* and *in vitro* for useful metabolites have been conducted in our laboratory. Rapid velopment of Plant Tissue Culture and growing fear that the earth is losing plant species at an rming rate, has focused attention on alternatives to whole plants for production of natural oducts. Plant tissue culture is potentially valuable for studying the biosynthesis of secondary etabolites and may also eventually provide an efficient means of producing important plant oducts (Bull, 1981). There has been considerable interest growing on investigation of the tential of plant tissue culture to traditional methods for the industrial production of plant oducts. The ability of plant cell cultures to produce natural products has been reported in several stems (Zenk, 1978). Hence the present paper deals with the production, isolation and

ddress for correspondence: Plant tissue culture & Biotech Lab., Dept. of Botany, Dungar College, Bikaner-334001.

identification of secondary metabolites from callus tissues of some arid zone plants viz. *Capparis decidua, Vigna aconitifolia* cultivars–RMO-257, RMO-40, IPCMO-880, Jwala, *Zizyphus mauritiana* cultivars–Gola, Seb and *Withania somnifera.*

MATERIALS AND METHODS

Tissue Culture Studies

Tissue cultures of *Capparis decidua* were raised from nodal explant and those of moth bean cultivars (RMO-257, RMO-40, IPCMO-880 and Jwala) from leaf explant excised from 5-7 days old *in vitro* grown seedling. Tissue cultures of *Zizyphus mauritiana* cultivars (Gola and Seb) and *Withania somnifera* were raised from leaves excised from mature plants. Tissues thus obtained were established for 18 months by frequent subculturing on Murashige and Skoog's (1962) medium supplemented with 1.5 mg/l 2,4-D and 0.5/l mg kinetin for the moth bean cultivars, 5.0 mg/l NAA, 1.5 mg/l 2,4-D and 1.0 mg/l kinetin in case of *C. decidua* and *Z. mauritiana* cultivars and 0.1 mg/l IAA and 0.1 mg/l BAP for unorganized callus of *W. somnifera*. The MS media was further supplemented with 0.1 mg/l, 0.3 mg/l and 0.5 mg/l IAA for organized cultures of *W.somnifera*. The tissues were maintained for 12-18 months by frequent subculturing at 26 ± 1°C, 55% relative humidity and diffused light conditions (300 lux), harvested at 2, 4, 6, 8 and 10 weeks and growth indices were calculated. The tissue cultures of all the plants studied were analyzed for flavonoids and steroidal compounds. Tropane alkaloids and withanolides were studied in cultures of *W. somnifera*.

Estimation of Flavonoids

The dried samples were separately soxhlet extracted (Subramanian & Nagrajan, 1969) in 80% ethanol (100 ml/g.d.w.) on a water bath for twenty-four hours. Each of the extract was concentrated and re-extracted in petroleum ether (Fraction first), ethyl ether (Fraction second) and ethyl acetate (Fraction third) in succession. Fraction first was rejected due to its richness in fatty substances whereas fraction second was analyzed for free flavonoids and fraction third for bound flavonoids in each of the samples. Fraction third of each of the test samples was hydrolyzed by refluxing with 7% sulphuric acid (10 ml/gm residue) for two hours. The mixture was extracted with ethyl acetate and the extract neutralized with distilled water. Concentrated ether and ethyl acetate fractions were applied on TLC plates along with standard reference compounds, visualized under UV light, placed in a chamber saturated with NH_3, developed in solvent system of n-butanol, acetic acid and water (4:1:5, upper layer) and sprayed with ethanolic $FeCl_3$. Isolates were eluted with ethyl acetate, crystallized from chloroform and subjected to mp, mmp and IR spectral studies for confirmation. The identified flavonoids were quantitated colorimetrically adopting the procedure of Kariyone *et al*., 1953 and Naghski *et al*., 1975 in case of quercetin and Mabry *et al*, 1970 in case of kaempferol.

Estimation of Steroids

The dried samples were refluxed with 30% HCl for four hours and filtered. The hydrolyzed tissue sample was washed with cold distilled water, dried and extracted (Nag *et al*., 1979) with benzene for twenty-four hours. The extract was concentrated *in vacuo* and finally taken up in chloroform. The chloroform extract along with reference compounds was applied on activated silica gel 'G' coated plates, developed in solvent system of hexane and acetone (80:20) and sprayed with 50% sulphuric acid. UV light was used to mark various steroidal compounds on unsprayed plates. The marked bands separately collected along with silica gel, eluted with chloroform, dried *in vacuo,* crystallized and subjected to mp, and IR studies along with their respective standard reference sterols. Quantitative estimation of identified sterols in each of the test samples was carried out using the method of Das and Banerjee, 1980. Diosgenin was quantitatively estimated by following the procedure of Sanchez *et al.*, 1972.

Estimation of Alkaloids

Dried tissues of *W. somnifera* treated with 0.75 ml ammonium hydroxide (28%), 1 ml ethyl alcohol (95%) and 2 ml ethyl ether, macerated for 12 hours, dried, extracted with chloroform for 24 hours, dried *in vacuo*, taken up in chloroform, evaporated, mixed with 0.25 ml 0.1 N ethanolic (90%) hydrochloric acid, centrifuged and analyzed for tropane alkaloids. The extract along with reference standard compounds was applied on activated silica gel 'G' coated plates, developed in solvent system of chloroform, methanol and 28% aqueous ammonia (30:60:2), dried, visualized under UV light and the spots corresponding with standard reference were marked. These plates were sprayed with Dragendraff's reagent (Munier & Macheboeuf, 1949) and heated in oven at 100°C for 15 min. The marked band from unsprayed plate collected separately eluted with chloroform, dried, crystallized in 50% ethanol and subjected to mp, mmp and infrared spectral studies along with standard reference samples for confirmation. Isolated compounds were quantitatively estimated by adopting the procedure of Feldman and Robb (1970) and the absorbance was measured on a spectrophotometer (Carl Zeiss Jena, DDR-VSU-2P).

Estimation of Withanolides

The dried tissues of *W. somnifera* were extracted with methanol (3 × 100 ml) overnight at room temperature. The extract pooled, concentrated to 25 ml, diluted with 100 ml of distilled water, extracted first with petroleum ether followed by $CHCl_3$, dried over Na_2SO_4 and taken for further analysis (Roja *et al.*, 1991). The extract was applied separately on Silica gel 'G' coated, activated glass plates along with standard reference, developed in solvent system of chloroform, ethyl acetate, methanol and benzene (72:4:8:16 v/v) and sprayed with freshly prepared Liebermann Burchard reagent (Ray *et al.*, 1996), heated at 100°C for 10 minutes and examined. The extract was further analyzed by HPLC. The absorbance selector was set at 254 nm but at a higher flow rate (1.2 ml min^{-1}) and confirmed on the basis of TLC mobility, colour reaction and HPLC retention comparable with authentic samples. For quantitative analysis, peak areas of HPLC were used to calculate the amount of withanolides present

RESULTS AND DISCUSSION

The tissues of all the eight plants were unorganized. The tissues of *C. decidua* were creamy brown and compact in nature while that of *V. aconitifolia* cultivars were creamy with greenish tinge and fragile. *Z. mauritiana* cultivars were green and fragile and of *W. somnifera* were greenish yellow brown in colour and fragile. Growth of tissues increased upto eight weeks in case of *C. decidua* and *V. aconitifolia* cultivars and upto six weeks in the tissues of *Z. mauritiana* cultivars and *W. somnifera.*

Flavonoids

The presence of quercetin and kaempferol was observed in the tissue cultures of all the plants studied. Presence of quercetin Rf 0.82; uv fluorescent- bluish yellow, NH_3 – yellow, $FeCl_3$ – bluish grey; melting point 309-311°C; uv max 258, 303, 375 nm in ethanol and kaempferol Rf 0.93; bright yellowish blue; NH_3 – deep yellow; $FeCl_3$ – brownish; melting point 271- 272°C, uv max 263, 324, 368 nm in ethanol was confirmed in all the samples. The characteristic IR spectral peaks were found to be super-imposable with those of their respective standard reference compounds of quercetin and kaempferol in all the samples tested. The amount of isolated kaempferol was higher in comparison to quercetin except *W. somnifera.*

The kaempferol content was found to be maximum (5.20 mg/gdw) in tissue cultures of *C. decidua* and IPCMO-880 cultivar of *V. aconitifolia* and minimum (3.59 mg/gdw) in Gola cultivar.

Flavonoids are secondary metabolites that are present at high levels in most plant seeds and grains. These compounds appear to play vital roles in defence against pathogens and predators and contribute to physiological functions such as seed maturation and dormancy (Brenda Winkel Shirley, 1998). Flavonoids have comprehensively been discussed (Skinner, 1955; Ebi & Ofoefule 2000; Ogunwande *et al.* 2001). Quercetin has been reported in static cultures of *Crotolaria juncia* (Jain & Khanna, 1974), *Calendula officinalis, Crotolaria burhia* and *Papaver rhoeas* (Khanna *et al.* 1980) and kaempferol from callus cultures of *Dolichas lablab, Glycinemax, Pisum sativum* (Rao & Raju, 1999) while both quercetin and kaempferol have been isolated from tissue culture of *Peganum harmala* (Harsh & Nag, 1984).

Steroids

Three spots of sterols corresponding with the standard reference compounds of β–sitosterol (purple, Rf 0.85), stigmasterol (purple, Rf 0.91) and lanosterol (light brown, Rf 0.89) were observed by TLC in all the tissue samples studied except *W. somnifera* in which stigmasterol was not detected. β–sitosterol, lanosterol and stigmasterol were confirmed by Co-TLC, melting point (β–sitosterol, 139-144°C; stigmasterol, 143-144°C; lanosterol, 141-142°C), mmp (undepressed) and superimposable IR spectra of the isolated and authentic samples of the steroids.

Phytosterols are naturally occurring 3-monohydroxy steroids of the C_{27}, C_{28} and C_{29} series, known to be the starting materials for a number of plant steroids used in pharmaceutical industry and have been reported from a number of plant species growing *in vivo* and *in vitro* (Sahoo *et al.* 1999; Binutu & Cordell, 2000; Bishop & Yokota, 2001). Santos *et al.* (1999) isolated campesterol, stigmasterol and β–sitosterol from callus obtained from *M. glomerata* leaf explants. Anjaneyulu and Radhika (2000) reviewed the steroid constituents of *Mangifera indica* in different plant parts.

Steroidal sapogenins are very useful to the medical field as they are precursors for the production of several steroidal hormones. Diosgenin is the chief raw material for the commercially useful drugs. Sahoo *et al*., 1999 studied the effect of growth regulator on diosgenin production in callus tissues of *Solanum jasminoides* grown on RT medium. Considerable progress has been made on diosgenin-bearing plants and in the steroid industry (Mandal & Chatterjee, 1985; Ravishankar & Grewal, 1986; Singh *et al.,* 1988; Dixit and Banerjee, 2000).

In the present paper diosgenin investigated from seven plants (Table 1) have been reviewed. Maximum (1.99%) amount of diosgenin was found in Seb cultivar of *Z.mauritiana* and minimum (0.31 mg/gdw) in IPCMO-880 cultivar of *V. aconitifolia*

Table 1 Secondary metabolites (mg/gdw) from tissue cultures of arid zone plants

Compounds identified	*C. decidua*	*V. acontifolia Cultivars*			*Z. mauritiana Cultivars*			*W. somnifera*
		RMO-257	*RMO-40*	*IPCMO-880*	*JWALA*	*GOLA*	*SEB*	
Quercetin	1.12	1.20	0.99	0.87	0.84	1.12	1.19	3.97
Kaempferol	5.2	4.35	3.80	5.20	4.90	3.59	3.72	2.79
Stigmasterol	5.55	8.50	8.28	8.82	8.26	5.39	5.48	–
Lanosterol	2.99	5.08	4.28	5.26	5.06	2.09	2.15	2.78
β-sitosterol	5.14	8.42	7.94	8.44	8.90	5.06	5.09	1.86
Diosgenin	1.40	0.44	0.39	0.31	0.34	1.89	1.99	

Alkaloids

The presence of tropane alkaloids (atropine and hyocyamine) was observed in *in vitro* cultures of *W. somnifera*. In MS medium (1 ppm IAA + 1 ppm BAP) the maximum concentration of atropine (0.45 mg/gdw) and hyocyamine (0.13 mg/gdw) was observed in six weeks old cultures and minimum in 2 weeks old cultures. The amount of tropane alkaloids was found to be directly proportional to the growth indices (Table 2).

Studies on the production of tropane alkaloids through tissue culture have been actively carried out because of shortage of alkaloids producing plants. Tropane alkaloids make up one distinct group of secondary metabolites. Tropane alkaloids have been reported from tissue cultures of *Datura* species (Khanna & Nag, 1972; Nag, 1981).

Effect of hormones on the alkaloids production has been reported in *Catharanthus roseus* (Deus-Neumann & Zenk, 1984), *Thalictrum minus* (Nakagawa *et al*., 1986), *Rauwolfia serpentina* (Yamamoto & Yamada, 1987) and *Lycium barbarum* (Nag, 1995).

Table 2 Alkaloids (mg/gdw) from tissue cultures of *W. somnifera*

Growth Age	*Gi*	*Atropine*	*Hyocyamine*
2	3.6	0.23	0.09
4	5.4	0.36	0.11
6	7.8	0.45	0.13
8	6.2	0.42	0.10

Withanolides

Three major coloured spots, coinciding with those of standard samples of withaferin 'A' (light brown Rf 0.26), withanolide 'D' (dark brown Rf 0.51) and one unknown (blue Rf 0.70) were observed. Higher levels of IAA (1.0 and 3.0 ppm) in the medium caused a marked increase in withanolide 'D' synthesis (24 µg/gdw and 37 µg/gdw respectively) which decreases (29 µg/gdw) with further increase in IAA concentration (5 ppm). However withaferin 'A' occurred in traces at lower concentration of IAA (1 ppm) but increased (0.72 mg/gdw and 0.96 mg/gdw) in its concentration at higher levels of IAA (3 and 5 ppm respectively). Unorganized cultures grown on MS (1 ppm IAA + 1 ppm BAP) medium did not contain any withanolides. HPLC analysis at flow rate 1.2 ml min^{-1} revealed retention time (RT) 11.80 for withanoloide 'D' and one more RT 7.51 showed a markedıy big area for an unknown compound (Table 3).

Table 3 Withanolide contents (µg/gdw) in unorganized and organized cultures influenced by various concentrations of IAA in *W. somnifera* at six weeks age.

Tissue Cultures	*Medium*	*Withaferin 'A'*	*Withanolide*
Unorganized	MS + 1 ppm IAA + 1 ppm BAP (control)	–	–
Organized	MS (control) + 1 ppm IAA	0.06	24
Organized	MS (control) + 3 ppm IAA	0.72	37
Organized	MS (control) + 5 ppm IAA	0.96	29

In vitro culture of different plant parts of *W. somnifera* has been studied by various workers for their withanolide content (Banerjee *et al.*, 1994; Ray *et al.*, 1996; Roja *et al.*, 1991), *W. somnifera* has wide ranging pharmacological activity and plant to plant variation therefore not a single withanolide observed in the suspension and callus culture (Roja *et al.*, 1991). However, *Agrobacterium* rhizo genes incorporated cultures induced hairy roots and showed the presence of withanolides (Banerjee *et al.*, 1994; Ray *et al.*, 1996). A major constraint in using the natural flora in the existence of a great deal of plant to plant chemovariability several chemotypes of withania have been investigated in detail (Kirson *et al.*, 1971), multiple shoot cultures could be an answer to overcome these constraints since they are known to exhibit a stable chemical profile over a prolonged period of subculture.

The main objective of this review is to speculate the biosynthetic potential of arid zone plants especially of cultivars developed as the investigations made by plant breeders are based only on agronomic features, yield capacity and disease resistance parameters.

Thus, the present study favours that the tissue cultures of Indian arid zone plants have the potentialities to synthesize the secondary products in good amounts. Much of the work done in the area of hormonal influence on product biosynthesis is confined to the effect of the hormones in terms of product formation only. More critical studies are needed to understand the hormonal regulations of product biosynthesis particularly in view of the fact that the hormones can exercise their influence at their nuclear, enzymatic, cellular, tissues and organ levels and bring about chromosomal aberration, cytodifferentiation and organ differentiation. These events may have a direct effect on the biosynthetic potential of the cultured tissues.

Even though potential of plants as a natural research has clearly been demonstrated, industrial scaling up has still many limitations. However, recent advances has shifted the focus from descriptive methodology to manipulation and regulation. Recent advances in the technology of scaling up of natural products have encouraged attempts to commercially produce plant products from plants specially from plant cell cultures.

REFERENCES

Anjaneyulu V. & Radhika P. (2000) The triterpenoides and steroids from *Mangifera indica* Linn. *Indian Journal of Chemistry.* 39(12): 883-893.

Banerjee S., Naqvi A.A., Mandal S. & Ahuja P.S. (1994) Transformation of *Withania somnifera* (L.) Dunal by *Agrobacterium* rhizogenes: Infectivity and Phytochemical styudies. *Phytotherapy Research* 8: 452-455.

Binutu O.A. & Cordell G.A. (2000) Constituents of *Zanthoxylum sprucci. Pharmaceutical Biology* 38(3): 210-213.

Bishop G.J. & Yokota T. (2001) Plants steroid hormones, brassinosteroids: Current highlights of molecular aspects on their synthesis/metabolism, transport, perception and response. *Plant and cell Physiology.* 42(2):114-120.

Brenda Winkel Shirley. (1998) Flavonoids in seeds and grains: Physiological function, agronomic importance and the genetics of biosynthesis. *Seed Science Research* 8: 415-422.

Bull E.A. (1981) The physiological role(s) of seconadry (natural) products. In: *The Biochemistry of Plants A Comprehensive Treatise* (Eds.) Stumpf P.K. and Conn E.E., Academic Press, pp. 1-19.

Das S.K. & Banerjee A.B. (1980) A rapid method for quantification of sterols after thin layer chromatography. *Indian J. Exp. Biol.* 18: 969-971.

Deus-Neumann B. & Zenk M.H. (1984) Instability of indole alkaloid production in *Catharanthus roseus* cell suspension cultures. *Planta Med.* 50: 427-431.

Dixit B.S. & Banerjee R. (2000) High diosgenin content in some selected clones of *Dioscorea floribunda. Indian Journal of Pharm. Sci.* 62(1): 53-54.

Ebi G.C. & Ofeofule S.I. (2000). Antimicrobial activity of *Pterocarpus osun* stems. *Fitoterapia* 71(4): 433-435.

Feldman J.A. & Robb B.J. (1970) Colorimetric determination of atropine, homoatropine, scopolamine and their derivatives by ferric hydroxamate. *J. Pharm. Sci.* 59: 1646-1647.

Harsh M.L. & Nag T.N. (1984) Antimicrobial principles from *in vitro* tissue culture. *Lloydia.* 47: 365-368.

Jain S.C. & Khanna P. (1974) Quercetin from *Crotolaria juncea* L. Tissue cultures. *Indian J. Exp. Biol.*, 12: 466.

Kariyone T., Hashimoto Y. & Kimura M. (1953) Microbial studies on plant components. IX. Distribution of flavonoids in plants by paper chromatography. *J. Pharma. Soc.* (Japan). 73: 253-256.

Khanna P. & Nag T.N. (1972) Isolation, identification and screening of phyllembin from *Embica officinalis* Gaertn Tissue Culture.

Khanna P., Sharma O.P., Sehgal M., Bhargava C., Jain M., Goswami A., Singhvi S., Gupta V., Agarwal R., Sharma P. & Jain S.C. (1980) Antimicrobial Principles from *in vitro* tissue culture of some plant species. *Indian J. Pharm. Sci.* 42: 113-117.

Kirson I., Glotter E., Lavie D. & Abraham A. (1971) Constituents of *Withania somnifera* Dunal part XII. The withanolides of an Indian chemotype. *J. Chem. Soc.* (C), 2032-2044.

Mabry T.J., Markham K.R. & Thomas M.B. (1970) (Eds.). In: *The Systematic Identification of Flavonoids.* Springer-Verlag, Berlin, pp. 119.

Mandal A. & Chatterjee S.K. (1985) Histochemical studies of *Dioscorea composita* (Hemsl.) Yams with special reference to starch grain in relation to its diosgenin content. *72nd Indian Sci. Cong., Abstr.* 45.

Munier R. & Macheboeuf M. (1949) Preparation of modified Dragendorff's reagent. *Bull. Soc. Chem. Biol.* 31: 1144-1150.

Murashige T. & Skoog F. (1962) A revised medium for rapid growth and bioassays with tobacco tissue cultures. *Physiol. Plantarum.* 15: 473-497.

Nag T.N. (1981) Action of tyrosine as precursor of tropane alkaloids in *Datura metel* L. tissue culture. *Comp. Physiol. Ecol.* 6(4): 239-240.

Nag T.N. (1995) Product production studies in tissue cultures of arid zone plants. *J. Indian Bot. Soc.* 74A: 311-316.

Nag T.N., Mathur C.S. & Goyal S.C. (1979) Phytochemical studies of *Tribulus alatus, T. terrestris* and *Agave wightii* contents of primary and secondary products. *Comp. Physiol. Ecol.* 4: 157-160.

Naghski J., Fenske C.S. (Jr.) & Couch I.F. (1975) Use of paper chromatography for the estimation of quercetin in rutin. *J. Am. Pharm. ssoc.* 40: 613-616.

Nakagawa K., Fukui H. & Tabata M. (1986) Hormonal regulation of berberine production in cell suspension cultures of *Thalictrum* minus. *Plant Cell Reports.* 5: 69-71.

Ogunwande I.A., Bello M.O., Olaware O.N. & Muili K.A. (2001) Phytochemical and antimicrobial studies on *Butyrospermum paradoxum Fitoterapia* 72(1): 54-56.

Rao S. & Raju B.G. (1999) Role of Biotechnology in production of secondary metabolites. In: *Role of Biotechnology in medicinal and aromatic plants.* Khan I.A. & Khanum A. (Eds) 392-409. Ukaaz Publications, Hyderabad.

Ravishankar G.A. & Grewal S. (1986) Development of two-stage culture methods and biochemical markers for growth and diosgenin production in *in vitro* grown cultures of *Dioscorea deltoidea.* VI IAPTC Minneapolis, 3rd-8th August, Absrtacts, p. 251.

Ray S., Ghosh B., Sen S. & Jha S. 1996. Withanolide production by root cultures of *Withania somnifera* transformed with *Agrobacterium* rhizogenes. *Planta Medica.* 62: 571-573.

Roja G., Heble M.R. & Sipahimalani A.T. (1991) Tissue cultures of *Withania somnifera*: morphogenesis and withanolide synthesis. *Phytother. Res.* 5: 185-187.

Sahoo S., Das R. & Jain S.C. (1999) Effect of growth regulators on steroidal sapogenin and glycoalkaloid in tissue culture of *Solanum jasminoides. Hamdard Medicus* 42(3): 99-103.

Sanchez G.L., Medina Acevedo J.C. & Soto R.R. (1972) Spectrophotometric determination of diosgenin in *Dioscorea composita* following thin layer chromatography. *Analyst.* 97: 973.

Santos P.A., Pereira A.M.S., Franca S.C., Lopes N.P. & Dos Santos P.A. (1999) Steroids and coumarin from callus of *Mikania glomerata* Sprengal. *Revista Brasileira de Ciencia Farmaceuticas* 35(2): 231-235.

Singh V., Sethia M., Mathur K. & Nag T.N. (1988) Desert medicinal plants: A source of steroids. Oikoassay, 5(2): 41-42.

Skinner F.A. (1955) Antibiotics. In: K. peach and MV Tracey, (Eds.) *Modern Mithoden der Pflanzenanslyse,* 3: 626-744. Springer verlag, Berlin.

Subramanian S.S. & Nagrajan S. (1969) Flavonoids of the seeds of *Crotolaria retusa* and *C. striata. Curr. Sci.* 38: 65.

Yamamoto O. & Yamada Y. (1987) Selection of reserpine producing cell strain using UV light and optimization of reserpine production in the selected cell strain. *Plant Cell Tissue and Organ Culture* 8: 125-133.

Zenk M.H. (1978) Introduction to plant tissue culture. In: *Frontiers of Plant Tissue Culture*, International Association for Plant Tissue Culture. Thorpe, T.A. (Ed.) Calagary, Canada, pp. 1-13.

CHAPTER

16

Steroidal Sapogenin in Moth Bean Cultivars *in vivo* and Tissue Culture

Sapna Tyagi and T.N. Nag

ABSTRACT

Diosgenin, a major raw material for the commercial production of steroidal contraceptives and sex hormones is useful in several medical therapies. Considerable progress has been made on diosgenin-bearing plants and in the steroid industry. Besides *Dioscoreas*, alternative sources of diosgenin yielding plants have been searched out. But so far the moth bean cultivars RMO-257, RMO-40, IPCMO-880 and Jwala have not been worked out for diosgenin production. Seeds and leaves of four cultivars of moth bean were collected from ARS, RAU, Bikaner and their cultures were established on Murashige and Skoog's (1962) medium supplemented with 0.5 mg/l kinetin and 1.5 mg/l 2,4-D for 18 months by frequent subculturing at 26 ± 1^0C, 55% relative humidity and diffused light conditions (300 lux). Seeds, leaves and tissue cultures (at maximum GI) were dried and analyzed.

INTRODUCTION

Diosgenin, a major raw material for the commercial production of steroidal contraceptives and sex hormones is useful in several medical therapies. Considerable progress has been made on diosgenin-bearing plants and in the steroid industry (Mandal and Chatterjee, 1985; Ravishankar and Grewal, 1986; Singh *et al.*, 1988; Dixit and Banerjee, 2000). Besides *Dioscoreas*, alternative sources of diosgenin yielding plants have been searched out. All these resources result in the depletion of forest resources and in some areas resulted complete eradication of wild plants. To produce diosgenin synthetically is also uneconomical. Therefore, emphasis is likely to shift to the recovery of diosgenin from newer natural resources. So far the moth bean cultivars RMO-257, RMO-40, IPCMO-880 and Jwala have not been worked out for diosgenin production. Further, considerable interest has been growing on investigation of the potential of plant tissue culture to the

*Address for correspondence: Plant Tissue Culture & Biotech Lab., Dept. of Botany, Dungar College, Bikaner-334001

We are grateful to Dr. R.C. Sharma, Associate Director Research, Rajasthan Agriculture University, Bikaner for providing experimental plant material and necessary facilities during the course of investigation.

traditional methods for the industrial production of plant products. This prompted the present investigation on the production of diosgenin from four moth bean cultivars.

MATERIAL AND METHODS

Seeds and leaves of four cultivars of moth bean (RMO-257, RMO-40, IPCMO-880 and Jwala) were collected from experimental field of Agriculture Research Station, Bikaner. Unorganized cultures were raised from leaf explant excised from aseptically grown *in vitro* seedlings and maintained on Murashige and Skoog's (1962) medium supplemented with 0.5 mg/l kinetin and 1.5 mg/l 2,4-D for 18 months by frequent subculturing at 26 ± 1°C, 55% relative humidity and diffused light conditions (300 lux), harvested at 2, 4, 6, 8 and 10 weeks and growth indices were calculated. The seeds, leaves and unorganized cultures (at maximum GI) were dried and analyzed.

Extraction Procedure

The dried samples (leaves, seeds and unorganized cultures) were hydrolyzed with 30% (v/v) hydrochloric acid (2 gm/20 ml) for four hours on water bath, washed separately with distilled water till the filtrate attained pH 7.0, dried at 60°C for eight hours and soxhlet extracted in benzene for 24 hours (Nag *et al.*, 1979) separately. Benzene extracts of various test samples were dried *in vacuo* and taken up in chloroform for further analysis.

Qualitative Analysis

Each of the extract along with reference sapogenins (diosgenin, gitogenin, hecogenin, kryptogenin, smilogenin, tigogenin and yamogenin) was applied separately on Silica gel 'G' coated and activated glass plates and developed in hexane and acetone (80:20, v/v). The plates were sprayed with 50% sulphuric acid heated at 100°C for 10 min. Examination of chromato plates revealed one spot corresponding to reference diosgenin (brown, Rf–0.43) in all the samples. The diosgenin spots were marked, scrapped from developed unsprayed plates, eluted with chloroform, rechromatographed in order to ascertain the purity of compound, crystallized from methanol-acetone (Kaul and Staba, 1968) separately and subjected to mp and IR (Perkin Elmer 337 grating Infra-red spectrophotometer) spectral studies along with standard reference.

Quantitative Estimation

Diosgenin was quantitatively determined by following the method of Sanchez *et al.* (1972). Standard stock solution of diosgenin was separately prepared in chloroform, out of which various concentrations were made ranging from 10 mg to 120 mg and applied separately on silica gel 'G' coated and activated glass plates, developed in a solvent system of hexane: acetone (80:20), air dried and exposed to iodine vapours, marked, heated to evaporate excess of iodine, scrapped along with silica gel, eluted with 5 ml methanol, centrifuged, supernatants collected, dried and 4 ml of 80% methanolic sulphuric acid was added and kept for 2 hours. Absorbance was determined using a spectronic-20 colorimeter (Bausch and Lomb) set at 405 nm against a blank solution (4 ml 80% methanolic sulphuric acid). Absorbance was plotted against their respective concentrations to compute a regression curve which followed Beer's law. The concentration of various samples was calculated (in %) by comparing the absorbance with the regression curve of diosgenin.

RESULTS AND DISCUSSION

Presence of diosgenin in seeds, leaves and eight-week-old tissues of four cultivars of moth bean (RMO-257, RMO-40, IPCMO-880 and Jwala) was confirmed by co-chromatography (diosgenin, Rf. 0.42) mp 207-208°C. The presence of diosgenin was further confirmed by superimposable IR spectrum of the isolated and reference compounds.

Leaves showed comparatively more amount than seeds in all the four cultivars. Leaves of cultivar RMO-257 showed maximum amount (0.40%) of diosgenin whereas minimum amount (0.12%) was observed in seeds of Jwala. Among the unorganized cultures, maximum diosgenin content (0.44%) in RMO-257 and minimum (0.31%) in IPCMO-880 was observed (Table 1). However tissue cultures of all the four cultivars showed more amount of diosgenin than seeds and leaves. The reason for more concentration of diosgenin might be due to the supplementation of phytohormones (2,4-D, IAA and kinetin) in the medium. Kaur (1997) also observed higher amount of diosgenin in six-week-old callus than seeds in *Gossypium* varieties.

Table 1 Diosgenin (%) in moth bean cultivars *in vivo* and in tissue cultures

Vigna aconitifolia cultivar	*Diosgenin content (%)*		
	Leaves	*Seeds*	*Unorganized cultures*
RMO-257	0.40	0.21	0.44
RMO-40	0.32	0.17	0.39
IPCMO-880	0.26	0.19	0.31
Jwala	0.22	0.12	0.34

Quantification of diosgenin revealed that unorganized tissue cultures have more potentiality to synthesize diosgenin than seeds and leaves.

The steroidal sapogenins of *Dioscorea tokaro* were found mainly in the organelles (mitochondria) and only traces in the supernatant (Akahori *et al.*, 1971). The sapogenins are not released into the medium by somatic disintegration. It seems likely that the sapogenins bound to the membranes of the organelles are not present as solutions in their matrix. The work of Behnke (1965) with *Dioscorea* and that of others with *Cucurbita* (Esau and Cheadle, 1962; Kallmann, 1964) indicates that with increased differentiation, there is a reduction or loss in the number of mitochondria. Assuming that mitochondria are intimately associated with steroid biosynthesis in plant cells, their observations may explain the marked difference in diosgenin production in differentiated and undifferentiated *D. detoidea* tissue cultures (Kaul and Staba, 1968). Chaturvedi and Chowdhary (1980) reported that 2,4-D or IAA, more or less, equally stimulated diosgenin biosynthesis. Khanna (1987) has also reported a number of steroidal sapogenins from tissue cultures of various plants. Aminuddin (1998) reported roots and callus cultures of *P. betle* as alternative source of diosgenin. This report also presents evidence that diosgenin can be isolated from the seeds, leaves and especially from tissue cultures of moth bean cultivars in significant amount and if properly utilized, may serve as an additional source for the production of diosgenin.

REFERENCES

Akahori A., Yasuda F. & Okanishi T. (1971) Studies on the steroidal components of domestic plants. LXV. Part LXIV. *Chem. Pharm. Bull.*, (Japan). 19: 2409.

Aminuddin (1998) Biosynthesis of diosgenin in tissue cultures of medicinal and aromatic plants. In: Role of Biotehnology in Medicinal and Aromatic Plants. Khan I.A. and Khanum A. (Eds.) Ukaaz Publications, Hyderabad, India.

Behnke H.D. (1965) Effect of growth hormones and some nitrogen sources on diosgenin biosynthesis by tuber-callus of *Dioscorea deltoidea. Indian J. Exp. Biol.*, 18: 913-915.

Chaturvedi H.C. & Chowdhary A.R. (1980) Effect of growth hormones and some nitrogen sources on diosgenin biosynthesis by tuber-callus of *Dioscorea deltoidea. Indian J. Exp. Biol.* 18: 913-915.

Dixit B.S. & Banerjee R. (2000) High diosgenin content in some selected clones of *Dioscorea floribunda. Indian Journal of Pharm. Sci.* 62(1): 53-54.

Esau K. & Cheadle V.I. (1962) Mitochondria in the phloem of *cucurbita. Bot. Gaz.* 124: 79-85.

Kallman R. (1964) On the fine structure of the sieve element proto plant. *Planta Med.* 21: 426.

Kaul B. & Staba E.J. (1968) *Dioscorea* tissue cultures. I. Biosynthesis and isolation of diosgenin from *Dioscorea deltoidea* callus and suspension cells. *Lloydia* 31: 171-179.

Kaur A. 1997. Influence of plant growth regulators on the production of primary and secondary products in cotton varieties growing *in vitro.* Ph.D. Thesis, M.D.S. University, Ajmer, India.

Khanna P. (1987) Useful metabolites from seventeen plant species: A review. AISABR at Bikaner, 1st-3rd Nov., Abstr. pp. 126-127.

Mandal A. & Chatterjee S.K. (1985) Histochemical studies of *Dioscorea composita* (Hemsl.) Yams with special reference to starch grain in relation to its diosgenin content. *72nd Indian Sci. Cong., Abstr.* 45.

Murashige T. & Skoog F. (1962) A revised medium for rapid growth and bioassays with tobacco tissue cultures. *Physiol. Plantarum.* 15: 473-497.

Nag T.N., Mathur C.S. & Goyal S.C. 1979. Phytochemical studies of *Tribulus alatus, T. terrestris* and *Agave wightii* contents of primary and secondary products. *Comp. Physiol. Ecol.* 4: 157-160.

Ravishankar G.A. & Grewal S. 1986. Development of twostage culture methods and biochemical markers for growth and diosgenin production in *in vitro* grown cultures of *Dioscorea deltoidea.* VI IAPTC Minneapolis, 3rd-8th August, Absrtacts, pp. 251.

Sanchez G.L., Medina Acevedo J.C. & Soto R.R. (1972) Spectrophotometric determination of diosgenin in *Dioscorea composita* following thin layer chromatography. *Analyst.* 97: 973.

Singh V., Sethia M., Mathur K. & Nag T.N. (1988) Desert Medicinal Plants: A Source of Steroids. Oikoassay, 5(2): 41-42.

CHAPTER

17

Production of Bacoside by Multiple Shoot Cultures and *in vitro* Regenerated Plantlets of Selected Cultivar of *Bacopa monnieri* (L.) Wettst.

Ashok Ahuja, K.K. Gupta, R.K. Khajuria, Aradhana Sharma, Arun Kumar, M. Sharada and M.K. Kaul

ABSTRACT

The use of plant cell cultures for the production of bioactive secondary metabolites is an attractive alternative approach. Multiple shoot cultures of medicinal plants are a good source of known and new biologically active metabolites. Bacosides, tetracyclic triterpenoid saponins have received considerable attention in recent years due to their potential memory enhancing activities. During the course of present studies, multiple shoot cultures and *in vitro* regenerated plants of *Bacopa monnieri* (L) Wettst were established utilizing defoliated nodal explants on MS medium supplemented with BAP (1.0 mg l^{-1}) and analyzed for bacoside contents using HPLC and LC-MS. Bacoside A_3 and A_2 were identified as the major bacoside components. Their structure and preferred confirmation were determined on the basis of spectroscopic data and molecular calculations. The total bacoside content was comparable to those detected for donor tissues. The total bacosides ranged between 1.76 to 2.70 % on dry weight basis. The potential to produce bacosides by multiple shoot cultures and *in vitro* miniature plantlets represent an attractive alternate source for production of these pharmaceutically important compounds.

Key Words: *Bacopa monnieri*, Brahmi, Bacoside, shoot cultures, plantlets, tissue cultures

INTRODUCTION

Medicinal plants provide mankind invaluable products and bioactive molecules. Since time immemorial man has been using a variety of herbs and herbal preparations in traditional system of medicines. *Bacopa monnieri* (L) Wettst (Scrophulariaceae) is an important medicinal herb

*Address for correspondence: Regional Research Laboratory, CSIR, Jammu Tawi, 180 001, India.
E-mail: ashoksahuja@ rediffmail.com

commonly known as Brahmi or 'Jalnimba' classified as 'Medhya Rasayan' in Indian traditional system. It is widely distributed throughout tropical and subtropical areas; used to relieve mental stress and improves memory function (Handa, 1994). Besides, it is also used to cure various ailments including bronchitis, inflammation, and asthma (Dey & Dutta, 1966; Singh & Dhawan, 1982; Kapoor, 1990; Rastogi *et al.*, 1994; Khanna *et al.*, 1995; Tripathi *et al.*, 1996). The alcohol extract of *B. monnieri* is reported to possess a relaxant effect on the trachea, pulmonary artery, and aorta of rabbit and guinea pig. Recent studies also suggest that the extract of *B. monnieri* acts as a non-selective calcium antagonist (Hou *et al.*, 2002). The medicinal properties of *B. monnieri* have been attributed to the presence of different tetracyclic triterpenoid saponins amongst which, bacosides are proving to be biologically active (Rastogi & Kulshershtra, 1998). The bacosides help repair damaged neurons by augmenting kinase, the protein involved in the synthesis of new neurons to replace old ones. Depleted synaptic activity is restored which leads to enhanced memory function (Vaidya, 1997). Owing to memory enhancing property of the bacosides the demand of *B. monnieri* is rapidly growing and more products are sold in the market utilizing this herb. A sufficient supply of this herb is rather limited and presently, commercial requirement is met solely from the wild natural populations causing it to be listed as a threatened plant (Tiwari *et al.*, 2001). Since both native and cultivated plants are in limited supply an alternative source of bacosides is of great interest. As such, tissue cultures could be considered a viable alternative for production of bacosides. Encouraging reports are available in the literature utilizing organ cultures for exploitation of secondary metabolite production in various medicinal plants (Rout *et al.*, 2000). Although, *in vitro* regeneration studies in *B. monnieri* have been attempted (Thakur *et al.*, 1976; Thakur & Ganapathy, 1978; Tiwari *et al.*, 1998; Tiwari *et al.*, 2000) no information on production of bacosides by the neoformations is available. Most recently, production of bacosides A and B have been reported in cell suspension cultures (Rahman *et al.*, 2002). The present communication describes production, isolation and identification of bacosides from multiple shoot cultures and tissue cultured plantlets of this important medicinal plant.

MATERIALS AND METHODS

Defoliated nodal segments (~0.5 cm) with single axillary buds collected from field grown plants of *Bacopa monnieri* (of high yielding strain RRL - 001BM developed by RRL, Jammu) were utilized as explant source material for establishment of multiple shoot cultures and subsequent plantlet regeneration therefrom. The selected clone utilized for initiation of cultures is consistently stable and suitable for cultivation in a wide range of agro-climatic conditions having sufficient moisture levels with uniform growth in the experimental field beds (Gupta, 2000). The explants were cultured on MS medium (Murashige and Skoog, 1962) containing 3% (w/v) sucrose supplemented with BAP + IBA (1.0 mg l^{-1} each). The pH of the medium was adjusted to 5.8 before autoclaving (15 lb in^{-2}, 121°C). The cultures were incubated at 25 ± 2°C with 16 hr/day photoperiod provided by cool white tube lights. Proliferated axillary shoots were excised and cut into propagules (3-4 shoots/cluster), cultured on the same medium, termed as first generation shoots. Subsequently, these were cut into nodal segments with single axillary buds and regularly sub-cultured at 4 week intervals. These were termed as second generation shoots. The number of shoots derived from a single axillary bud at the end of each subculture (4 week) was recorded as the rate of multiplication. For rooting of the *in vitro* microshoots (2.5-4 cm length), MS medium supplemented with IBA (1.0 mg l^{-1}) was used. Two week old plantlets regenerated *in vitro* were acclimatized under controlled greenhouse conditions in small pots containing a 1:1 sand:garden soil potting mixture, with RH of 70-80%, temperature of 25 ± 2°C for 2 week before transfer to field conditions during the month of February (Fig. 1A-E). The culture samples were subsequently dried and analyzed for bacoside contents.

Fig. 1A-E Regeneration of plantlets from defoliated nodal explants of *Bacopa monnieri.* **(A)** First generation multiple shoots on MS medium supplemented with BAP (1.0 mgl^{-1}) + IBA (1.0 mgl^{-1}). **(B/C)** Second generation well-elongated multiple shoots on MS medium supplemented with BAP (1.0 mgl^{-1}) + IBA (1.0 mgl^{-1}). **(D)** *In vitro* rooted plantlet. **(E)** 4 week old acclimatized plantlets ready for transfer to field beds.

Table 1 Bacoside production from *in vitro* regenerated and donor tissues of *Bacopa monnieri* (L.) wettst.

Culture type	*Age (weeks)*	*Bacoside contents (% dwb)**		
		Total	A_3	A_2
Multiple shoot cultures (First generation)	4	1.83	0.93	0.22
Multiple shoot cultures (Second generation)	4	1.76	1.08	0.39
Plantlets regenerated *in vitro*	4	2.51	1.62	0.34
Micropropagated plants (greenhouse)	8**	2.61	1.66	0.36
Micropropagated plants (field- grown)	0**	2.70	2.02	0.44
Donor parent plant (RRL-01BM)	24	3.36	2.28	0.51

*All values expressed as a percentage on dry wt basis are mean of three replicates.
**After transfer to green house/field

Extraction and isolation of bacosides

The multiple shoot cultures and *in vitro* regenerants from first and second generation were harvested and dried at 50 ± 2°C overnight. Powdered plant material (10 g) was extracted in a soxhlet with methanol (150 ml) for 4 hrs. The extract was concentrated to 60 ml under *vacuo* and made upto 90 ml with water and successively extracted with n-hexane (100 ml × 3) and n-butanol (50 ml × 4). Butanol extracts were dried under *vacuo* to obtain total bacoside contents. The procedure was repeated for *in vitro* raised regenerated plants grown in the green house and those in the field beds.

Quantitative determination

Bacosides were quantified by LC-MS carried out on Agilent - 1100 series HPLC system hyphenated to Ion Trap Mass Spectrometer (Esquire 3000, Bruker). HPLC analysis was conducted using Acetonitrile:water (48:52) as the mobile phase with a flow rate of 0.5 ml min^{-1} at wavelength of 210 nm on a Merck RP – 18, 5 μm, 4.6 × 250 mm column. The column temperature was maintained at 30°C for maximum peak efficiency. The analysis was carried out using the positive electro spray mode of ionization. The MS data obtained was in agreement with that reported in literature (Rastogi *et al.*, 1994; Rastogi & Kulshershta, 1998). Bacoside A_3 and A_2 eluted at retention times of 8.452 and 9.470 minutes respectively which exhibited molecular ion peaks at m/z 951 $[M+Na]^+$ and 921 $[M+Na]^+$ in the positive mode (Fig. 2A-C). HPLC grade acetonitrile and methanol (Rankem make) purchased from Ranbaxy Chemicals Ltd. (Mohali, Punjab, India) were used without further purifications. The samples and standards were prepared in HPLC grade methanol and filtered through 0.45 μm millipore filter. Stock solutions of the standard reference mixture were prepared and working solutions in the concentration range of 4.77 to 20.32 μg were utilized for plotting of calibration curves for standard markers–bacoside A_3 and A_2. Concentrations

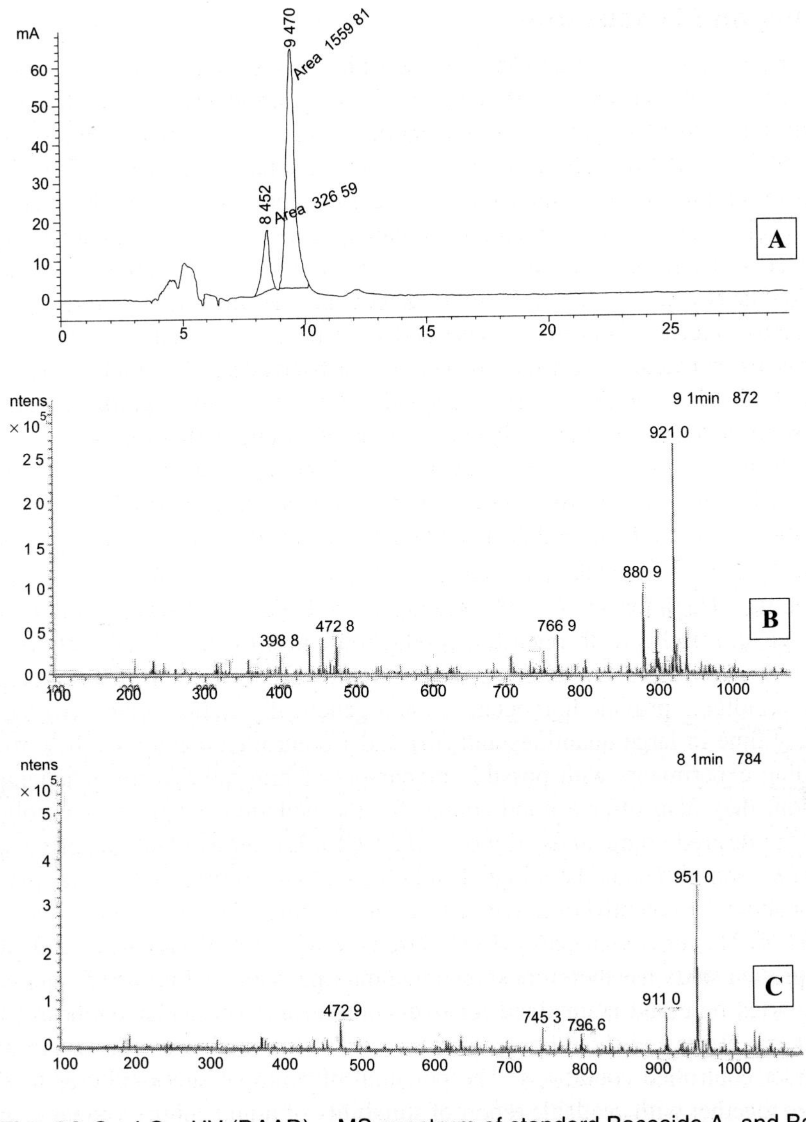

Fig. 2A-C – LC – UV (DAAD) – MS spectrum of standard Bacoside A_2 and Bacoside A_3. **A.** HPLC chromatogram of standard Bacosides; **B.** MS spectrum of Bacoside A_2. **C.** MS spectrum of Bacoside A_3.

of bacoside A_3 and A_2 were found to be in the ratio of 41:9 as determined by HPLC at 210 nm with a purity of 0.99985 and 0.99782 respectively. On the basis of the standard curves obtained, the percentage of bacoside A_2 and A_3 were determined in samples of intact plant, shoot cultures, and regenerated plants.

Observations and Discussion

Present studies have examined the contents of bacosides in selected strain of *Bacopa monnieri* (RRL-001BM) with a total bacoside yield of 3.36% on dry weight basis. Qualitative as well as quantitative data for bacosides synthesized and produced in multiple shoot cultures and *in vitro* regenerants established utilizing defoliated nodal segments are tabulated in Table 1. On an average 18.35 ± 2.15 shoots/culture were produced after 4 weeks of inoculation. The shoot cultures grew consistently to reach rapid growth with maximal doubling time of 2 weeks (15 days). Regenerated shoots were successfully rooted with ease to get plantlets which were further acclimatized and transferred to field beds with > 90% survivability. Shoot cultures, miniature plantlets (*in vitro*) and acclimatized plants collected from greenhouse and field beds were analyzed for total bacoside contents. All tissues analyzed were found to synthesize bacosides. The total bacoside contents ranged between 1.76-2.70% on dry weight basis and did not show any significant variation with respect to bacoside yield among the analyzed samples of multiple shoot cultures and plantlets (Table 1). Identification of bacosides was confirmed by direct comparison with reference marker compounds on TLC. Further confirmation by LC-MS profile showing qualitative similarity of bacosides identified as bacoside A_3 and A_2 from donor plants and tissue culture regenerants.

Multiple shoot cultures of medicinal plants provide a good approach for production of several bioactive metabolites (Hagimori *et al.*, 1984; Heble, 1985; Roja *et al.*, 1987; Miura *et al.*, 1988). Higher product yield, stability with regard to production and growth of shoot cultures offers an attractive approach over the intact plant for production of pharmaceutically important compounds. Besides, (i) shoot cultures provide homogeneous and genetically stable tissues, (ii) they may be propagated at any time in large quantities and (iii) under controlled conditions they offer a good system for feeding experiments with possible precursors of biosynthesis for enhancing product yield. In addition, they also offer a good source for the isolation of enzymes involved in the biosynthesis of the desired compounds (Endo *et al.*, 1988). Earlier available reports suggest that shoot cultures in *B. monnieri* could be initiated and utilized to micropropagate 'elite' plants. These studies are silent about the potential of tissue cultures or regenerated plants for bacoside production (Thakur *et al.*, 1976; Thakur & Ganapathy, 1978; Tiwari *et al.*, 1998; Tiwari *et al.*, 2000; Mathur *et al.*, 2001). The present study has therefore shown that multiple shoot cultures of *B. monnieri* could be grown as reported here and maintained for over three years. These can synthesize bacosides comparable to donor plants and should prove useful as stable alternative source for bio-production of bacosides under controlled conditions. The potential of multiple shoot cultures to synthesize bacosides *in vitro* together with available report of suitability of liquid culture system to grow these multiple shoots as shake cultures recently reported by Tiwari *et al.* (2000) can be exploited by cultivation of multiple shoot cultures in bioreactors and production of bacosides. The feasibility to cultivate organ cultures in bioreactors and production of metabolites are quite promising (Park *et al.*, 1989; Ales & Tomas *et al.*, 1998). Thus, it may be concluded that *in vitro* grown shoot cultures of *B. monnieri* can also be effectively utilized as an alternative source to the natural cultivated herb and for bio-production of these important memory-enhancing compounds.

Acknowledgements

We are grateful to Director, Regional Research Laboratory, Jammu, for providing facilities.

REFERENCES

Ales N. & Tomas V. (1998) *In vitro* propagation of *Stevia rebaudiana* plants using multiple shoot cultures, *Planta Med.* 64: 775-776.

Dey P.K. & Dutta C. (1966) Effect of psychotropic phytochemicals on cerebral amino acid level in mice, *Indian J. Exp. Biol.* 4: 216-218.

Endo T., Goodbody A., Vukovic J. & Misawa M. (1988) Enzymes from *Catharanthus roseus* cell suspension cultures that couple Vindoline and Catharanthine to form 3', 4' – Anhydroxyvinblastine, *Phytochemistry.* 7: 2147-2149.

Gupta S. (2000) Project completion report, Agrotechnology & Improvement of *Bacopa monnieri* (pp 1-27) RRL, Jammu.

Hagimori M., Matsumoto T. & Mikami Y. (1984) Jar fermentor culture of shoot forming cultures of *Digitalis purpurea* L. using a revised medium, *Agric. Biol. Chem.* 48: 965-970.

Handa S.S. (1994) Rasayana Drugs, *Pharma Times.* 26: 17-25.

Heble M.R. (1985) Multiple shoot cultures: A viable alternative *in vitro* system for the production of known and new biologically active plant constituents. In: Newman H.K., Barz W. & Reinhard E. (eds) *Primary and Secondary Metabolism of plant Cell Cultures* (pp 281-289) Springer-Verlag, Heidelberg, Germany.

Hou C.-C., Lin S.-J. ,Cheng J.-T. & Hsu F.-L. (2002) Bacopside III Bacopasanin G and Bacopasides A, B and C from *Bacopa monnieri, J Nat. Prod.* 65: 1759-1763.

Kapoor L.D. (1990) Hand Book of Ayurvedic Medicinal Plants (pp 61-63) CRC Press, Florida.

Khanna N.K., Jain P. & Godhawari J.L. (1995) Studies on anti-inflammatory effects of Brahmi Rasayan in rodents, *Indian J. Pharmacol.* 27: 49-50.

Mathur S., Darokar M.P., Dwivedi S., Khanuja S.P.S. & Kumar S. (2001) Descriptors for Brahmi *Bacopa monnieri, J. Med and Aromat Plant Sci.* 23: 91-96.

Miura Y., Hirata K., Kurano N., Miyamoto K. & Uchida K. (1988) Formation of vinblastine in multiple shoot cultures of *Catharanthus roseus., Planta Med.* 54: 18-20.

Murashige T. & Skoog F. (1962) A revised medium for rapid growth and bioassays with tobacco tissue cultures, *Physiol. Plant.* 15: 473-497.

Park J.M., Hu W.S. & Staba E.J. (1989) Cultivation of *Artemisia annua* L. plantlets in a bioreactor containing a single carbon and a single nitrogen source, *Biotech. Bioeng.* 34: 1209-1213.

Rahaman L.U.,Verma P.C., Singh D.,Gupta M.M. & Banerjee S. (2002) Bacoside production by suspension cultures of *Bacopa monnieri. Biotechnology Lett.* 24: 1427-1429.

Rastogi S., Pal R. & Kulshershta D.K (1994) Bacoside A_3 – A triterpenoid saponin from *Bacopa monnieri, Phytochemistry* 36: 133-137.

Rastogi S. & Kulshreshta D.K. (1998) Bacoside A_2 – A triterpenoid saponin from *Bacopa monnieri, Indian J. Chem.* 38B: 353-356.

Roja P.C., Sipahimalani A.T., Heble M.R. & Chadha M.S. (1987) Multiple shoot cultures of *Rauwolfia serpentina, J. Nat. Prod.* 50: 872-875.

Rout G.R., Samantaray S. & Das P. (2000) *In vitro* manipulations and propagation of medicinal plants, *Biotechnology Advance.* 18: 91-120.

Singh H.K., Dhawan B.N. (1982) Effect of *Bacopa monniera* Linn. (brahmi) extract on avoidance responses in rat, *J. Ethnopharmacol.* 5: 205-214.

Thakur S. & Ganapathy P.S. (1978) Morphogenesis of organ differentiation in *Bacopa monnieri* stem cultures, *Indian J. Exp. Biol.* 16: 514-516.

Thakur S., Ganapathy P.S. & Johri B.N. (1976) Differentiation of abnormal plantlets in *Bacopa monnieri, Phytomorphology.* 26: 422-424.

Tiwari V., Singh B.D. & Tiwari K.N. (1998) Shoot regeneration and somatic embryogenesis from different explants of Brahmi (*Bacopa monniera* (L.) Wettst), *Plant Cell Rep.* 17: 538-543.

Tiwari V., Tiwari K.N. & Singh B.D. (2000) Suitability of liquid cultures for *in vitro* multiplication of *Bacopa monniera* (L.) Wettst., *Phytomorphology.* 50: 337-342.

Tiwari V., Tiwari K.N. & Singh B.D. (2001) Comparative studies of cytokinins on in vitropropagation of *Bacopa monniera* (L.), *Plant Cell and Tissue Culture.* 66: 9-16.

Tripathi Y.B., Chaurasia S., Tripathi E., Upadhyay A. & Dubey G.P. (1996) *Bacopa monniera* Linn. as an antioxidant: mechanism of action, *Indian J. Exp. Biol.* 34: 523-526.

Vaidya A.D.B. (1997) The status and scope of Indian medicinal plants acting on central nervous system, *Indian Journal of Pharmacology.* 29: 340-343.

CHAPTER

18

Biotechnology of Medicinal Plants

Alka Narula, Sanjeev Kumar*, M.Z. Abdin and P.S. Srivastava

INTRODUCTION

Almost 50% of the commercial drugs available in the market even today are of plant origin. Because of ruthless exploitation, the availability of these herbs is threatened. The problem is further aggravated due to increased anthroprogenic activities coupled with rapidly eroding ecosystem. Strategies are therefore required to produce propagules on a large scale and conserve the fast depleting drug-yielding plants. *In vitro* propagation of medicinal plants has already been achieved by using various explants [6-16; see Table 1]. It was believed that enhancement in the yield of secondary metabolites was dependent on prolonged tissue cultures or organogenesis [18]. However, it has now been demonstrated that the ability of product biosynthesis continues throughout the culture regime and can be detected at various stages of growth and differentiation [19].

Enhancement in the Yield of Secondary Metabolites

Effect of chemical milieu

Yield of secondary metabolites can be enhanced by modifying the chemical milieu and culture conditions. The composition of culture medium not only affects growth and production of metabolites but also plays a critical role in initiation of morphogenic events [20].

Abiotic stress-use of metal ions

Trace elements have indeed been considered abiotic elicitors or inducing factors [22] that can trigger the biosynthesis of secondary metabolites. Studies on the effect of divalent ions such as, Co^{2+} and Cu^{2+} seem to have received more attention because of their positive effects on the production of secondary metabolites [21, 23, 24]. Increase in Co^{2+} from 1 to 5 mM caused enhanced production of betalains in *Beta vulgaris* [25]. Enhanced shikonin production in the

*Address for correspondence: Centre for Biotechnology, Faculty of Science, Jamia Hamdard, New Delhi 110062.
*Amity Institute of Biotechnology, Amity IT University Campus, Sector 125, Noida 201 303.

Table 1 Examples of micropropagation of some medicinal plants

Plant	Therapeutic use	Cultured explant	Medium*	Response	Ref
Abrus precatorius	Abortifacient	Epicotyl segments	MS + NAA (0.1) + BAP (0.5) MS + NAA (0.1) + Kn (0.5)	Plantlets Plantlets	15
Ammi majus	Leucoderma	Nodal segments Isolated shoot	MS + IAA (0.5) + Kn (2.0) + CH (1000) MS + IAA (0.5) + Kn (5.0) + CH (500) + Ad (40) MS + IBA (0.2) + Glu (100)	Multiple shoots *In vitro* flowering and fruiting Plantlets	1,2, 81
Artemisia annua	Antimalarial, Anti-HIV	Cotyledonary leaves, stem segments Hypocotyl segments Immature inflorescence segments	MS + (in μM) NAA (0.5) + BAP (13.0) + GA_3 (0.3) + Glu (700) + Asp (300) + Arg (300) + Cyst Hcl (30) MS + (in μM) NAA (1.0) + BAP (13) + CM (2%) MS + (in μM) NAA (0.5) + BAP (13.0)	Multiple shoots Multiple shoots	4
Allium wallichii	Tuberculosis, nerve defects, cold, cough	Seedlings (without the root portion)	MS + Zt (20 μM)	Multiple shoots	82
Arnica montana	Anti-inflammatory	Nodal segments	MS + (in μM) NAA (5.3) + 2iP (5.0) + Phloroglucinal (0.6 mM) + Ad (0.2 μM)	Plantlets	83
Atropa acuminata	Antispasmodic, narcotic, analgesic, Antiasthamatic	Shoot tips, nodal segments Isolated shoots	MS + BAP (1.0) + IBA (1.0) RT + IAA (1.0) RT + IBA (1.0)	Shoot buds Elongated shoots Plantlets	84
Bacopa monniera	Memory vitalizer	Nodal segments	MS + NAA (0.2) + BAP (5.0) + CH (500) MS + NAA (0.1) + BAP (0.5) + CH (500)	Callus, multiple shoots Plantlets	3
Catharanthus roseus	Antileukaemic	Juvenile explants and mature stem segments Callus	MS + NAA (2.0) + BAP (5.0) + Asp (100) + CH (1000) MS + NAA (0.1) + BAP (5.0) + Zt (1.0) + Asp (100) + Glu (100)	Callus Multiple shoots	8
Clerodendrum inerme	Substitute for Quinine	Leaf segments	MS + BAP (2.0) + NAA (0.5) MS + IAA (2.0)	Callus, multiple shoots Plantlets	85

(Contd.)

(Contd.)

Coleus forskohlii	Anti-inflammatory	Shoot tips	MS + (in µM) IAA (0.57) + Kn (0.46)	Plantlets	86
Datura innoxia	Anticholinergic	Anthers	WB + NAA (1.0) + BAP (2.0) + CH (500) WB + IBA (0.1)	Multiple shoots Plantlets	11
Dioscorea bulbifera	Antifertility	Nodal segments	MS + IAA (0.1) + Kn (5.0) + CH (500) + Charcoal (2000)	Plantlets	87
Holorrhena antidysentrica	Dysentry, colic pain	Node with axillary buds	MS + BAP (15 µM) MS + IBA (35 µM)	Multiple shoots Plantlets	88
Isoplexis canariensis	Source of cardiac glycosides	Seedling explants Isolated shoots	MS + BAP (5.0 µM) MS + (in µM) BAP (2.25) + IAA (0.17) MS + (in µM) BAP (0.22) + IAA (0.17)	Multiple shoots Plantlets	16
Nothapodytes foetida	Antitumour	Leaf segments	MS +TDZ (1.36 µM) MS (1/2)) + (in µM) BAP (2.22) + IBA (0.49)	Multiple shoots Plantlets	89
Peganum harmala	Abortifacient	Cotyledonary node with shoot tips	MS + (in µM) NAA (0.1) + BAP (5.0) MS + IBA (8.0 µM)	Multiple shoots Plantlets	90
Phyllanthus caroliniensis	Antidiabetic and used against Hepatitis B virus	Nodal segments Isolated shoots	MS + (in µM) BAP (5.0) + Kn (1.25-5.0) MS	Multiple shoots Plantlets	91
Plantago ovata	Laxative	Shoot buds Callus	MS + (in µM) 2,4-D (4.5) + Kn (2.3) MS + (in µM) BAP (4.4) + NAA (2.7) + CM (10%)	Callus Somatic embryogenesis	92
Plumbago rosea	Anticancerous	Nodal segments	MS + IAA or IBA (0.1) + BAP (1.5) + Ad (50) MS (1/2) + IBA (0.25)	Multiple shoots Plantlets	93
P. zeylanica	Leprosy, skin diseases	Nodal segments	MS + (in µM) Ad (27.2) + IBA (2.46) MS + IBA (4.92 µM)	Multiple shoots Plantlets	94
Psoralea corylifolia	Psoriasis	Nodal segments	MS + BAP (0.5) MS + IAA (5.7)	Multiple shoots Plantlets	95

(Contd.)

(Contd.)

Rauwolfia serpentina	Cardiovascular diseases, sedative or tranquilizer	Shoot apices	MS + BAP (2.0)	Multiple shoots	96
		Nodal segments	MS + BAP (1.0) + NAA (0.1)	Multiple shoots	
			MS + NAA (2.0) + BAP (1.5)	Plantlets	
Sterculia foetida	Abortifacient, diuretic	Cotyledonary nodes	MS + BAP (4.0)	Multiple shoots	97
			MS + IAA (2.0)	Plantlets	
Stevia rebaudiana	Antidiabetic	Nodal segments	MS + (in μM) BAP (8.87) + IAA (5.71)	Multiple shoots	98
			MS (1/2) + IBA (4.90 μM)	Plantlets	
Silybum marianum	Antihepatotoxic	Nodal segments	MS + IAA (0.1) + Kn (5.0)	Multiple shoots	99
			MS + NAA (0.1) + Zt (0.5)	Plantlets	
Solanum khasianum	Steroidal drugs	Leaf segments	MS + 2,4- D (3.0) + Kn (1.0)	Callus	100
		Callus	MS + BAP (3.0)	Multiple shoots	
		Shoots	MS + NAA (2.0)	Plantlets	
Tylophora indica	Asthma, bronchitis anti-tumorous	Leaf segments	MS + (in μM) 2,4–D (9.04) + Kn (0.05)	Callus	101
		Callus	MS + 2iP (9.84 μM)	Somatic embryos	
		Somatic embryos	MS	Plantlets	
Typhonium flagelliforme	Anticancerous	Shoot buds (from rhizome)	MS + (in mM) IBA (2.46) + BAP (1.33)	Plantlets	5
Valeriana jatamansi	Tranquillizer	Shoots buds	MS + BAP (4.44 μM)	Multiple shoots	9
			MS + (in μM) NAA (4.03) + BAP (4.44)	Plantlets	
Vitex negundo	Anticancerous	Nodal segments	MS + NAA (0.1) + BAP (2.0)	Multiple shoots and flowering	102
		Shoots with immature flowers	MS + NAA (0.5) + BAP (0.1)	Flower maturation, plantlets	
Withania somnifera	Antistress, antitumor,	Leaf segments	MS + BAP (1.0)	Multiple shoots	103
	Anti-inflammatory	MS + IBA (10.0)		Plantlets	

MS = Murashige and Skoog medium, WB = Wood and Braun medium, RT = Revised Tobacco medium

Ad = Adenine, Arg = Arginine, Asp = Asparagine, BAP = Benzyl amino purine, CH = Casein hydrolysate, CM = Coconut milk,

2,4-D = 2,4 dichlorophenoxyacetic acid, Glu = Glutamine, IAA = Indole-3-acetic acid, IBA = Indole-3-butyric acid, 2iP = 2 isopentenyladenine, Kn = Kinetin, NAA = α-naphthaleneacetic acid, TDZ = Thidiazuron, Zt = Zeatin,

*Concentration of growth hormones are in mgl $^{-1}$ unless mentioned otherwise

cultures of *Lithospermum erythrorhizon* have been attributed to the concentration of both copper and sulphate [21].

Tissue culture techniques have helped not only in raising metal tolerant plants but have also demonstrated that subjecting the cultures of medicinal plants to abiotic stress can be crucial in increasing the yield of secondary metabolites [26]. Several investigations have indeed demonstrated the possibility to raise metal tolerant plants *in vitro* [see 27]. Heavy metals have different role in metabolic functions. Some of them including Cu and Zn are required as micronutrients in biological systems to act as cofactor and/or as part of prosthetic groups of enzymes in a wide variety of developmental pathways [28]. Addition of Cu in the medium promoted somatic embryogenesis as well as its subsequent development in *Citrus* [29]. Cu stimulated regeneration in wheat, *Nicotiana tabacum, Bacopa monniera* [30] and *Dioscorea bulbifera* (Narula *et al.*, unpublished). Cu has proved to be more effective than Zn in enhancing the yield of xanthotoxin in *Ammi majus* and lepidine in in *vitro cultures* of *Lepidium sativum* [1, 2, 26]. Heavy metals and others are reported to have a positive effect on alkaloid production in *Catharanthus roseus* [31]. Fe^{2+} [32] and Cu^{2+} [30] proved beneficial for the synthesis of shikonin. Endress [33] and Obrenovic [34] have already demonstrated profitable role of Cu^{2+} on the accumulation of betacyanins in callus cultures of *Portulacca grandiflora* and *Amaranthus caudatus* seedlings. Higher concentrations of Cu^{2+} in the media supported increased accumulation of sapogenin steroid in *Agave amaniensis* [35, 36].

Plant growth regulators

Plant growth regulators (auxins and cytokinins) are also effective triggers of secondary metabolites. An optimum concentration of 2,4-D (25 mgl^{-1}) favoured the production of L-Dopa in cell cultures of *Mucuna* while lower concentration (0.1 ppm) proved favourable for alkaloid production in *Cinchona ledgeriana*. In addition to concentration, the type of auxin used also exerts a strong influence on secondary product formation. According to Zenk *et al.* [37] in *Morinda citrifolia* presence of 2,4-D reduced the production of anthraquinones but NAA enhanced the accumulation. In *Nothapodytes foetida,* alkaloid synthesis and biomass accumulation increased in nutrient media containing NAA than the media with 2,4-D [38].

Like auxins, cytokinins also influence secondary metabolite production. BAP enhanced shikonin production in *Lithospermum erythrorhizon* and kinetin promoted L-Dopa synthesis in callus cultures of *Stizolobium hassjoo*. Likewise, in the presence of BAP maximum accumulation of withanolide occurred in the cultures of *Withania somnifera* [39].

Higher concentrations of NAA + Kn or IAA + Kn promoted the yield of diosgenin in *D. bulbifera;* NAA + Kn induced much higher content (Narula *et al.*, unpublished). Corroborative results have been recorded in *D. deltoidea.* The medium with 2,4-D favoured diosgenin production most consistently. GA and high BAP concentrations proved toxic [40]. GA or kinetin otherwise increase the steroid content in *Phaseolus aureus* and *Corylus avellana* and doubled production of diosgenin in *Solanum xanthocarpum* [40]. Zhao *et al.* [41] observed that an increase in jaceosidin production was accomplished by increased concentration of NAA. This concurs with the results of Matsumoto *et al.* [42].

Addition of precursors

Addition of precursors of desired compounds to the culture medium also enhances the yield of secondary products. Ajmalicine production in *Catharanthus roseus* could be stimulated to 10-fold by the addition of secologanin [43]. Quinine in *Cinchona* cultures, rosemarinic acid in *Coleus blumei* and capsaicin production in *Capsicum frutescens* [44] are some examples where precursor addition induced enhanced yield of related metabolites.

Use of elicitors

Elicitors can be of biotic or abiotic origin [45]. Biotic elicitors are products of fungal, yeast or bacteria cultures, fungal mycelial extracts, culture filtrates, fractions or compounds obtained from microbial cell walls. Autoclaved fungal mycelia induced the accumulation of diosgenin in *Dioscorea deltoidea* [46]. Berberine and shikonin production enhanced in the cultured cells subjected to fungal extracts [47, 48]. A beta-glucan elicitor from yeast promoted alkaloid production in *Eschscholtzia*. In *Tabernaemontana divaricata*, reserpine accumulation increased by an elicitor prepared from *Candida albicans* [49]. Purified fractions of bacteria also elicited diosgenin and capsaicin production [46]. Elicitation of capsaicin in *Capsicum frutescens* occurred with chitosan, curdlan and xanthan.

Synthesis of alkaloids can be similarly elicited with Jasmonic acid and its esters [50]. In fact, fungal cell wall elicitors and methyl jasmonate (MeJa) can induce secondary metabolism in soybean cell cultures. Treatments with exogenous MeJa can elicit the accumulation of several classes of alkaloids in a wide range of plant species [51]. Methyl jasmonate therefore could be an useful tool for the enhancement of lignan production.

Effect of physical conditions

Even physical conditions of the medium have proved crucial for the production of secondary metabolites in cultures. Cell suspension cultures have been favourites for the production of valuable medicinal compounds.

Production of Synthetic Seeds

The first successful example of synthetic seed technology is in alfalfa [52] and celery [53]. Various vegetative propagules like axillary buds, shoot tips, bulbs, protocoms have been used [54]. The production of 'Syn' seeds has been reported in *Atropa belladonna, Hyoscyamus muticus, Mentha arvensis, Picrorhiza kuroa* [55], *Dioscorea alata*, *D. floribunda*, [56], *Clitoria ternatea* [57], *Guazuma crinita* [58] and others.

Role of Cryopreservation

Cryopreservation can be a preferred tool for long-term conservation of germplasms with genetic stability and retention of biosynthetic potential [59]. Cryopreservation has been achieved by using various explants.

Because meristems are genetically more stable they have been preferred over cell and callus cultures. Shoot tips of medicinal plants such as *Cichorium* sp. [60], *Dioscorea deltoidea, D. floribunda* [59], *Holostemma annulare* [61], *Mentha* sp. [62] and others have been cryopreserved successfully. The technique also avoids genetic erosion due to periodic subculture. The storage problem can be overcome by freeze preservation of callus and cell suspensions in liquid nitrogen. Cell suspensions of medicinal plants, *Atropa belladonna, Datura innoxia, Nicotiana tabacum, Panax ginseng,* etc, are reported to retain their biosynthetic potential after freezing. Cryopreservation of somatic embryos helps in storage at appropriate stage that can be used whenever required. The somatic embryos of carrot, orange and asparagus frozen in liguid nitrogen yield high viability and regenerate complete plants.

Cryopreservation of pollen can ensure mass production of haploid plants, genetic stability as well as conservation of genetic resources. Anther segments and pollen embryos of *Atropa belladonna, Brassica sps, Nicotiana tabacum* and *Primula obconica* have been successfully frozen and entire plants have been regenerated after one year of storage [63].

Use of Molecular Markers

Maintenance of the genetic fidelity of micropropagated plants is of utmost importance to retain the qualities of high yield, uniform quality, shorter rotation period, etc. Only then the advantage of use of elite genotypes can be demonstrated [64, 65]. It has been argued that micropropagation can be rewarding if complete genetic fidelity is maintained [65]. For this reason, for harvesting true-to-type plants, axillary branching or somatic embryogenesis have mostly been preferred. These strategies have been considered to prevent genetic changes that may result in culture regimen during differentiation in tissue cultures [66].

The best method for detecting genetic changes by randomly surveying the genome is the AFLP markers. These markers do not depend on prior sequence information. AFLP markers have wide genome coverage as compared to other DNA-based markers [67]. Therefore AFLP markers are considered an ideal tool for evaluating genetic variation. The technique has been employed for analyzing genetic variation in somatic embryoids of pecan [68], for ascertaining clonal fidelity in tissue culture raised progenies of a medicinally important plant, *Azadirachta indica* [69], and for assessment of genetic variation in economically important plant species such as, *Cichorium* [70], *Withania* sps [71], etc.

Genetic Engineering in Medicinal Plants through *Agrobacterium*

Genetic engineering enjoys many advantages over routine tissue culture technology that may ensure fast growth and stable high level production of secondary metabolites. Genetic engineering therefore is a favourable tool for biotechnological exploitation. *Agrobacterium tumefaciens* and *A. rhizogenes* have been used for introduction of genes into the desired plant genome. Transformation through Ti plasmids has become possible in plants, the transformed tissues and hairy roots have proved useful and have been profitable. The transformed tissues have become potential sources for stable production of metabolites.

Hairy root cultures of *Trigonella-foenum-graecum* L. produced twice the amounts of diosgenin [72]. Studies indicate that *Agrobacterium rhizogenes* affects the levels of polyamines in transformed plants [73] which may influence growth as well as production of secondary metabolites. *Atropa baetica* hairy roots synthesized and accumulated a conspicuously high amount of tropane alkaloids

[74]. In *Hyoscyamus albus*, hyoscyamine content was more in the transformed roots [75]. Transformed roots of *Catharanthus roseus* yielded higher ajmalicine and catharanthine than the non-transformed cultures. Growth of cultures also get affected due to transformation. Usually transformed tissues show higher growth. Transformation stimulated increased biomass and tropane alkaloid production in axenic root cultures of *Calystegia sepium* and *Atropa belladonna* [77]. Compared to transformed plants, non-transformed plants contained low amounts of tropane alkaloids, especially 6 β-hydroxy hyoscyamine and scopolamine in the roots [78]. The ginsenoside production, however, varied during different stages of growth in *Panax ginseng*. The capacity to produce alkaloids also varied in *Datura* [76]. Farnesyl diphosphate synthase gene promoted artemisinin 3-4 times higher in hairy roots [79] in *Artemisia*. High STR (strictocidine synthase) activity enhanced the flow of metabolites through the indole pathway in *Catharanthus roseus* [80]. The *str* gene is required for increased terpenoid indole alkaloid (TIA) production.

Conclusions

In addition to crop plants, medicinal plants have become an important group of natural resources that have acquired importance. Because of the resurgence of the herbal system of medicine, the natural herbs are now required in much more higher quantities than before putting additional pressure on demand. Strategies therefore are required to not only to micropropagate the plants but also to enhance the yield. Biotechnological approaches described above will prove to be of tremendous help in achieving the goal.

REFERENCES

[1] Purohit M., Pande D., Datta A. & Srivastava P.S. (1995a) Enhanced xanthotoxin content in regenerating cultures of *Ammi majus* and micropropagation, *Planta Med.* 61: 481-482.

[2] Purohit M., Pande D., Datta A. & Srivastava P.S. (1995b) *In vitro* flowering and high xanthotoxin content in *Ammi majus* L., *J. Plant Biochem. Biotechnol.* 4: 73-76.

[3] Ali G., Purohit M., Mughal M.H., Iqbal M. & Srivastava P.S. (1996) A rapid protocol for micropropagation of *Bacopa monniera* L Wettst. a medicinal plant, *Plant Tiss. Cult. Biotech.* 2: 208-211.

[4] Gulati A., Bharel S., Abdin M.Z., Jain S.K. & Srivastava P.S. (1996) In vitro micropropagation and flowering in *Artemisia annua*, *J. Plant Biochem. Biotechnol.* 5: 31-35.

[5] Sai S.T., Keng C.L., Pargini N. & Teo Chris K.H. (2000) In vitro propagation of *Typhonium flagelliforme* (Lodd) Blume, *In Vitro Cell. Dev. Biol. Plant.* 36: 402-406.

[6] Malabadi R.B. & Nataraja K. (2001) Shoot regeneration in leaf explants of *Clitorea ternatea* L. cultured in vitro, *Phytomorphology* 51: 169-172.

[7] Uander D.W. (1991) Callus induction in *Phyllanthus* species and inhibition of viral DNA polymerase and reverse transcriptase by callus extracts, *Plant Cell Rep.* 10: 461-466.

[8] Datta A. & Srivastava P.S. (1997) Variation in vinblastin production by *catharanthus roseus* during in vivo and in vitro differentiation, *Phytochemistry* 46: 135-137.

[9] Kaur R., Sood M., Chander S., Mahajan R., Kumar V. & Sharma D.R. (1999) In vitro propagation of *Valeriana jatamansi*, *Plant Cell Tiss. Org. Cult.* 59: 227-229.

[10] Fracaro F. & Echeverrigaray S. (2001) Micropropagation of *Cunila galioides*, a popular medicinal plant of south Brazil, *Plant Cell Tiss. Org. Cult.* 64: 1-4.

[11] Srivastava P.S., Purohit M, Pande D. & Datta A. (1993) Phenotypic variation and alkaloid content in the androgenic plantlets of *Datura innoxia, Phytomorphology* 43: 209-216.

[12] Sudha C.G., Krishnan P.N., Seeni S. & Pushpangadan (2000) Regeneration of plants from in vitro root segments of *Holostemma annulare* (Roxb.) K Schum., A rare medicinal plant, *Curr. Sci.* 78: 1532-1535.

[13] Butiuc-Keul A.L. & Deliu C. (2001) Clonal micropropagation of *Arnica montana* L., a medicinal plant, *In Vitro Cell. Dev. Biol. Plant.* 37: 581-585.

[14] Selvakumar V., Anbudurai P.R. & Balakumar T. (2001) In vitro propagation of the medicinal plant *Plumbago zeylanica* L. through nodal explants, *In Vitro Cell. Dev. Biol. Plant.* 37: 280-284.

[15] Narula A., Kumar S. & Srivastava P.S. (2001) In vitro response of explants of *Abrus precatorius* in different media, *J. Trop. Med. Plants.* 2: 57-65.

[16] Perez–Bermudez P., Seitz H.U. & Gavidia I. (2002) A protocol for rapid micropropagation of endangered *Isoplexis, In Vitro Cell. Dev. Biol. Plant.* 38: 178-182.

[17] Fowler M.W. (1982) Plant cell cultures: fact and fantasy, *Biochem. Soc. Trans.* 11: 23-28.

[18] O' Neill C.M., Arthur A.E. & Mathias R.J. (1996) The effect of proline thioproline and methyl glyoxalbis (guanylhydrazone) on shoot regeneration frequencies from stem explants of *B. napus, Plant Cell Rep.* 15: 695-698.

[19] Parr A.J., Payne J., Eagles J., Chapman B.T., Robins R.J. & Rhodes M.J.C. (1990) Variation in tropane alkaloid accumulation within the solanaceae and strategies for its exploitation, *Phytochemistry* 29: 2545-2550.

[20] Zenk M.H., El-shagi H., Arens H., Stockigt J., Weiler E.W. & Deus B. (1975) Formation of indole alkaloids serpentine and ajmalicine in cell suspension cultures of *Catharanthus roseus*, In: W. Barz, E. Reinhard, M.H. Zenk (Eds), *Plant Tissue Culture and its Biotechnological Applications,* Springer, Berlin Heidelberg New York, pp. 27-43.

[21] Fujita Y., Hara Y., Suga C. & Morimoto T. (1981) Production of shikonin derivatives by cell suspension cultures of *Lithospermum erythrorhizon* II. A new medium for the production of shikonin derivatives, *Plant Cell Rep.* 1: 61-63.

[22] Verpoorte R., Van der Heijden R., Ten Hoopen H.J.G. & Memelink J. (1999) Metabolic engineering of plant secondary metabolites pathway for the production of fine chemicals, *Biotechnol. Lett.* 21:467-479.

[23] Furze J.M., Rhodes M.J.C., Parr A.J., Robins R.J., Whitehead I.M. & Threlfall D.R. (1991) Abiotic factors elicit sesquiterpenoid phytoalexin production but not alkaloid production in transformed roots of *Datura stramonium, Plant Cell Rep.* 10: 111-114.

[24] Sri Andrijany V., Indrayanto G. & Sochono L.A. (1999) Simultaneous effect of calcium, magnesium, copper and cobalt ions on sapogenin steroids content in callus cultures of *Agave amaniensis, Plant Cell Tiss. Org. Cult.* 55: 103-108.

[25] Trejo-Tapia G., Jimenez-Aparicio A., Rodriguez-Monroy M., De. Jesus-Sanchez A. & Gutierrez-Lopez G. (2001) Influence of cobalt and other microelements on the production of betalains and the growth of suspension cultures of *Beta vulgaris, Plant Cell Tiss. Org. Cult.* 67: 19-23.

[26] Saba, Pande D., Iqbal M., Srivastava P.S. (2000) Effect of $ZnSO_4$ and $CuSO_4$ on regeneration and lepidine content in *Lepidium sativum, Biol. Plant.* 43: 253-256.

[27] Chakarvarty B. & Srivastava S. (1997) Effect of cadmium and zinc on metal uptake and regeneration of tolerant plants in linseed, *Agric. Ecosyst. Environ.* 61: 45-50.

[28] Ali G., Srivastava P.S. & Iqbal M. (1998) Morphogenic response and proline content in *Bacopa monniera* cultures grown under copper stress, *Plant Sci.* 138: 191-195.

[29] Jumin H.B. (1995) Plant regeneration via somatic embryogenesis in *Citrus* and its relatives. *Phytomorphology* 45: 1-8.

[30] Srivastava P.S., Ali G., Iqbal M., Narula A. & Bharti N. (2002) Micropropagation of *Bacopa* and effect of heavy metals on growth performance. In: Nandi S.K., Palni L.M.S., Kumar A. (Eds), *Role of Plant Tissue Culture in Biodiversity Conservation and Economic Development,* Gyanodaya Prakashan, Nanital, India, pp. 325-344.

[31] De Backer-Royer C., Vannereau A., Duret S., Villavigens R. & Cosson L. (1990) Action de metaux lourds Cu et Cd surdes cellules de *Catharanthus roseus cultivees* in vitro adaption et production alcaloidique. *Col. Inst. Natl. Res. Alim.* 51: 247-251.

[32] Mizukami H., Konoshima M. & Tabata M. (1977) Effect of nutritional factors on shikonin derivative formation in *Lithospermum* callus cultures, *Phytochemistry* 16: 1183-1186.

[33] Endress R. (1976) Betacyanin akkumulation in kallus von *Portuglacca grandiflora* var JR unter dem eintluss v Vonphytohormonem und Cu^{2+} ionen auf unterschiedlichen grund medien, *Biochem. Physiol. Pflanz.* 169: 87-98.

[34] Obrenovic S. (1990) Effect of Cu (11) D-penicillanine on phytochrome mediated beta-cyanin formation in *Amaranthus caudatus* seedlings, *Plant Physiol. Biochem.* 28 639-646.

[35] Indrayanto G., Rahayu L., Rahman A. & Noerani P.E. (1993) Effect of calcium, strontium and magnesium ions on the formation of phytosteroids in callus cultures of *Agave amaniensis, Planta Med.* 59: 97-98.

[36] Indrayanto G., Pratnaningsih B. & Santosa M.H. (1996) The influence of molybdate and calcium ions on the sapogenin content in callus cultures of *Agave amaniensis. Indonesian J. Pharm.* 7: 21-27.

[37] Zenk M.H., El-Shagi H. & Schulte U. (1975) Anthraquinone production by cell suspension culture of *Morinda citrifolia, Planta Med. Suppl.* 79: 101.

[38] Fulzele D.P., Satdive R.K. & Pol B.B. (2001) Growth and production of camptothecin by cell suspension cultures of *Nothapodytes foetida, Planta Med.* 67: 150-152.

[39] Ray S. & Jha S. (2001) Production of Withaferin A in shoot cultures of *Withania somnifera, Planta Med.* 67: 432-436.

[40] Marshall J.G. & Staba E.J. (1976) Hormonal effects on diosgenin biosynthesis and growth in *Dioscorea deltoidea* tissue cultures, *Phytochemistry* 15: 53-55.

[41] Zhao D., Xing J., Maoyin Li, Dongping Lu & Qiao Zhao. (2001) Optimization of growth and jaceosidin production in callus and cell suspension cultures of *Saussurea medusa, Plant Cell Tiss. Org. Cult.* 67: 227-234.

[42] Matsumoto T., Nishida K., Noguchi M. & Tamaki E. (1973) Some factors affecting the anthocyanin formation by *Populus* cells in suspension culture, *Arg. Biol. Chem.* 37: 561-567.

[43] Hay C.A., Anderson L.A., Roberts M.F. & Phillipson J.D. (1986) Effect of precursors on the yield of secondary products, *Plant Cell Rep.* 5: 1-4.

[44] Lindsey K. (1986) Precursor caused higher capsaicin production by cell cultures of *Capsicum frutescens, Phytochemistry* 25: 2793.

[45] Whitehead I.M. & Threlfall D.R. (1992) Production of phytoalexins by plant tissue cultures, *J. Biotech.* 26: 63-81.

[46] Rokem J.S., Schwarzberg J. & Goldberg J. (1984) Autoclaved fungal mycelia increase diosgenin production in cell suspension cultures of *Dioscorea deltoidea*, *Plant Cell Rep.* 3: 159-160

[47] Funk C., Gugler K. & Brodelius P. (1987) Increased secondary product formation in plant cell suspension cultures after treatment with yeast carbohydrate preparation (Elicitor), *Phytochemistry* 26: 401-405.

[48] Kim D.J. & Chang H.N. (1990) Increased shikonin production in *Lithospermum erythrorhizon* suspension cultures with in situ extraction and fungal cell treatment (Elicitor), *Biotech. Lett.* 12: 443-446.

[49] Wijsnama R., Go JTK A., Van Weerden I.N., Harkes P.A., Veerpoorte B. & Barheim S.A. (1985) Anthraquinones as phytoalexins in cell and tissue cultures of *Cinchona* sp, *Plant Cell Rep.* 4: 241-244.

[50] Hashimoto T. & Yamada Y. (1994) Alkaloid biogenesis: Molecular aspects, *Ann. Rev. Plant Physiol. Plant. Mol. Biol.* 329: 257-285.

[51] Aerts R.J., Gisi D., De Carolis E., De Luca V. & Baumann T.W. (1994) Methyl jasmonate vapor increases the developmentally controlled synthesis of alkaloids in *Catharanthus* and *Cinchona* seedlings, *Plant J.* 50: 635-642.

[52] Fujii J., Slade D. & Redenbaugh K. (1989) Maturation and green house planting of alfalfa synthetic seeds, *In Vitro Cell. Dev. Biol. Plant.* 25: 1179-1182.

[53] Sanada M., Sakamoto Y., Hayashi M., Mashiko T., Okamoto A., Ohnishi N., Celery & Lettuce. (1993) In: Redenbaugh K. (Ed.), *Synthetic Seeds: Application of Synthetic Seeds to Crop Improvement,* CRC Press, Boca Raton, USA, pp. 305-327.

[54] Rao P.S., Ganapathi T.R., Suprasana P. & Bapat V.A. (1996) Synthetic seed technology as a method of plant propagation and delivery of tissue cultured plants, In: Pareek L.K. (Ed.), *Trends in Plant Tissue Culture and Biotechnology,* Agrobotanical Publishers, Jaipur, India, pp. 47-52.

[55] Ahuja P.S., Mathur J., Lal N., Mathur A., Mathur A.K. & Kukreja A.K. (1989) Towards developing artificial seeds by shoot bud encapsulation, In: Kukreja A.K., Mathur A.K., Ahuja P.S. & Thakur R.S. (Eds) *Tissue Culture and Biotechnology of Medicinal and Aromatic Plants,* Paramount Publishing House, New Delhi, India, pp. 22-28.

[56] Mandal B.B., Chandel K.P.S. & Dwivedi S. (1996) Cryopreservation of yam (*Dioscorea* spp.) shoot apices by encapsulation-dehydration, *Cryo Lett.* 17: 165-174.

[57] Malabadi R.B. & Nataraja K. (2002) In vitro storage of synthetic seeds in *Clitoria ternatea* Linn., *Phytomorphology.* 52: 231-243.

[58] Maruyama E., Kinoshita I., Ishii K., Ohba K. & Saito A. (1997) Germplasm conservation of the tropical trees, *Cedrela odorata* L., *Guazuma crinita* Mort. and *Jacaranda mimosaefolia* D. Don by shoot tip encapsulation in calcium alginate and storage at 12-25°C, *Plant Cell Rep.* 16: 393-396.

[59] Ahuja S., Dixit S., Mandal B.B. & Srivastava P.S. (2002) Cryopreservation of medicinal plants: prospects and potentials, *J. Trop. Med. Plants.* 3: 113-120.

[60] Vanadenbussche B., Demeulemeester M.A.C. & De Proft M.P. (1993) Cryopreservation of alginate coated in vitro grown shoot tips of chicory (*Cichorium intybus* L.) using rapid freezing, *Cryo Lett.* 14: 259-266.

[61] Decruse S.W., Seeni S. & Pushpagadan P. (1999) Cryopreservation of alginate coated shoot tips of in vitro grown *Holostemme annulare* (Roxb.) K. Schum, and endangered medicinal plant: Influence of pre-culture and DMSO treatment on survival and regeneration, *Cryo Lett.* 20: 243-250.

[62] Towill L.E. (1990) Cryopreservation of isolated mint shoot tips by vitrification, *Plant Cell Rep.* 9: 178-180.

[63] Bajaj Y.P.S. (1983) Regeneration of plants from pollen embryos of *Arachis*, *Brassica* and *Triticum* spp. Cryopreserved for one year, *Curr. Sci.* 52: 484-485.

[64] Debergh P.C. & Read P.E. (1990) Micropropagation, In: P.C. Deberg, R.H. Zimmerman (Eds), *Micropropagation, Technology and Application,* Kluwer Academic, Dordrecht, Boston, London, pp. 1-13

[65] Rani V. & Raina S.N. (2000) Genetic fidelity of organized meristem-derived micropropagated plants: a critical reappraisal, *In Vitro Cell. Dev. Biol. Plant.* 36: 319-330.

[66] Shenoy V.B. & Vasil I.K. (1992) Biochemical and molecular analysis of plants derived from embryogenic cultures of napier grass (*Pennisetum purpureum* K. Schum.), *Theor. Appl. Genet.* 83: 947-954.

[67] Breyne P., Boerjan W., Gerats T., Van-Montagu M. & Van-Gysel A. (1997) Applications of AFLP in plant breeding molecular biology and genetics, *Belg. J. Bot.* 129: 107-117.

[68] Vendrame W.A., Kochert G. & Wetzstein H.Y. (1999) AFLP analysis of variation in pecan somatic embryos, *Plant Cell Rep.* 18: 853-857.

[69] Singh A., Negi M.S., Moses V.K., Venkateswarlu B., Srivastava P.S. & Lakshmikumaran M. (2002) Molecular analysis of micropropagated neem plants using AFLP markers for ascertaining clonal fidelity, *In Vitro Cell. Dev. Biol. Plant.* 38: 519-524.

[70] Kiers A.M., Mes T.H.M., van der Meijden R. & Bachmann K. (2000) A search for diagnostic AFLP markers in *Cichorium* species with emphasis on endive and chicory cultivar groups, *Genome* 43: 470-476.

[71] Negi M.S., Singh A. & Lakshmikumaran M. (2000) Genetic variation and relationship among and within *Withania* species as revealed by AFLP markers, *Genome* 43: 975-980.

[72] Merkli A., Christen P. & Kapetaniddis I. (1997) Production of diosgenin by hairy root cultures of *Trigonella foenum-graecum* L., *Plant Cell Rep.* 16: 632-636.

[73] Martin-Tanguy J., Sun L.Y., Burtin D., Vernoy R., Rossin N. & Tepfer D. (1996) Attenuation of the phenotype caused by the root inducing, left-hand, transferred DNA and its *rol* gene-correlations with changes in polyamine metabolism and DNA methylation, *Plant Physiol.* 111: 259-267.

[74] Zárate R. (1999) Tropane alkaloid production by *Agrobacterium rhizogenes* transformed hairy root cultures of *Atropa bactica* Willk (Solanaceae), *Plant Cell Rep.* 18: 418-423.

[75] Doerk Schmitz K., Witte L. & Alfermann A.W. (1994) Tropane alkaloid patterns in plants and hairy roots of *Hyoscyamus albus, Phytochemistry* 35: 107-110.

[76] Mallol A., Cusido R.M., Palazon J., Bonfill M., Morales C. & Pinol M.T. (2001) Ginsenoside production in different phenotypes of *Panax ginseng* transformed roots, *Phytochemistry* 57: 365-371.

[77] Jung & Tepfer D. (1987) Use of genetic transformation by the Ri T-DNA of *Agrobacterium rhizogenes* to stimulate biomass and tropane alkaloid production in *Atropa belladona* and *Calystegia sepium* roots grown in vitro, *Plant Sci.* 50: 145-151.

[78] Aoki T., Matsumoto H., Asako Y., Matsunaga & Shimomura K. (1997) Variation of alkaloid productivity among several clones of hairy roots and regenerated plants of *Atropa belladona* transformed with *Agrobacterium rhizogenes* 15834, *Plant Cell Rep.* 16: 282-286.

[79] Chen Da-Hua, Liu Chang-Jun, Ye He-Chun, Li Guo-Feng, Liu Ben-ye, Meng Yu-Ling & Chen Xiao-Ya. (1999) Ri-mediated transformation of *Artemisia annua* with a recombinant farnesyl diphosphate synthase gene for artemisinin production, *Plant Cell Tiss.Org. Cult.* 57: 157-162.

[80] Canel C., Lopes Cardoso M.I., Whitmer S., Fits L., Pasquali G., Van der Heijden R., Hoge J.H.C. & Verpoorte R. (1998) Effect of over expression of strictosidine synthase and tryptophan decarboxylase on alkaloid production by cell cultures of *Catharanthus roseus, Planta.* 205: 414-419.

[81] Pande D., Purohit M. & Srivastava P.S. (2002) Variation in xanthotoxin content in *Ammi majus* L. cultures during in vitro flowering and fruiting, *Plant Sci.* 162: 583-587.

[82] Wawrosch C., Malla P.R. & Kopp B. (2001) Micropropagation of *Allium wallichii* kunth, a threatened medicinal plant of Nepal, *In Vitro Cell Dev. Bio. Plant.* 37: 555-557.

[83] Butiuc-Keul A.L. & Deliu C. (2001) Clonal propagation of *Arnica montana* L., a medicinal plant, *In Vitro Cell. Dev. Biol. Plant.* 37: 581-585.

[84] Ahuja A., Sambyal M. & Koul S. (2002) In vitro propagation and conservation of *Atropa acuminata* Royle ex Lindl—An indigenous threatened medicinal plant, *J. Plant Biochem. Biotech.* 11: 121-124.

[85] Baburaj S., Ravichandran P. & Selvapandian M. (2000) In vitro adventitious shoot formation from leaf cultures of *Clerodendrum inerme* (L) Gaertn., *Indian J. Exptl. Biol.* 38: 1274-1276.

[86] Bhattacharyya R. & Bhattacharya S. (2001) In vitro multiplication of *Coleus forskohlii* Briq.: An approach towards shortening the protocol, *In Vitro Cell. Dev. Biol. Plant.* 37: 572-575.

[87] Narula A., Srivastava P.S. & Rangaswamy N.S. (2000) In vitro culture studies on *Dioscorea* species, *J. Trop. Med. Plants* 1: 60-74.

[88] Raha S. & Roy S.C. (2001) In vitro plant regeneration in *Holorrhena antidysenterica* Wall., through high frequency axillary shoot proliferation, *In Vitro Cell. Dev. Biol. Plant.* 37: 232-236.

[89] Thengane S.R., Kulkarni D.K., Shrikhande V.A. & Krishnamurthy K.V. (2001) Effect of thidiazuron on adventitious shoot regeneration from seedling explants of *Nothapodytes foetida*, *In Vitro Cell. Dev. Biol. Plant.* 37: 206-210.

[90] Saini R. & Jaiwal P.K. (2000) In vitro multiplication of *Peganum harmala*—an improtant medicinal plant, *Indian J. Exptl. Biol.* 38: 499-503.

[91] Catapan E., Otuki M.F. & Viana A.M. (2000) In vitro culture of *Phyllanthus caroliniensis* (Euphorbiaceae), *Plant Cell Tiss. Org. Cult.* 62: 195-202.

[92] Das M. & Raychaudhuri S.S. (2001) Enhanced development of somatic embryos of *Plantago ovata* Forsk. by additives, *In Vitro Cell. Dev. Biol. Plant.* 37: 568-571.

[93] Das G., Rashmi M.N. & Rout G.R. (2001) In vitro clonal propagation of *Plumbago rosea* L.: a potential medicinal plant, *Plant Cell Biotech. Mol. Biol.* 2: 105-112.

[94] Selvakumar V., Anbudurai P.R. & Balakumar T. (2001) In vitro propagation of the medicinal plant *Plumbago* zeylanica L. through nodal explants, *In Vitro Cell. Dev. Biol. Plant.* 37: 280-284.

[95] Jeyakumar M. & Jayabalan N. (2000) An efficient method for regeneration of plantlets from nodal explants of *Psoralea corylifolia* Linn., *Plant Cell Biotech. Mol. Biol.* 1: 37-40.

[96] Sehrawat A.R., Sanjogta Uppal & Chowdhury J.B. (2002) Establishment of plantlets and evaluation of differentiated roots for alkaloids in *Rauwolfia serpentina*, *J. Plant Biochem. Biotech.* 11: 105-108.

[97] Anitha S. & Pullaiah T. (2001) In vitro propagation of *Sterculia foetida* Linn (Sterculiaceae), *Plant Cell Biotech. Mol. Biol.* 2: 139-144.

[98] Sivaram L. & Mukundan U. (2002) Feasibility of commercial micropropagation of *Stevia rebaudiana* in India, *J. Trop. Med. Plants.* 3: 97-103.

[99] Saba, Iqbal M. & Srivastava P.S. (2000) In vitro micropropagation of *Silybum marianum* L. from various explants and silybin content, *J. Plant Biochem. Biotech.* 9: 81-87.

[100] Bhalsing S.R. & Maheshwari V.L. (1997) In vitro culture and regeneration of *Solanum khasianum* and extraction of solasodine, *J. Plant Biochem. Biotech.* 6: 39-40.

[101] Jayanthi M. & Mandal P.K. (2001) Plant regeneration through somatic embryogenesis and RAPD analysis of regenerated plants in *Tylophora indica* (Burm.F. Merrill), *In Vitro Cell. Dev. Biol. Plant.* 37: 576-580.

[102] Thiruvengadam M., Jayabalan N. (2001) In vitro flowering of *Vitex negundo* L. a medicinal plant, *Plant Cell Biotech. Mol. Biol.* 2: 67-70.

[103] Ray S. & Jha S. (2002) Regeneration of *Withania somnifera* plants, *J. Trop. Med.* Plants 3: 89-95.

CHAPTER

19

Somaclonal Variants for Salt Tolerance and *in vitro* Propagation of Peanut

V.P.S. Shekhawat and Ashwani Kumar

ABSTRACT

Plant regeneration from mature zygotic embryo of *Arachis hypogaea* through direct differentiation as well as through de-differentiation via callus formation has been achieved on Murashige and Skoog's medium supplemented with different growth regulators. The optimum concentration of 2-4,Dichlorophenoxy acetic acid for callus induction was 5 mgl^{-1}. Shoot buds were produced upon transferring this callus on medium supplemented with 6-benzylaminopurine (BAP) where the optimum concentration was 6 mgl^{-1}. However addition of 1 mgl^{-1} kinetin caused enhancement in number of shoot buds produced. Direct differentiation in shoot buds from mature zygotic embryo was observed on medium supplemented with 6 $mg.l^{-1}$ 6-benzylaminopurine (BAP) and 2 $mg.l^{-1}$ alpha naphthalene acetic acid. Histological studies have been carried out to study the origin of roots and shoots from callus. Salt tolerant callus of *Arachis hypogaea* has also been developed. Mature zygotic embryos were subjected to MS medium supplemented with NaCl (5.0, 7.5, 10.0, 12.5 and 15 gl^{-1}) and 6 mgl^{-1} 6-benzylaminopurine (BAP) and 2 mgl^{-1} alpha naphthalene acetic acid (NAA). Leaves from shoots regenerated on MS medium supplemented with 7.5 gl^{-1} NaCl were plucked and cultured on MS medium supplemented with 5 gl^{-1} NaCl and 1 mgl^{-1} 2,4-Dichlorophenoxy acetic acid. Whereas leaves taken from control could not produce callus on MS medium supplemented with the same concentration and died within three to four days. Proline content was twofold more in salt tolerant callus. Growth of salt tolerant callus was less as compared to control, cultured on salt free medium. Present investigation demonstrates the development of salt tolerant callus without going through lengthy procedure of callus induction and then selection.

Key Words: Growth regulator, Micropropagation, Multiple shoot, Organogenesis, Peanut, Salt tolerance callus

Oilseed crops in agricultural economy of India occupy an important place. In terms of acreage, tonnage and value, oilseed is major agricultural crop, just next to food grains. Among the cultivated edible oil seed crops *Arachis hypogaea* L. is an important one. India occupies the first position both with regards to area and production in the world.

*Address for correspondence: Department of Botany, University of Rajasthan, Jaipur, Rajasthan-302004, India
E-mail: msku4@hotmail.com

In the present context salinity has been considered a major determinant for crop production (Boyer, 1982) and productivity of *A. hypogaea* is also badly affected by salinity. Besides conventional selection and breeding methods of crop improvement, plant tissue culture technique, which is an important component of biotechnology, offers efficient means for crop improvement (Gupta *et al.*, 1998). Plant tissue culture techniques that make use of somaclonal variation (Lee & Phillips, 1988) do not require previous knowledge of the genetic basis of resistance and facilitates *in vitro* selection of cell and tissue for any particular trait. *In vitro* selection, besides its use as a tool for obtaining salt tolerant plants, may offer potential means for quick evaluation of germplasm for salt tolerance. The phenomenon of obtaining NaCl tolerant cell lines tissue culture has been extensively reviewed (Staverek & Rains, 1984). NaCl tolerance has also been demonstrated for callus of *Medicago sativa* (McCoy, 1987), canola (Chandler & Thorpe, 1987) and for callus of maize (Ram Gopal, 1986). Salt tolerant plants using tissue culture technique have also been regenerated from tobacco (Nabors *et al.*, 1975); oat (Nabors *et al.*, 1982), sugar beet (Freytag *et al.*, 1990), tomato (Cano *et al.*, 1998), potato (Ochatt *et al.*, 1999) and citrus (Agustin and Eduardo, 1995). Although *A. hypogaea* cultures have been exploited for physiological studies *in vitro* (Kumar, 1974) but no efforts have been made to obtain salt tolerant plants of *A. hypogaea* using tissue culture techniques. Further lack of a reliable and quick regeneration method for *A. hypogaea* in tissue culture has proven to be one of the major obstructions in the application of biotechnology based protocols for crop improvement. Consequently, extensive research in recent years has produced numerous reports on regeneration in *A. hypogaea* via organogenesis and somatic embryogenesis (Kanyand *et al.*, 1994; Baker and Wetzstein 1995; Chengalrayan *et al.*1995; Heatly & Smith 1996; Venkatachalam & Jayabalan 1997; Gupta & Kumar, 1998; Ponsamual *et al.*, 1998; Pestana *et al.*, 1999; Victor *et al.*, 1999). Plant regeneration has been reported from various explants like hypocotyledonary notch region (Murch *et al.*, 1999), leaf let (Baker & Wetzstein, 1992), immature cotyledons (Ozias-Akins, 1989), hypocotyl (Mcently, 1994; Li *et al.*, 1994) and epicotyl. Although immature embryos or cotyledonary explants collected from field shows promising results but the contamination rates are very high (Venkatachalam *et al.*,1998) further availability of such explants may be another hindrances in applications of such protocols.

The present investigation was undertaken with an object to obtain somaclonal variants for production of salt tolerant plants. Regeneration protocol from mature zygotic embryo of *A. hypogaea* has also been optimized.

MATERIALS AND METHODS

Explant

Mature zygotic embryo of dry pods of *A. hypogaea* (cv local) was taken as explant. Seeds were washed with lab detergent (Labolene - Glaxo) and soaked overnight in sterilized distilled water. Zygotic embryos dissected mechanically were used as explant. Surface sterilization was carried out with 70% (v/v) ethanol for 1 min. Then the explants were treated with 0.1% (w/v) for 90 sec followed by three washes in sterilized distilled water. All the sterilization procedures were carried out under aseptic conditions.

Culture medium and conditions

The excised zygotic embryos were cultured on MS medium (Murashige & Skoog, 1962) supplemented with 3% (w/v) sucrose and different concentration of different growth regulators. 0.8% agar was used as gelling agent. pH of medium was adjusted to 5.86 with 0.5 N hydrochloric acid or 1N sodium hydroxide prior to autoclaving. Cultures were incubated at 25° ± 2°C in growth chambers under continuous illumination from cool white fluorescent tubes.

Histological studies

Orgenogenic callus was fixed in formalin-acetic acid-alcohol (FAA) consisting of formalin, acetic acid and 70% ethanol in a proportion of 1:1:18. The fixed materials were preserved in 70% alcohol, dehydrated through tertiary butyl alcohol (TBA) series and embedded in paraffin for 2-3 days. Sections were cut on a rotary microtome at 7-9 µm. Haupt's adhesive (Johansen, 1940) was used for affixing the paraffin ribbons to the slides, stained with morphological stains, a method commonly used for normal anatomy (Johansen, 1940).

Primary screening and selection for salt tolerance and callus induction

Mature zygotic embryos of *Arachis hypogaea* cv local were inoculated on MS medium (Murashige and Skoog, 1962) supplemented with BAP (6 mgl^{-1}) and NAA (2 mgl^{-1}), 3% sucrose and 0.8% w/v Agar was used as gelling agent. Different quantity of NaCl i.e. 5.0, 7.5, 10.0, 12.5, 15.0 gl^{-1} was added in the medium separately. Control was without supplement of NaCl. Cultures were incubated for 4 weeks at 25° + 2ºC under continuous light. After 4 weeks of salt challenge, the regenerated shoots surviving at the highest concentration were selected. Leaves were plucked from these shoots and taken as explants for callus induction and transferred on MS medium supplemented with 2,4-D (1 mgl^{-1}) and NaCl (5 gl^{-1}) and sucrose (3% w/v). Figure (1) shows the flow chart diagram for experimental scheme.

Growth

Initial fresh weight was determined after four weeks of culture from the time of inoculation of leaf and final fresh weight of callus was determined after 4 weeks of culture. Callus growth was expressed as fresh weight gain over initial fresh weight.

Proline assay

500 mg callus tissue was homogenized in 10 ml of 3% (w/v) sulfosalicylic acid and the homogenate filtered through whatmann-2 filter paper. Two ml of filtrate was reacted with 2 ml acid ninhydrin and 2 ml of glacial acetic acid in a test tube for 1 hour at 100ºC. After one hour tubes were transferred on ice bath to terminate the reaction. The reaction mixture was agitated with 4 ml toluene in a separating funnel and mixed vigorously for 15-25 seconds. The chromophore containing toluene was aspirated from the aqueous phase, warmed to room temperature and the absorbance was read at 520 nm using toluene as blank. The proline concentration was determined from a standard curve and calculated on fresh weight basis.

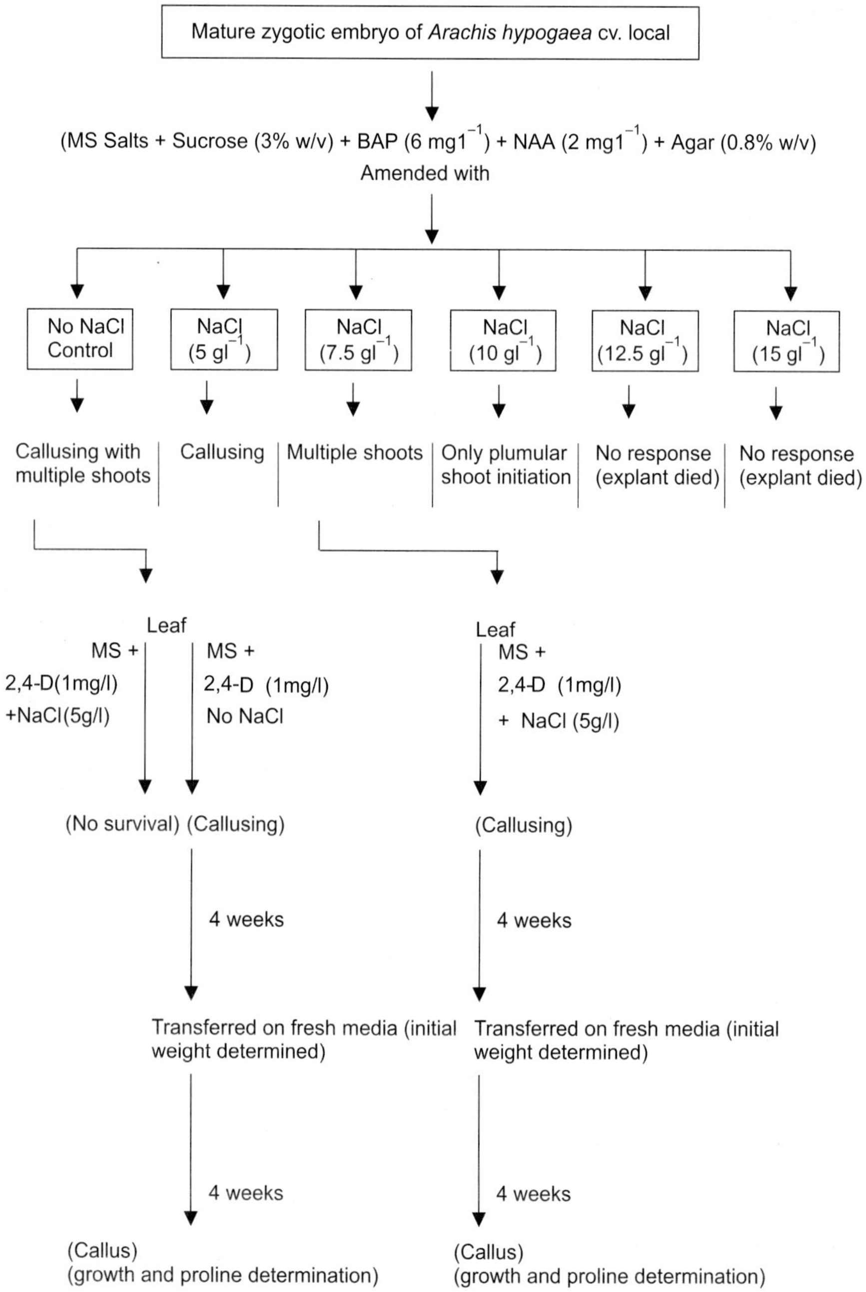

*Percent response

Fig. 1 Flow chart for experimental procedure for development of salt tolerant callus of *Arachis hypogaea* L.

Observations

Callus induction

IAA, NAA and 2, 4-D in 1-10 $mg l^{-1}$ concentration were tried for callus induction from the excised embryo of *A. hypogaea*. All these auxins exerted different effects on callus formation (Table 1). IAA produced brown friable non-regenerative callus together with plumular shoot proliferation. The best response in terms of callusing was observed at 5.0 $mg l^{-1}$ concentration of IAA. NAA had better response as compared to IAA and 2,4-D. However non-regenerative poor whitish callus, which was associated with friable green callus, was obtained from excised embryo cultured on medium supplemented with NAA. The excised embryo cultured on medium supplemented with 2,4-D produced good amount of regenerative compact green callus (Fig. 2A). The optimum concentration of 2,4-D for callusing was 5 $mg l^{-1}$. The 2,4-D supplementation to the medium was the best for callusing from the excised embryo as compared to other auxins i.e. IAA and NAA.

Direct differentiation from excised embryo via multiple shoots

Multiple shoot production was achieved in the present investigation. Cytokinin BAP alone and in combination with auxins, i.e. IAA, NAA, 2,4-D were tried for the induction of multiple shoots from the excised zygotic embryo. Extent of direct differentiation from the excised zygotic embryos was variable in different growth regulators treatments (Table 2). The excised embryos cultured on MS medium supplemented with BAP (2-10 $mg l^{-1}$) were differentiated into plumular shoot together

Table 1 Effect of IAA, NAA, and 2,4-D on Callus formation from the excised zygotic embryo of *Arachis hypogaea*.

Growth regulator	*Response*	*% Response*	*Remarks*
IAA (mgl^{-1})			
1	C^{+} + S	55%	Friable callus brownish
2.5	C^{++} + S	60%	in colour with shoot
5	C^{++} + S	80%	
7.5	C^{+} + S	73%	
10	C^{+} + S^{+}	53%	
NAA (mgl^{-1})			
1	C^{+} + S^{+}	93%	Profuse whitish callus
2.5	C^{++} + S^{+}	93%	associated with brown
5	C^{++}	83%	friable callus with poor
7.5	C^{++}	80%	growth
10	C^{+}	73%	
2,4-D (mgl^{-1})			
1	C^{+} + S	66%	Regenerative compact
2.5	C^{++} + S	86%	green callus together
5	C^{+++}	60%	with yellowish brown
7.5	C^{++}	45%	friable callus at margins.
10	C^{+}	30%	

S = Shoot : C = Callus

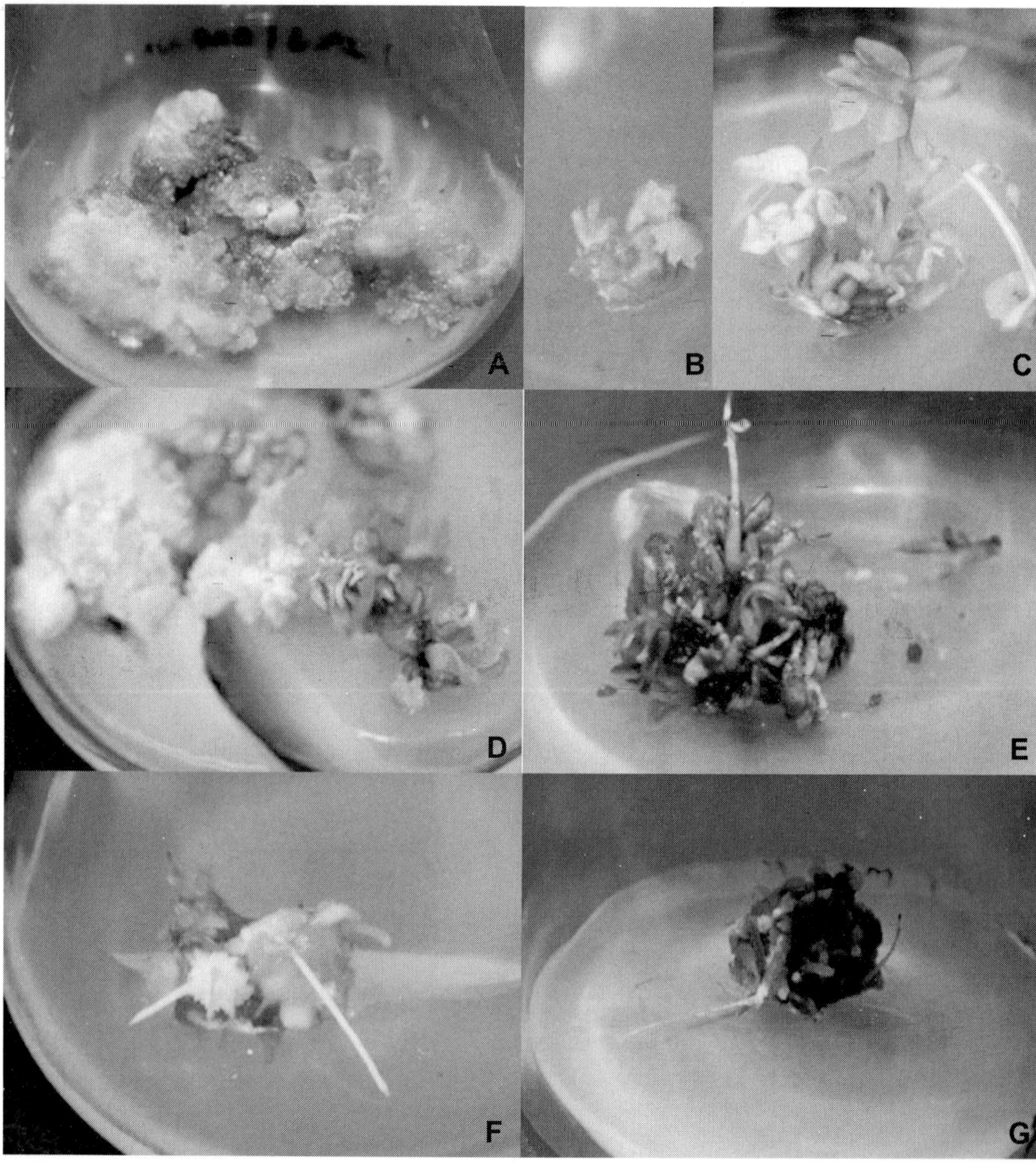

Fig. 2 *In Vitro* propagation of *Arachis hypogaea* from mature zygotic embryo. **(A)** compact green regenerative callus maintained on MS medium containing 5 mgl^{-1} 2,4-D. **(B/C)** multiple shoot induction from mature zygotic embryo. **(D)** Shoot bud induction from callus on MS medium containing BAP 6 mgl^{-1} **(E)** shoot bud induction from callus on MS medium supplemented with kinetin 1 mgl^{-1}. **(F)** Shoot buds and roots proliferation on medium containing IBA 2 mgl^{-1}. **(G)** typical rhizogenesis on growth regulator free medium.

Table 2 Effect of different growth regulators on differentiation of excised zygotic embryo of *Arachis hypogaea.*

Growth regulator		*Nature of response*	*Remarks*
BAP (mgl^{-1})			
2		S^+	Shoot together with friable callus
4		S^+	
6		$S^{++} + C^+$	
8		$S^+ + C^+$	
10		$S^+ + C^+$	
BAP (mgl^{-1})	*IAA (mgl^{-1})*		
2	2	$S^+ + C^+$	Plumular shoot proliferation together with callusing
4	2	$S^+ + C^+$	
6	2	$S^+ + C^+$	
8	2	$S^+ + C^+$	
10	2	$S^+ + C^+$	
BAP (mgl^{-1})	*NAA (mgl^{-1})*		
2	2	$S^{++} + C^+$	Differentiation into multiple shoot together with callusing
4	2	$S^{++} + C^+$	
6	2	$S^{+++} + C^+$	
8	2	$S^{++} + C^+$	
10	2	$S^{++} + C^+$	
BAP (mgl^{-1})	*2,4-D (mgl^{-1})*		
2	2	$S^+ + C^+$	Plumular shoot proliferation together with callusing
4	2	$S^+ + C^+$	
6	2	$S^{++} + C^+$	
8	2	$S^{++} + C^{++}$	
10	2	$S^+ + C^+$	

(S = Shoot : C = Callus)

with scanty callusing. Differentiation into multiple shoots was observed from the excised embryo when NAA was provided into medium together with BAP (Fig. 2B-C). BAP (6 mgl^{-1}) in combination with NAA (2 mgl^{-1}) proved to be the best among all the combinations of BAP + NAA tried here for the induction of multiple shoots. Only plumular shoot proliferation together with scanty callus was observed in the medium supplemented with different combinations of BAP + IAA. The excised embryos cultured on medium supplemented with 2,4-D (2 mgl^{-1}) with different concentrations of BAP were differentiated into plumular shoot together with good amount of green callus.

Oraganogenesis via callus formation

The callus obtained from excised embryo cultured on MS medium supplemented with 2,4-D (5 mgl^{-1}) was subjected to different growth regulators in order to induce organogenesis. They exhibited varying influence on organogenesis from callus tissue (Table 3). Callus, upon transfer on hormone free MS medium did not produce any shoot bud and only typical rhizogenesis was observed (Fig. 2G). Well-developed roots were observed after 4 weeks of culture. This rhizogenic callus did not produce any shoots even upon transfer on medium containing hormones. The callus produced from the excised embryo cultured on medium supplemented with 2,4-D (5 mgl^{-1}) was

Table 3 Effect of different growth regulators on differentiation of callus originated from mature zygotic embryo of *Arachis hypogaea*.

Growth regulator		*Nature of response*	*Remarks*
BAP (mgl^{-1})			
1	–	S^{+}	Differentiation into shoot buds
2	–	S^{+}	
3	–	S^{+}	
4	–	S^{+}	
5	–	S^{++}	
6	–	S^{+++}	
BAP (mgl^{-1})	*K (mgl^{-1})*		
	1	S^{++++}	Differentiation into shoot buds
	2	S^{+}	
6	3	S^{+}	
	4	S^{+}	
	5	S^{+}	
BAP (mgl^{-1})	*IAA (mgl^{-1})*		
	1	S^{+}	Shoot buds and callus formation
	2	S^{+}	
6	3	C^{+}	
	4	C^{+}	
	5	C^{+}	
BAP (mgl^{-1})	*NAA (mgl^{-1})*		
	1	S^{+}	Differentiation into shoot buds
	2	S^{+}	
	3	S^{+}	
	4	S^{+}	
	5	S^{+}	
BAP (mgl^{-1})	*IBA (mgl^{-1})*		
	1	$S^{+} + C^{+} + R^{+}$	Differentiation into shoot buds + Roots, together with callusing
	2	$S^{+} + C^{+} + R^{++}$	
	3	S^{+}	
	4	S^{+}	
	5	S^{+}	
No growth regulator		R^{+++}	Rhizogenesis

(S = Shoot Buds : C = Callus : R = Roots)

transferred to MS medium supplemented with various concentrations of BAP (1-5 mgl^{-1}). Numerous shoot buds (up to 10) were produced by callus after 4 weeks of culture (Fig. 2D). BAP (6 mgl^{-1}) concentration was found better among all the concentrations tried with regards to number of shoot buds produced. Growth and number of shoot buds produced were greatly enhanced when Kinetin (1-5 mgl^{-1}) was provided in the medium. 1 mgl^{-1} Kinetin along with BAP (6 mgl^{-1}) supported maximum growth and differentiation of shoot buds (Fig. 2E). Different auxins i.e. IAA, NAA and IBA were tried in combination with BAP (6 mgl^{-1}) for the enhancement of production of shoot buds. There was no considerable increase in number of shoot buds produced in medium supplemented with IAA and IBA. No shoot buds differentiation from callus was observed beyond 2 mgl^{-1} concentrations of IAA. Organogenic callus cultured on medium supplemented with NAA (1-5 mgl^{-1}) together with BAP (6 mgl^{-1}), differentiated in to shoot buds. Although no enhancement in number of shoots produced was observed but growth of regenerated shoots was greatly enhanced. In medium supplemented with 1 mgl^{-1} and 2 mgl^{-1}, IBA rooting was also observed together with shoot bud production (Fig. 2F)

Rooting

The *in vitro* regenerated shoots either via direct differentiation or via organogenesis were rooted on MS medium without growth regulators and supplemented with IBA (0.5-2 mgl^{-1}). IBA (1.5 mgl^{-1}) concentration was the best for rooting. No rooting was observed on growth regulator free medium or MS medium supplemented with 0.5 mgl^{-1} IBA. Rooting together with callus formation was observed in 2 mgl^{-1} IBA concentration.

Histological studies

The longitudinal and transverse sections of the organogenic callus showed that the initiation of shoot occurred endogenously in the inner region of the callus (Fig. 3A-B). The meristematized cells of this region become long, flattened and smaller in size. Cells outer to the meristematized region became larger and took less stain and this meristematized tissue grew out to give rise to the shoot apices. Transverse sections of the basal region of the callus regenerating shoots showed presence of 5-8 vascular bundles. The root primordia also originated endogenously from very deep region of the callus tissue. Root primordia can be clearly distinguished from the surrounding tissue with cells arranged longitudinally and having central dark stained stele (Fig. 3C-D). Transverse section of the developing root primordium showed clearly defined endodermis and a tetrarch conditions differentiating from four different regions (Fig. 3E-F).

Primary selection for salt tolerance

Exposure to increasing concentration of NaCl resulted in a progression of reduced response from zygotic embryos cultured on MS medium (Fig. 4). Further they exhibited variable response on MS medium supplemented with different concentrations of NaCl. In control where no NaCl was added to the medium, multiple shoot proliferation together with callusing was observed. On MS medium supplemented with 5 gl^{-1} NaCl scanty callusing was observed. However multiple shoot proliferation was observed on MS medium supplemented with 7.5 gl^{-1} NaCl. Only plumular shoot initiation was observed which did not proliferate further, in media supplemented with 10 gl^{1} NaCl. Beyond 10 gl^{1} NaCl concentration zygotic embryo did not show any response and died within 1-2 weeks.

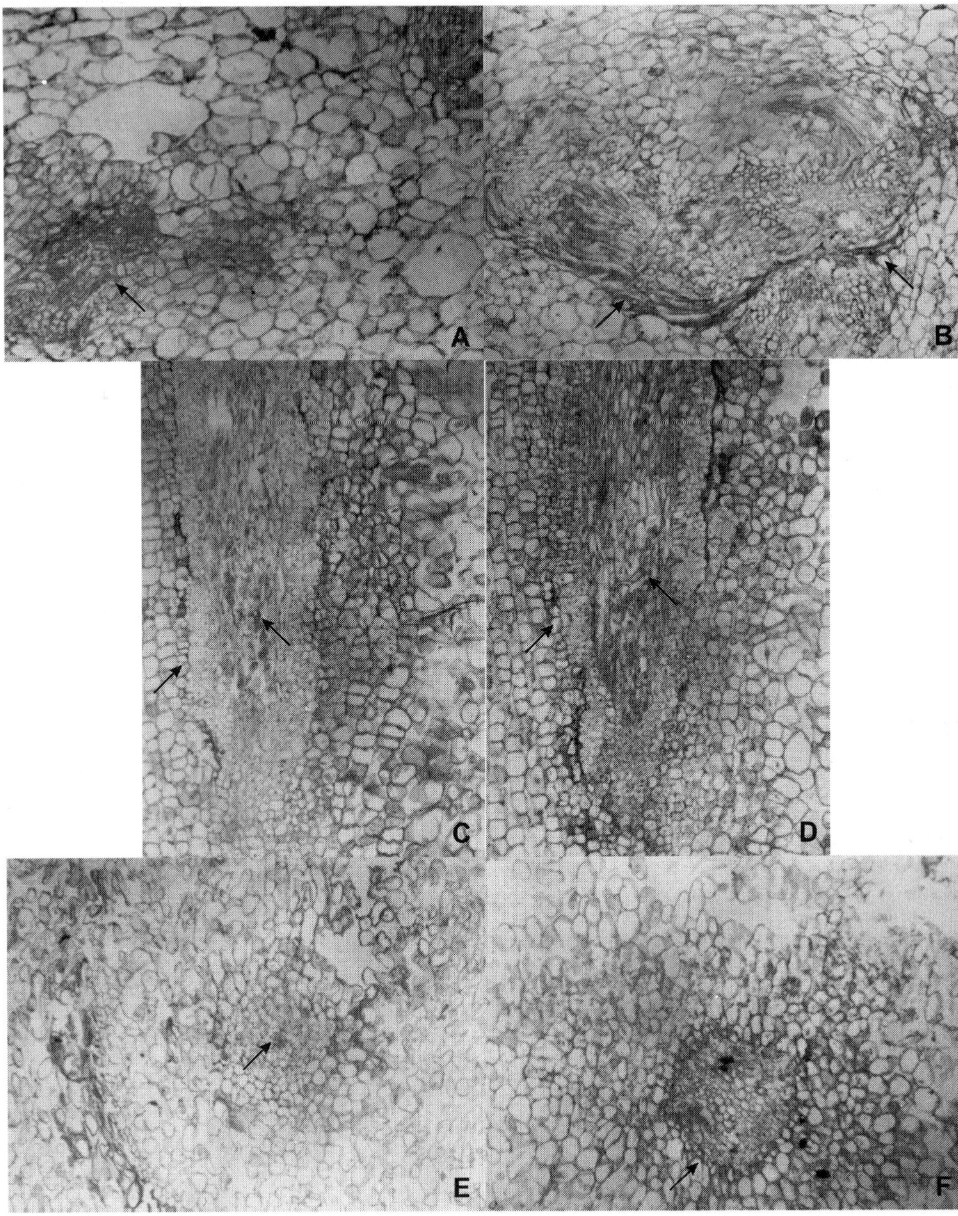

Fig. 3 Histological examination of regenerating callus. (A/B) section showing dark stained meristematized regions and origin of shoot primordia. (B/D). Root primordia originated endogenously, section shows cells arranged longitudianlly and central dark stained stelar region. (E/F). transverse section of developing root primordium.

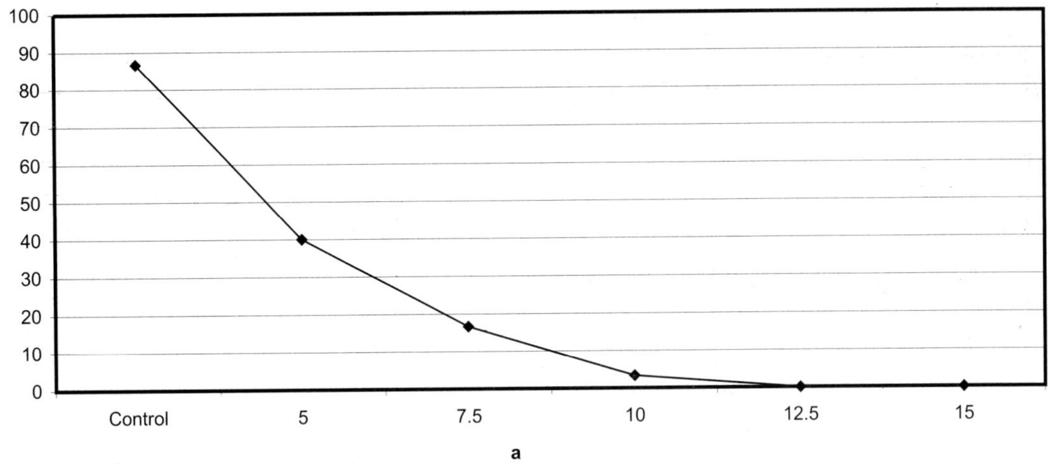

Fig. 4 Percent response of excised zygotic embryo of *A. hypogaea* on MS medium supplemented with different concentrations of NaCl.

Salt tolerant callus

Leaves plucked from *in vitro* regenerated shoots from excised embryo cultured on MS media supplemented with 7.5 gl^{-1} NaCl showed good callusing on medium supplemented with 2,4-D (1 mgl^{-1}) and NaCl (5 gl^{-1}). Callus was originated from petiolar cut end. Leaves from control flasks where no NaCl was provided in the medium failed to survive on medium containing 2,4-D (1 mgl^{-1}) and NaCl (5 gl^{-1}). However good callusing was observed on salt free medium, here also callusing was initiated from petiolar cut end.

Callus growth

The gain in fresh weight was more in callus originated from control leaves taken from regenerated shoots in control flasks where no NaCl was provided in the medium. The relative gain in fresh weight was 333% over initial fresh weight. Growth in callus originated from leaves of regenerated shoots of salt challenged embryo cultured on MS medium supplemented with NaCl (5 gl^{-1}) was poor as compared to control. The relative gain in fresh weight was 209% over initial fresh weight.

Proline content

Proline content in control and salt tolerant callus was also variable. Callus originated from regenerated salt tolerant shoot had about 2-fold more proline as compared to callus originated from leaves taken from regenerated shoots in control flasks where no NaCl was provided in the medium.

Discussion

Dependence of success of *in vitro* morphogenesis on explant, physiological status of explant, media factors and different growth regulators is well recognized. Concentrations of different growth regulators in the media are critical to regulate the *in vitro* growth and morphogenesis (Skoog & Miller, 1957; Davies, 1995). The dedifferentiation route or indirect organogenesis (Bhojwani & Razdan, 1989) for multiplication *in vitro* involves induction of callus by whatever means and then induction of shoot buds or roots (organogenesis). Ratio of cytokinin and auxin is pertinent. If the system is synthesizing adequate auxin then only cytokinin may be necessary for induction of shoot primordia (Krikorian, 1995). In the present investigation, callus derived on media supplemented with 2,4-D produced shoot buds upon transfer on media containing BAP and number was enhanced when kinetin was provided in medium. Higher concentration of IAA in the medium resulted in inhibition of production of shoot buds from the callus and only rhizogenesis was observed on medium without any growth regulator. Low auxin and high cytokinin concentrations in the medium resulted in the induction of shoot morphogenesis. Multiple shoot formation from cotyledonary node explant of peanut, on the medium supplemented with high cytokinin and low cytokinin has been reported (Venkatachalam & Jayabalan, 1997). During the present study higher concentration of cytokinin together with low auxin resulted in differentiation of zygotic embryo in multiple shoots. The number of shoots produced from cotyledonary node explant was the maximum on medium containing BAP and IBA (Venkatachalam & Jayabalan, 1997). However in the present investigation number of shoots produced from excised zygotic embryo was the maximum on medium supplemented with BAP and NAA. Similarly, induction of caluogenic buds on medium supplemented with BAP and NAA has been reported in *A. hypogaea* cultures (Chengalrayan *et al.*, 1995). Auxin at low concentration has promotive effect on root induction from shoots produced *in vitro*. During the present investigation the best rooting was observed on medium supplemented with IBA. Similar findings have been reported by Venkatachalam & Jayabalan (1997). It is a generally accepted view that detrimental organogenic factors are endogenous growth hormones themselves, and they are under the control of exogenously applied growth regulators and they may not be necessarily of same categories (Centeno *et al.*, 1996). The non-purine type cytokinin "1-phennyl-3-(1,2,3- thiadizol-5-yl) urea" (Thidiazuron) has a promotive effect on multiple shoot induction from pea nut. Higher concentration of TDZ had positive influence on multiple shoot formation from hypocotyl and cotyledon explants (Kanyand, 1994). Victor *et al.* (1999) suggested that TDZ-induced accumulation of purine cytokinin (e.g. BAP) could be key component of TDZ-induced morphogenic responses. It has been commonly observed that roots appear endogenously whereas shoot originate exogenously (Bhojwani & Razdan, 1989). Meristemoids i.e. cambium like cells which at later stages may give rise to roots or shoots and have been considered site of organ formation were observed in the deep inner region of callus and root and shoot primordia originated endogenously. Similar pattern of shoot formation have also been reported in some other plants (Bhojwani & Razdan, 1989). Plant genotype can be critical factor in morphogenetic responses in peanut (McKently, 1995). In *Daucus carota*, Random amplified polymorphic DNA (RAPD) analysis of embryogenic and non-embryogenic species revealed that presence or absence of two specific bands was detrimental for the occurrence of somatic embryogenesis (Imani *et al.*, 2001). Similar DNA sequences may also be detrimental and regulatory in morphogenesis of *A. hypogaea*. Development of cultivars tolerant to salinity can contribute significantly in agricultural economy. In the present investigation we demonstrated development of NaCl tolerant callus, which showed survival and satisfactory growth on salt amended media. Tolerance to salts is a polygenic trait and involves highly complex osmo-regulatory functions. Plant tissue culture provides an option to obtain plant with identical genotype to the parents. *In vitro* selection from tissue culture has resulted in production of plants tolerant to salt (Croughan *et al.*, 1981, Staverek & Rains, 1984).

Selection of variant cell lines for tolerance to elevated external NaCl concentrations from the population of cultured somatic cells is one important alternative for the production of salt tolerant plants. In the present work, at the first, selection of explants was carried out through inoculating the zygotic embryo on the medium supplemented with NaCl and plant growth regulators. This could be a better system for testing and selection for salt tolerance. Advantage of this method could be that during selection, very little new variability will be generated. Shoot apex culture method, rather than callus culture, for the selection and screening for salt tolerance, has been suggested by Martinez *et al.* (1996). The results obtained in the present study suggest that NaCl challenge to mature zygotic embryo may be useful for rapid evaluation and screening for salt tolerance in *A. hypogaea. In vitro* selected callus of potato and tomato had elevated levels of proline, which is a key metabolite for osmotic adjustment in response to salinity stress (Martinez *et al.,* 1996, Cano *et al.,* 1998). Proline, which may act as a protein compatible hydrotrope (Srinivas and Balasubramaniam, 1995) protects membranes and proteins against the adverse effects of high concentraton of inorganic ions (Pollard and Wyn Jones, 1979). Proline may be accumulated to balance inorganic ions like Na^+ and Cl^- accumulated by callus from salt amended media. In the present investigation about 2-fold more proline was accumulated by salt tolerant callus as compared to control. The physiological and biochemical responses of the callus were also similar to that found in whole plant (Pèrez Alfocea *et al.*, 1994 and Cano *et al.*, 1996). Efforts are in progress to regenerate salt tolerant plants of *A. hypogaea.*

REFERENCES

Agustin P.G. & Eduardo P.M. (1995) Selection of a NaCl tolerant Citrus plant, *Plant Cell Rep.* 14: 314-318.

Bates L.S., Waldren R.P. & Teare I.D. (1973) Rapid determination of free proline for water stress studies. *Plant Soil,* 39: 205-207

Baker C.M. & Wetzstin H.Y. (1995) Repetitive somatic embryogenesis in peanut cotyledon cultures by continual exposure to 2,4-D, *Plant Cell Tissue Organ Cult.* 40: 249-254.

Bhojwani S.S & Razdan M.K. (1989) *Plant Tissue Culture: Theory and Practice* (Elsevier Science Publ. Amsterdam, Oxford, NewYork, Tokyo).

Boyer J.S. (1982) Plant productivity and environment, *Science.* 218: 443-448.

Cano E.A, Pèrez-Alfocea F., Moreno V. & Bolarin M.C. (1996) Responses to NaCl stress of cultivated and wild tomato species and their hybrids in callus cultures, *Plant Cell Rep.* 15: 791-794.

Cano E.A., Pèrez-Alfocea F., Moreno V., Caro M. & Bolarin M.C. (1998) Evaluation of salt tolerance in cultivated and wild tomato species through *in vitro* shoot apex culture, *Plant Cell Tissue Organ Cult.* 53: 19-26.

Chandler S.F. & Thorpe T.A. (1987) Characterization of growth, water relations and proline accumulation in sodium sulphate tolerant callus of *Brassica napus* L. cv. Westar (Canola), *Plant Physiol.* 84: 106-111.

Centeno M.J., Rodriguez A., Feito J. & Fernandez B. (1996) Relationship between endogenous auxin and cytokinin levels and morphogenic responses in *Actinida deliciosa* tissue cultures, *Plant Cell Rep.* 16: 58-62.

Chengalrayan K., Mhaske V.B. & Hazra S. (1995) *In vitro* regulation of morphogenesis in peanut (*Arachis hypogaea* L.), *Plant Sci.* 110: 259-268.

Croughan T.P., Stavereks S.J. & Rains D.W. (1981) *In vitro* development of salt resistant plants, *Env. Exp. Bot.* 21: 317-324.

Davies P.J. (1995) *Plant Hormones, Physiology, Biochemistry and Molecular Biology* (Kluwer Academic Publ. Dodrecht, Boston, London).

Freytag A.H., Wrather J.A. & Erichsen A.W. (1990) Salt tolerant sugar beet progeny from tissue cultures challenged with multiple salts, *Plant Cell Rep.* 8: 647-650.

Gupta S. & Kumar A. (1998) Influence of nitrogen sources on growth and differentiation of *Arachis hypogaea* L. *Int. J. Mendel.* 15: 55-56.

Gupta S., Roy S. & Kumar A. (1998) Improvement of peanut & sunflower using *in vitro* technique, *Int. J. Mendel.* 15: 41-42.

Heatley M.E. & Smith R.H. (1996) Whole plant regeneration from the shoot apex of *Arachis hypogaea L. In Vitro Cell. Dev. Biol. Plant.* 32(2): 115-118.

Imani J., Thi L.T., Langen G., Arnholdt Schmitt B., Roy S., Lein C., Kumar A. & Neumann K.H. (2001) Somatic embryogenesis and DNA organization of genomes from selected *Daucus* species, *Plant Cell Rep.* 20: 537-541.

Johansen D.A. (1940) *Plant Micro Technique* (Tata Mcgraw Hill Publ. Co. Ltd. Bombay, New Delhi).

Kanyand M., Dessai A.P. & Prakash C.S. (1994) Thidiazuron promotes high frequency regeneration of peanut (*Arachis hypogaea*) plants *in vitro, Plant Cell Rep.* 14(1): 1-5.

Krikorian A.D. (1995) Hormones in tissue culture and micropropagation, In *Plant Hormones: Physiology Biochemistry and Molecular Biology,* pp.774-796 ed P J Davies (Kluwer Acadamic Publ.: Dordrecht, Boston, London).

Kumar A. (1974) Studies on growth, physiology, and biochemistry of plant tisuues in tissue culture, *Phytomorphology.* 24: 1-2.

Lee M. & Phillips R.L. (1988) The Chromosomal basis of somaclonal variation, *Annu. Rev. Plant Physiol. Plant Mol. Biol.* 39: 413-437.

Li Z., Jarret R.L., Pittman R.N. & Demski J.W. (1994) Shoot organogenesis from cultured seed explants of peanut (*Arachis hypogaea* L.) using thidiazuron, *In Vitro Plant.* 30: 187-191.

Martinez C.A., Maestri M., Lani E.G. (1996) *In vitro* salt tolerance and proline accumulation in Andean potato (*Solanum* spp.) differing in frost resistance, *Plant Sci.* 116: 177-184.

McCoy T.J. (1987) Characterization of alfalfa (*Medicago sativa* L.) plants regenerated from selected NaCl tolerant cell lines, *Plant Cell Rep.* 6: 417-422.

McKently A.H. (1995) Effect of genotype on somatic embryogenesis from axes of mature peanut embryos, *Plant Cell Tissue Organ Cult.* 42: 251-254.

Murashige T. & Skoog F. (1962) A revised medium for rapid growth and bioassay with tobacco tissue culture, *Pl. Physiol.* 15: 473-497.

Murch S.J., Victor J.M.R., Krishnaraj S. & Saxena P.K. (1999) The role of proline in thidiazuron-induced somatic embryogenesis of peanut, *In Vitro Cell. Dev. Biol. Plant.* 35: 102-105.

Nabors M.W., Kroskey C.S. & McHugh D.M. (1982) Green spots are predictors of high callus growth rates and shoot formation in normal and in salt stressed tissue cultures of oat (*Avena sativa* L.), *Z. Pflanjenphysiol.* 105: 341-349.

Nabors M.W., Daniels A., Nadolny L. & Brown C. (1975) Sodium chloride tolerant lines of tobacco cells, *Plant Sci. Lett.* 4: 155-159.

Ochatt S.J., Marconi P.L., Radice S., Arnozis P.A. & Caso O.H. (1999) *In vitro* recurrent selection of potato: production and characterization of salt tolerant cell lines and plants, *Plant Cell Tissue Organ Cult.* 55: 1-8.

Ozias-Akins P. (1989) Plant regeneration from immature embryos of peanut, *Plant Cell Rep.* 8: 217-218.

Perez-Alfocea F., Guerrier G., Estan M.T. & Bolarin M.C. (1994) Comparative salt responses at cell and whole plant levels of cultivated and wild tomato species and their hybrid, *J. Hort. Sci.* 69: 639-644.

Pestana M.C., Lacorte C., de-Freitas V.G., de-Oliveira D.E. & Masnsur V. (1999) *In vitro* regeneration of peanut (*Arachis hypogaea* L.) through organogenesis. Effect of culture temperature and silver nitrate, *In Vitro Cellu. Dev. Pl. Biol.* 35: 214-216.

Pollard A. & Wyn Jones R.G. (1979) Enzyme activities in concentrated solutions of glycinebetaines and other solutes, *Planta* 144: 291-298.

Ponsamual J., Huhnaan D.V., Cassidy B.G. & Postbeittenmiller D. (1998) *In vitro* regeneration via caulogeneis and brassin-induced shoot conversion of dormant buds from plumular explants of peanut (*Arachis hypogaea* L. cv okrun), *Plant Cell Rep.* 17: 373-378.

Ram Gopal S. (1986) Protein synthesis in a maize cell exposed to NaCl and mannitol, *Plant Cell Rep.* 5: 430-434.

Skoog F. & Miller C.O. (1957) Chemical regulation of growth and organ formation in plant tissues cultured *in vitro,* In *The Biological Action of Growth Substances,* pp 118-131 (Cambridge University Press: Cambridge).

Srinivas V. & Balasubramaniam D. (1995) Proline is a protein-compatible hydrotrope, *Langmuir.* 11: 2830-2833.

Stavarek S.J. & Rains D.W. (1984) Cell culture techniques selection and physiological studies of salt tolerance, In: *Salinity Tolerance in Plants: Strategies for Crop Improvement,* Staples R.C. & Toenniessen G.H. (Ed.) pp 321-334 (Wiley- Interscience, New York).

Venkatachalam P. & Jayabalan N. (1997) Effect of auxin and cytokinin on efficient plant regeneration and multiple shoot formation from cotyledons and cotyledonary node explant of groundnut (*Arachis hypogaea L.*) by *in vitro* culture technology, *App. Biochem. Biotech.* 67: 237-247.

Venkatachalam P., Geetha N. & Jayabalan N. (1998) Induction of somatic embryos and plantlet development in cell suspension cultures of *Arachis hypogaea* L., *Breeding Sci.* 48: 231-236.

Victor J.M.R., Murthy B.N.S., Murch S.J., Krishnaraj S. & Saxena P.K. (1999) Role of endogenous purine metabolism in thidiazuron induced somatic embryogenesis of peanut (*Arachis hypogaea* L.), *Plant Growth Regula.* 28: 41-47.

CHAPTER

20

Flow Cytometry: A Comparatively New Tool in Plant Biotechnology

Amita Pal

INTRODUCTION

Flow-cytometry (FCM), as the name implies, is the process of analysis of the optical properties of cells in a moving fluid stream. The optical and light scattering properties of cells are measured as single cell passes through the flow-cytometer. The fluorescence and light scattering properties depend on the size and granularity of the cells. The main advantage of flow-cytometry is the measurement of at least 3-5 parameters at a time and its swiftness in measuring number of cells/ nuclei. The method is very precise, accurate, less labour-intensive, less time consumable and reproducible. As a cell passes in front of the laser, several measurements are made based on the physical characteristics of the cells. Based on these characteristics, cells scatter the laser light and emit fluorescence, which provide information about the cell's size, internal complexity (granularity), base composition and relative fluorescence intensity.

To accomplish this analysis, flow-cytometer is composed of four independent modules: 1. Fluidics, 2. Optics, 3. Electronic device and 4. Data processing software.

The fluidics subsystem delivers the cells to the point of the light source/ measurement point. This is achieved by directing the suspended cells into another stream that acts as 'sheath' to the stream in which cells are flowing.

The optical system has a light source, which guides light from the excitation sources into the measurement area and a collection optics that collects the light scatter and emitted fluorescence and filters, which split the collected light into spectral bands (colour) for multiparameter analysis.

The electronic device converts the fluorescence and scatter signals into digital values for computer analysis of the data through a very complex electronic system comprising of PIN diodes (to measure forward angle scatter), photomultiplier tubes (for orthogonal scatter and fluorescence signals) and analog to digital converters.

*Address for correspondence: Plant Molecular & Cellular Genetics Section, Bose Institute, P 1/12 CIT Scheme VII M, Kankurgachi, Kolkata 700054. E-mail: amita@bic.boseinst.ernet.in

The flow cytometers are linked to Macintosh computer loaded with Cell Quest software for data acquisition and ploidy analysis and also with ModFit LT 3.0 cell cycle analysis software. Data from the analytic cytometers are stored on user managed 230 MB Macintosh-formatted optical disks or Zip drive 100 or 250 MB disks.

CLINICAL APPLICATIONS OF FLOW CYTOMETRY

FCM was originally developed as a method for rapid counting and analysis of human erythrocyte cells. With technical evolution and by the development of new fluorescent dyes it became a useful tool in clinical diagnosis and medical biology, such as, immuno-phenotypic analysis of leukaemias and lymphoma cells, detection of minimal residual diseases, stem cell enumeration, T-cell cross matching prior to solid organ transplantation and also for post-operative monitoring; detection of anti-antibodies to leucocytes, platelets, erythrocytes in anti-immune condition, and immunodeficiency diseases.

Current clinical use of FCM includes helper T-lymphocyte counts needed to monitor the course and treatment of HIV-infection, and to determine tumor cell DNA content and to monitor the proliferation of tumor which helps in assessing prognosis of the disease and to determine the accurate treatment for patients with malignant diseases including breast cancer. Recent application also includes isolation of animal and human chromosomes for the construction of genetic libraries; to separate sex chromosomes and sperms for animal breeding. The use of FCM also extends to identify and characterize unexplored marine bacteria.

The other common uses of flow-cytometry are to measure the genome size of plant and animal cells, to sort mammalian and plant cells, yeast and bacteria. The component of cells, viz. chromosomes, chloroplasts and mitochondria can also be sorted using flow-cytometry. In a cell various properties can be measured using flow-cytometer, such as DNA content of nucleus, determination of ploidy level, the expression of surface antigen, the activity of an intracellular enzyme or pH.

EQUIPMENT

In this article two equipment FACSCALIBUR and FACSVANTAGE of Becton and Dickinson are described, which are useful for various plant- biotechnological studies and are housed in Bose Institute, Kolkata.

(A) FACSCALIBUR is an analytical bench top instrument with a cuvette flow cell, 15 mW, and air-cooled 488-nM argon laser. The laser beam is focused and shaped using an elliptical lens system, which gives a spot size of about 20×60 μM. It can simultaneously measure fluorescence in 2 wavelength ranges, fixed in the region of maximum emission from fluorescein, phycoerythrin and FITC light scattering.

FACSCALIBUR analyzes cells as they pass one at a time through a focused laser beam. It can measure upto five parameters at a time-forward light scatter (FSC), side light scatter (SSC) and two fluorescence parameters and can measure the pulse height, area and width of any fluorescence parameter.

The following are some important functional components of both the equipment:

(i) Detectors: A solid-state detector is fitted to measure forward-scattered light and for measuring fluorescence and orthogonal (right angle) scatter, photo-multiplier is used. Detectors and amplifiers adjust the signals so that they appear in appropriate place on the data plots. Light signals are generated as particles pass through the laser beam in the flow cytometer. These light signals are converted to electrical signals (voltage) and assigned a channel number on a data plot. These channel numbers range from 0-255 or 0-1023, depending on the selection of resolution.

a. *Voltage:* The voltage parameter is set to adjust the photomultiplier tube (PMT) voltage for SSC or side scattering at different fluorescence range FLI to FL3 or upto FL5.
b. *Amplifier:* The Amp Gain allows users to set amplifier settings for each parameter. Both the linear and log amplification can be set as per requirement.
c. *Compensation:* Compensation allows adjustment for the spectral overlap between fluorochromes when the samples are stained with 2 or more fluorochromes. Fluorochromes emit light over whole length of spectra. Therefore, signal from one fluorochrome may overlap in a detector used for another fluorochrome. As for example, FITC appears primarily in the FL1 detector but some of its fluorescence overlap into the FL2 detector. Phycoerythrin appears primarily in the FL3 detector but some of its fluorescence overlaps into the FL2 and FL3 detectors. Proper setting of the compensation can eliminate such spectral overlaps electronically.

(ii) Signal processing: After pre-amplification, the signal from a photo-multiplier undergoes further processing. The instrument is set to respond to signals derived from the cells/nuclei and to ignore debris and 'spikes' from electronic noise. A threshold level is set on one or possibly two parameters in such a way that the electronics are only become responsive if the signal raises above this threshold. Amplification of the response can be done either in linear or in logarithmic mode.

(iii) Analog to digital conversion: The signal undergoes analog to digital conversion before being transmitted to computer. The resolution of the data will depend on the analog to digital converter (ADC) an eight-bit converter will yield 256 channels; a ten-bit converter gives 1024 channels and so on. The software we are using for genome size analysis, ploidy study etc. runs on a MAC platform.

(iv) Data collection: Most flow cytometers collect between 3 and 10 "parameters" or "variables". Each variable can be used to discriminate some component of the cell populations. The standard for data collection has been known as the "Flow Cytometry Standard" or FCS data file structure. Data can be collected in either histogram mode or list mode. Data can be collected in a gated or ungated mode. If the data are ungated, all events registered by the instruments electronic circuits will be saved in listmode files. If gated collection is chosen, it is possible to collect a list mode file of only those events that are set as per requirement of the operator.

(B) FACSVANTAGE (Fluorescence Activated Cell Sorter) has the analytical capabilities like that of the FACSCALIBUR and additionally has a fluidic sorter. With the sort option FACSVANTAGE can sort a subpopulation of the given sample. The equipment can use as many as three laser beams to illuminate two observation points spatially separated. There is a wide choice of laser type for the users including water and air-cooled argon, krypton or mixed gas ion-laser, He- Cd and He-Ne laser etc.

The laser beam in both the instruments, viz. FACS-CALIBUR and VANTAGE is focussed to a round or elliptical spot, which illuminates the cell stream after the stream emerges from a nozzle, through an orifice into air. Both sheath and sample fluid-streams are normally driven by air or gas pressure, microsample-handling module has a syringe pump.

Forward scatter signals are collected through a lens equipped with iris diaphragms, which users have to fix to define the required collection angle; the collection lens for fluorescence and orthogonal scatter is a Leitz microscope objective with a long working distance and a numerical aperture of 0.6. When two illuminated lasers of FACSVANTAGE are used at a time, the optics are arranged in such a way so that light from different observation points is reflected by mirrors and/or dichroics to different detectors.

Flow sorters are employed for droplet sorting, in which the fluidic stream is broken up into droplets, and the droplets containing the selected cells are electronically charged and deflected into a collection vessel by passage through an electrical field.

Some instrument uses mechanical actuators to collect cells from a continuous fluid stream. The rate of sorting is slower in such mechanical sorter as compared to the droplet sorter. However, their close fluidic systems are better adapted for work with potentially infectious or otherwise hazardous materials, which might be dispersed in the aerosols possibly generated by droplet sorter.

Flow sorting is primarily useful for the quantification of heterogeneity in a given cell population. First it requires identification and then subsequent characterization of subpopulation of cells. The selected homogeneous cell population was then processed further for definite purposes. Interested readers are requested to go through the book by Shapiro (1995) and a review article by Galbraith (1989).

Although the application of FCM in plant research has been delayed until late 1980s, presently it has several applications in plant breeding and seed production (reviewed by Dolezel, 1997).

APPLICATIONS IN BASIC AND APPLIED RESEARCH OF HIGHER PLANTS

(i) Genome size estimation: The information on nuclear genome size of most of the cultivated crops is still not known. However, the genome size is important as it helps to resolve phylogenetic relationships and to analyze the correlation between the genome size and agronomic traits and also this information is a prerequisite for the molecular biologists intend to map the genome of a particular taxa.

FCM is a convenient method that has been extensively used for estimation of nuclear genome size in plants due to its speed and precision (Dolezel et al., 1989, 1995 and Dolezel, 1997). Since fluorescent dye could not penetrate the cell wall conveniently, therefore, it is mandatory to isolate intact nuclei for the FCM estimation of DNA contents, cell cycle study etc. Guilfoyle (1995) has reviewed the nuclei isolation protocols. Later, Dolezel and coresearchers (1992b) compared three DNA fluorochromes for FCM estimation of nuclear DNA contents in plants. The various fluorochromes used in FCM analysis showed DNA bind differentially to the nucleic acids. Ethidium bromide (EB) and propidium iodide (PI) bind through intercalation with the nucleic acid. In order to measure DNA contents, treatment with RNase is required to eliminate fluorescence contributed by RNA through binding of the fluorochrome used. Hoechst 33258, 4′, 6-diamidine-2-phenylindole (DAPI) and mithramycin specifically bind with DNA and not with RNA, hence RNase treatment is not essential. Unlike EB and PI, these dyes exhibit base pair specificity while binding to DNA.

To start with Arumuganathan and Earle (1991) have measured 2C value for 100 plant species. Recently genome size analysis was carried out in *Vigna mungo* and *V radiata,* nuclei were isolated using Otto buffer I and II (Otto, 1990). FCM estimation of nuclear genome size in picogram amount or mega base pair number of the nucleotides can be obtained by the simultaneous measurement of nuclei from the specimen and reference standards. The genome size of the specimen is then calculated with the value obtained from the ratio of G 1 peak position of specimen standard by multiplying the known genome size value (pg) of the reference standard (Fig. 1).

The diploid genome size of three lines of *Vigna mungo* and eight cultivars of *V. radiata* were measured, using *Lycopersicon esculentum* as the internal standard, as shown below:

Lycopersicon esculentum cv. Stupicke (2C = 1.96 pg DNA)
Vigna mungo Cv. T9 (2C = 1.16 pg DNA)
V. mungo Cv. Line YMVHS (2C = 1.18 pg DNA)
V. mungo Cv. Line YMVHR (2C = 1.17 pg DNA)
V. radiata Cv. Bl (2C = 1.18 pg DNA)
V. radiata Cv. ML 337 (2C = 1.19 pg DNA)
V. radiata Cv. PDM 116 (2C = 1.23 pg DNA)
V. radiata Cv. PUSA 101 (2C = 1.2 pg DNA)
V. radiata Cv. PM-2 (2C = 1.15 pg DNA)
V. radiata Cv. PIMS-4 (2C = 1.17 pg DNA)
V. radiata Cv. TARM-2 (2C = 1.18 pg DNA)
V. radiata Cv. T-44 (2C = 1.2 pg DNA)

(ii) Cell cycle analysis: The cell cycle starts from M phase (point of mitosis) and cell division, followed by a period of cellular growth (Gl phase of interphase) preceding duplication of the nuclear genome (S phase), which in turn is followed by a second period of cellular growth (G2 phase) prior to the next M-phase.

Population of cells at a given time is either proliferating or quiescent. Qualitative and quantitative information on the proliferative characteristics of cells and cell populations are often used to monitor differentiation pathways during development, effect of mutagens and carcinogens on cell division, to measure the ratio of living Vs. apoptotic cells in a population etc. Major advancement in the FCM technique facilitated our knowledge about the progression of the cell

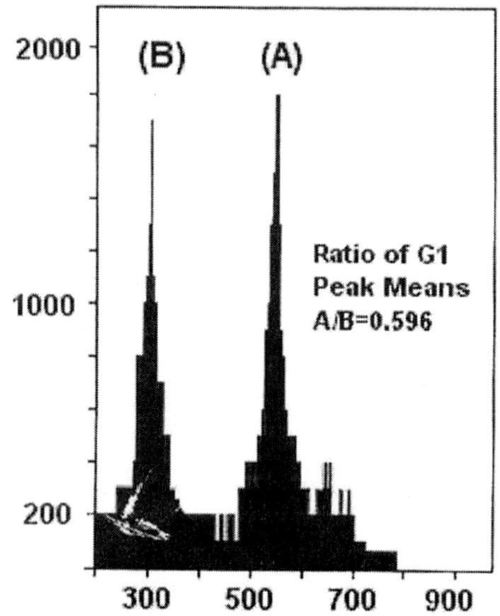

Fig. 1 Estimation of genome size of *Vigna mungo* Cv. T9 (B) and Lycopersicon *esculentum* (A, 2C = 1.96 pg DNA) was used as internal reference standards. The 2C DNA contents of *V. mungo* was calculated using the ratio of G 1 peak means: 0.596 X 1.96 = 1.16 pg.

cycle and its regulation. It provides estimation of cell numbers in the different phases of the cell cycle i.e. GI, S, G2 and M.

Monitoring DNA synthesis during cell cycle: In early days, the S-phase of the cell cycle was studied by the incorporation of radioactive nucleotide into cells but now use of the thymidine analogue 5-bromodeoxyuridine (BrdU) has become more prevalent. This BrdU was incorporated into DNA without affecting other cell cycle parameters and its incorporation was estimated by using a fluorescently tagged anti-BrdU antibody. BrdU was incorporated into cells by short exposure, the cells were then fixed and the DNA was denatured to expose the incorporated BrdU for antibody binding; the ssDNA was stained by fluorescently tagged anti-BrdU antibody and a DNA-specific fluorochrome, e.g. PI. Stained cells were then analysed in a flow cytometer and were excited at 488 nm and recorded as measure of relative BrdU incorporation and red fluorescence (620nm) for PI was recorded as measure of DNA content of that cell. These two parameters help measuring the fraction of DNA synthesizing-cells and correlating with the stage of the cell cycle, where the cells belong at the time of fixation (Moretti *et al.*, 1992).

(iii) Ploidy analysis: Haploid genome size (C), defined as the nuclear DNA content of the gamete. For somatic cells in Gl phase, the nuclei are defined as having a 2C DNA value. For those in G2 phase, the nuclei have a 4C DNA content and so on.

Analysis of the relative fluorescence intensity of nuclei isolated from young plant tissue yields a histogram showing a major peak corresponding to the nuclei in the G 1 phase of the cell cycle and

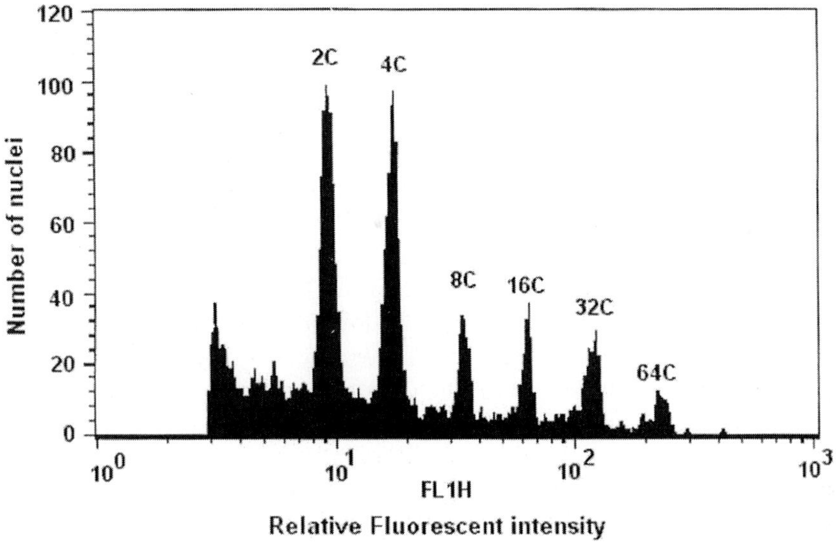

Fig. 2 Histogram showing distribution of ploidy level (2 C to 64 C) in the cotyledon of *V. mungo.*

a minor peak corresponding to G2 nuclei. To analyze ploidy level, the portion of the G 1 peak on a histogram is compared to that of a reference tissue with known ploidy. A more reliable approach involves comparison of G 1 peak positions within a single histogram of reference tissue (known genome size) and the sample tissue that are processed together (Fig. 2).

In banana-like hops and sugar-beet the commonly used cultivars are parthenocarpic triploids. Dolezel (1997) have measured genome size of a number of *Musa* genotypes and analysed their ploidy status using FCM.

This is a very reliable tool to analyze the purity of seed lot and can be used in seed banks and plant breeding station. Baert *et al.* (1992) used FCM to estimate the persistence of tetraploid in pasture, ploidy mixture of perennial ryegrass.

Since cotyledons of legumes are known to possess genomes with endoreduplicated DNA content, therefore we proposed to use FCM to measure the ploidy levels of cotyledons of *V mungo.* It is interesting to note that genomic endoreduplication occurs upto 64C in cotyledons of both the species of *Vigna* studied so far (Fig. 2). We reported for the first time the genome endoreduplication in cotyledons of *V. radiata* (Pal *et al.* 2004) and dynamics of cotyledons during ontogeny was shown by FCM analysis (Das & Pal, 2003).

(iv) Flow karyotyping and chromosome sorting: Genome analysis and mapping in most cultivated crops is neglected because of the large genome size. Dolezel and his associates (1992b) have circumvented this limitation by isolating pure metaphase chromosomes from root-tips of *Vicia faba* using flow-sorter. Later, this high yielding protocol was used for flow-karyotyping and sorting of *V. faba* chromosomes (Lucretti *et al.*, 1993).

The method consists of synchronization of cells at the metaphase and preparation of suspended intact chromosomes. The FACS-sorted chromosomes were then stained with a DNA-specific fluorochrome. The fluorescence was measured and used to quantitatively classify each chromosome / chromosome types. Resulting fluorescence distribution pattern is called as Flow karyotype (Lucretti & Dolezel, 1997). Flow karyogram consists of several peaks representing individual chromosome or chromosome groups and has proven to be a very useful tool for the detection of numerical and structural chromosome-aberrations (Boschman *et al.*, 1992).

Chromosomes those can be unambiguously discriminated on a flow karyotype can be sorted in high number by gating that particular fluorescent channel. The flow-sorted chromosomes are being used for subsequent molecular analysis, such as, for localization of specific DNA sequences with known function, isolation of complex probes for chromosome painting, construction of chromosome specific DNA libraries and isolation of chromosome specific molecular markers. Wang *et al.* (1992) have constructed a chromosome enriched *Hpa II* library from flow sorted wheat chromosomes. Arumuganathan *et al.* (1994) have developed flow cytogenetics for tomato and they have used flow sorted chromosomes for isolation of chromosome specific probes. Chromosomes of translocation lines of wheat, oat and barley are often sorted to locate the desirable gene/s having the agronomic trait/s. Marker assisted breeding of crop plants requires high density map in order to identify markers closely linked with the traits of interest. This is very difficult to obtain in species where either the genome size is large or for the species without having much genomic information, e.g. wheat. Recently, flow-sorted wheat chromosomes were isolated and were identified using PRINS and PCR amplification with specific DNA sequences (Vrana *et al.*, 2000).

Physical mapping of the genes on chromosome

Flow-sorted mammalian chromosomes were used for gene mapping (Lebo, 1982); construction of chromosome specific DNA libraries (Van Dilla & Deaven, 1990), including large-YAC-insert DNA libraries (Mc Connick *et al.* 1993); generation of chromosome painting probes (Ferguson-Smith, 1997) and targeted generation of molecular markers (Lan *et al.*, 1999).

Physical mapping of the gene/s is essential for the estimation of physical distances within the genetic maps and for the chromosomal localization of the genes. Information on physical gene location, especially for agronomically useful traits is of immense utility for genetic manipulations in basic and applied research. Physical mapping is often achieved by hybridizing labeled probes *in situ.* An alternate approach is PCR amplification of the target gene using sequence specific primers on flow sorted chromosomes (Cotter *et al.*, 1989). This method had limited application, as it requires sufficiently pure suspensions of metaphase chromosomes distinctive enough for successful sorting of all chromosomes of a plant genome, which is difficult to obtain (Wang *et al.*, 1992). Subsequently, Dolezel and co-researchers (1992a) developed a reliable procedure for the isolation of metaphase chromosome and sorting of desirable chromosome type at high numbers from root-tips of *Vicia faba.*

Macas and his associates (1993) has shown localization of seed protein genes on flow sorted *V. faba* chromosomes for the first time. The sorted chromosomes of *V. faba* were amplified with sequence-specific primers to localize seed-specific protein genes and pseudogenes. Sorted *V. faba* chromosomes were also used for physical mapping by Macas *et al.* (1996). Kubalakova *et al.* (1997) mapped repeated DNA sequence by PRINS and C-PRINS in flow-sorted plant chromosomes.

(v) Sex determination at an early developmental stage: Most of the flowering plants are hermophrodite, having male and female reproductive organs. However, 11% of plant species have unisexual flowers and 4% are dioecious with separate male and female individuals. Sex determination at an early stage of development for the dioecious-cum-economically important plant species has broad application in plant breeding programme. Dolezel & Gohde (1995) have demonstrated that high resolution FCM analysis of nuclei isolated from young leaves can discriminate between male and female plants of *Melandrium album.* Recently, Kejnovsky *et al.* (2001) have localized male-specifically expressed *MROS* genes of *Silene latifolia* (syn. *Melandrium album)* by PCR on flow-sorted sex chromosomes and autosomes. S. *latifolia* has heteromorphic sex chromosomes, XX in females and XY in males (Charlesworth & Guttman 1999, Moneger *et al.* 2000), that are larger than the autosomes and enable their separation by flow-sorting. Matsunga *et al.,* 1996, 1997) have discovered a group of four male- reproductive organ-specific (MROS) genes. However, from physical localization study Kejnovsky *et al.* (2001) have shown that all these genes are localized on the autosomes and at least two copies of *MROS3* are X-linked.

APPLICATION OF FCM IN *IN VITRO* STUDIES

Most of the above mentioned applications could be extended, in the *in vitro* studies. Genome size analysis and FCM analysis of the ploidy status are very reliable and efficient tools to determine clonal fidelity/genetic stability of the culture regenerants. FCM analysis does not require metaphase stage of dividing cells (which is a prerequisite for the chromosome count by classical method), sample preparation is easy, rapid (several hundreds of nuclei can be analysed per second, contrary to chromosome count), accurate, mixoploids can be detected (Dolezel, 1998). More so, as because, *in vitro* cultures are often characterized by low frequency of dividing cells. Wyman *et al.* (1992) has demonstrated ploidy level stability of adventitious shoots regenerated *in vitro* through FCM measurement. Kreuger and co-workers (1996) have successfully used FCM to investigate ploidy level stability of somatic embryos of cucumber cell line for artificial seed production. Brutovska *et al.* (1998) revealed cytogenetical variability amongst the *in vitro* regenerated *Hypericum perforatum* plants.

Ploidy analysis using FCM can be exploited to evaluate ploidy status of anther or pollen raised plants. Haploid plants can be developed *in vitro* from anther, ovule and microspore culture. Subsequently, haploid regenerants are treated with polyploidizing chemicals to produce fertile homozygous diploids. Possibility of screening potential-embryogenic- microspores through FCM has been demonstrated by Deslauriers *et al.* (1991). Tosca *et al.* (1995) optimized the conditions for chromosome doubling in Gerbera using colchicines and oryzalin and screening of diploids using FCM.

In protoplast-mediated hybridization, hybrid formation between two genetically distant species is possible circumventing sexual barriers. However, screening of heterokaryons is time consuming and laborious. Hammatt *et al.* (1990) have demonstrated an efficient application of FCM to identify heterokaryons after somatic hybridization following electrofusion of protoplasts. Determination of genome size can also be extended for the proper evaluation of the culture regenerants raised from somatic hybridization between *two* different species having considerable difference in genome size. Samoylov and Sink (1996) monitored chromosome elimination in asymmetric somatic hybrids using FCM analysis. Ramulu *et al.* (1996) employed FCM to investigate microplast-mediated chromosome transfer technology.

All these above applications indicate the immense scope of FCM in modern plant biotechnological research.

Acknowledgements

Author expressed her gratitude to the Indian National Science Academy for nominating in the INSA/ ASCR Scientist Exchange Programme to learn techniques of Flow-cytometry.

REFERENCES

Arumuganathan K. & Earle E.D. (1991) Nuclear DNA content of some important plant species *Plant Mol Biol. Rep.* 9: 208-218.

Arumuganathan K., Martin G.B., Telenius H., Tanksley S.D. & Earle E.D. (1994) Chromosome 2-specific DNA clones from flow-sorted chromosomes of tomato. *Mol. Gen Genet.* 24(2): 551-568.

Baert J., Reheul D., Van Bockstaelee E. & De Loof A. (1992) A rapid method by flow cytometry for estimating persistence of tetraploid perennial ryegrass in pasture mixtures with diploid perennial ryegrass *Biol. Plant.* 34: 381-385.

Boschman G.A., Manders E.M.M., Rens W., Slater R. & Aten J.A. (1992) Semi automated detection of aberrant chromosomes in bivariate flow karyotypes *Cytometry*. 13: 469-477.

Brutovska R., Cellarova E. & Dolezel J. (1998) Cytogenetic variability of *in vitro* regenerated *Hypericum perforatum* L. plants and their seed progenies. *Plant Sci.* 133: 221-229

Charlesworth D. & Guttman. (1999) The evolution of dioecy and plant sex chromosome systems, pp. 25-49 In: *Sex Determination in Plants.* Ainsworth C.C. (Ed.) Bios Scientific Publishers, Oxford.

Cotter F., Nasipuri S., Lam G. & Young B.D. (1989) Gene mapping by enzymatic amplification from flow sorted chromosomes. *Genomics*. 5: 470-474.

Das S. & Pal A. (2003) Differential DNA endoreduplication and protein profile during cotyledon ontogeny of *Vigna radiata. J. Plant Biochem. & Biotech.* 12: 1118.

Deslauriers C., Powell A.D., Fuchs K & Pauls K.P. (1991) Flow cytometric characterization and sorting of cultured *Brassica napus* microspores *Biochem. Biophys. Acta*. 1091: 165-172.

Dolezel J., Binarova P. & Lucretti S. (1989) Analysis of nuclear DNA content in plant cells by flow cytometry. *Biol. Plant.* 31: 113-120.

Dolezel J., Sgorbati S. & Lucretti S. (1992a) Comparison of three DNA fluorochromes for flow cytometric estimation of nuclear DNA content in plants. *Physiol. Plant.* 85: 625-631.

Dolezel J., Cihalikova J. & Lucretti S. (1992b) A high-yield procedure for isolation of metaphase chromosomes from root tips of *Vicia faba* L. *Planta*. 188: 93-98.

Dolezel J. & Gohde W. (1995) Sex determination in dioecious plants *Melandrium album* and *M. rubrum* using high-resolution flow cytometry. *Cytometry*. 19: 103-106.

Dolezel J., Lucretti S. & Macas J. (1995) Flow cytometric analysis and sorting of plant chromosomes pp 185-200 In: Brandham P. & Bennett M.D. (Eds) Kew Chromosome Conference Iv, Royal Botanic Gardens, Kew, U.K.

Dolezel J. (1997) Application of flow cytometry for the study of plant genomes. *J. Appl. Genet.* 38: 285-302.

Dolezel J., Lysak M.A., Van den Houwe I., Dolezelova M. & Roux N. (1997) Use of flow cytometry for rapid ploidy determination in *Musa* species *INFOMUSA* 6: 61.

Dolezel J. (1998) Flow cytometry, its application and potential for plant breeding. In: *Current Topics in Plant Cytogenetics Related to Plant Improvement,* pp. 80-90 Ed Lelley T. Universitatts Verlag, Vienna.

Ferguson-Smith M.A. (1997) Genetic analysis by chromosome sorting and painting: Phylogenetic and diagnostic applications *Eur. J. Hum. Genet.* 5: 253-265.

Galbraith D.W. (1989) Analysis of higher plants by flow cytometry and cell sorting. *Int. Rev. of Cytol.* 116: 165-228.

Guilfoyle T.J. (1995) Isolation and characterization of plant nuclei. In: *Methods in Cell Biology,* pp 101 Galbraith D.W., Baurque D.P. & Bohnert H.J. (Eds.) Academic Press, N.Y.

Hammatt N., Lister A., Blackhall N.W., Gartland J., Ghose T.K., Gilmour D.M., Power J.B., Davey M.R. & Cocking E.C. (1990) Selection of plant heterokaryon from diverse origins by flow cytometry. *Protoplasma* 154: 34-49.

Kejnovsky E., Vrana J., Matsunaga S., Soucek P., Siroky J., Dolezel J. & Vyskot B. (2001) Localization of male-specifically expressed *MROS* genes on *Silene latifolia* by PCR on flow-sorted sex chromosomes and autosomes. *Genetics*. 158: 1269-1277.

Kreuger M., Van Der Meer W., Postma E., Abbestee R., Raanmakers N. & Van Holst G.J. (1996) Genetically stable cell lines of cucumber for the large-scale production of diploid somatic embryos. *Physiol Plant.* 97: 303-310.

Kubalakova M., Macas J. and Dolezel J. (1997) Mapping of repeated DNA sequences in plant chromosomes by PRINS and C-PRINS. *Theor. Appl. Genet.* 94: 758-763.

Lan H., Shepel L.A., Haag J.D. & Gould M.N. (1999) Linkage mapping of rat chromosome 5 markers generated from chromosome specific libraries. *Mamm. Genome.* 10: 687-691.

Lebo R.V. (1982) Chromosome sorting and DNA sequence localization: A review. *Cytometry*. 3: 145-154.

Lucretti S., Dolezel J., Schubert I. & Fuchs J. (1993) Flow karyotyping and sorting of *Vida faba* chromosomes. *Theor. Appl. Genet.* 85: 665-672.

Lucretti S. & Dolezel J. (1997) Bivariate flow karyotyping in broad bean (*Vicia faba*) *Cytometry* 28: 236-242.

Macas J., Dolezel J., Lucretti S., Pich U., Meister A., Fuchs J. & Schubert I. (1993) Localization of seed protein genes on flow-sorted field bean chromosomes. *Chromosome Research* 1: 107-115.

Macas J., Gualberti G., Nouzova M., Samec P., Lucretti S. & Dolezel J. (1996) Construction of chromosome-specific DNA libraries covering the whole genome of field bean (*Vicia faba* L). *Chr. Res.* 4: 531-539.

Matsunaga S., Kawano S., Takano H., Uchida H. & Sakai A. (1996) Isolation and developmental expression of mal reproductive organ-specific genes in a dioecious campion, *Melandrium album* (*Silene latifolia*). *Plant J.* 10: 679-689.

Matsunaga S., Kawano S. & Kuroiwa T. (1997) MROS 1, a male reproductive organ specific genes in dioecious campion *Silene latifolia* is expressed in mature pollen. *Plant Cell Physiol*. 38: 499-502.

McCormick M.K., Campbell E., Deaven L. & Moyzis R. (1993) Low-frequency chimeric yeast artificial chromosome libraries from flow-sorted human chromosomes 16 and 21. *Proc. Natl. Acad. Sci.* USA 90: 1063-1067.

Moneger F., Barbacar N. & Negrutiu I. (2000) Dioecious *Silene* at the X-road: the reason Y-sex. *Plant Reprod.* 12: 245-249.

Moretti F., Lucretti S. & Dolezel J. (1992) Plant cell cycle analysis on isolated nuclei using a monoclonal antibody against BrdUrd. In: Proc. IX Natl. Meeting of the Italian Group of Cytometry. *Eur. J. Histochem.* 36: 367.

Otto F. (1990) DAPI staining of fixed cells for high-resolution flow cytometry of nuclear DNA. In *Methods in Cell Biology* pp. 105-110; Crissman H.A., & Darzynkiewicz Z., (Eds.) Academic Press, New York.

Pal, A., Vrana J. & Dolezel J. (2004). Flow cytometric analysis of variation in the level of nuclear DNA endore- duplication in the colyledons amongst *Vigna radiata* culivars. *Caryologia* 57: 262-266.

Ramulu K.S., Dijkhuis P., Rutgers E., Blass J., Krens F.A., Dons J.J.M, Cohjn.Hooymaus C.M. & Veerhoeven H.A. (1996) Microprotoplast-mediated transfer of single specific chromosomes between sexually incompatible plants. *Genome* 39: 921-933.

Samoylov V.M. & Sink K.C. (1996) The role of irradiation dose and DNA content of somatic hybrid calli in producing asymmetric plants between an interspecific tomato hybrid and eggplant. *Theor. Appl. Genet.* 92: 850-857.

Shapiro H.M. (1995) *Practical Flow Cytometry:* Third edition pp. 1-517, Wiley, New York, USA.

Tosca A., Pandolfi R., Citterio S., Fasoli A. & Sgorbati S. (1995) Determination by flow cytometry of the chromosome doubling capacity of colchicines and oryzylin in gynogenetic haploids of Gerbera. *Plant Cell Rep.* 14: 455-458.

Van Dilla M.A. & Deaven L.L. (1990) Construction of gene libraries for each human chromosome. *Cytometry.* 11: 208-218.

Vrana J., Kublaikova M., Simkova H., Cihalikova J., Lysak M.A. & Dolezel J. (2000) Flow-sorting of mitotic chromosomes in common wheat (*Triticum aestivum* L). *Genetics* 156: 2033-2041.

Wang M.L., Leitch A.R., Schwarzacher T., Heslop-Harrison J.S. & Moore G. (1992) Construction of a chromosome enriched *Hpall* library from flow-sorted wheat chromosomes. *Nucleic Acids Res.* 20: 1897-1901.

Wyman J., Brassard N., Fipo D. & Laliberte S. (1992) Ploidy level stability of callus tissue, axillary and adventitious shoots of *Larix* X *eurolepis* Henry regenerated *in vitro. Plant Sci.* 85: 189-196.

CHAPTER

21

Steroid Factories: Prospects and Practicalities in Effective Pharming of *Solanaum* Plants

Asha Jacob and Nutan Malpathak*

ABSTRACT

The established pharmaceutical products derived from phytosteroids include corticosteroids, contraceptives, sex hormones and anabolic agents. Solasodine is one of the important steroidal precursors, which have been detected and explored for pharmaceutical purposes. There is a need to enhance the production of the steroidal alkaloids, which is being done using the tools of biotechnology. This review article discusses the various steroidal precursors, their use and conversion to the steroidal molecule, with special reference to solasodine from various *Solanum* species and the biotechnological advances made in their scale up, extraction, pharming, its future prospects and application.

Key Words: Solasodine, Steroidal alkaloid, Solanaceae, Cell culture, Hairy root Culture, Bioreactor.

INTRODUCTION

The plant kingdom has been explored for an inexpensive and abundant source of steroidal alkaloids. Since 1947, diosgenin was used commercially for the production of corticosteroid hormones [1]. Diosgenin occurs as rhamno-rahmno-glucoside attached to dioscin, in the rhizome of several species of *Dioscorea* such as *D. deltoidea, D. floribunda, D. composita* and *D. prazeri* [1] but the commercial cultivation of *Dioscorea* is not encouraging, as it requires stringent climatic condition, it shows slow growth and delayed maturity of tubers. This has created an acute shortage of raw material and today it has become an important incentive to increase the present production by selecting an alternative commercially profitable source. Steroid alkaloids are present both in the animal and plant kingdoms. Steroid precursors other than diosgenin that are being considered are Solasodine ($C_{27}H_{43}O_2N$), Hecogenin ($C_{27}H_{42}O_4$), Stigmasterol ($C_{29}H_{48}O$), Sitosterol ($C_{29}H_{50}O$), Cholesterol ($C_{27}H_{46}O$), Deoxycholic acid ($C_{24}H_{40}O_5$), Ergosterol ($C_{28}H_{44}O$) [1]. Japan has recently

*Address for correspondence: Department of Botany, University of Pune, Pune-7, India.

entered the field with cholesterol owing to its easy biotransformation to other steroids. Over thirty new steroidal alkaloids have been isolated from various species of *Salamandra, Phyllobates* and *Bufo* [2]. These have been obtained from the skin gland evacuation via cholesterol. Steroidal alkaloids have also been obtained from the marine organisms such as *Cephalodiscus gilchristi, Zoanthid* species and from the marine sponge. *Corticium* species are also known to be source of steroidal alkaloids [3, 4]. Steroidal alkaloids are also present in plants of *Holarrhena, Solanum and Veratrum.* Solasodine present in *Solanum* plants has gained significant importance globally as both diosgenin and solasodine are degraded to 16-dehydro-pregnenolone acetate as the first step in steroid synthesis [5, 6, 7, 8]. It has been seen to be a good replacement for diosgenin [9, 10]. It is the substrate for the four groups of steroidal drugs manufactured [11], Fig 1. The steroidal drugs manufactured can be classified into 4 groups: Group A: *Corticosteroids:* Here the corticosteroids are derived basically from the hydrocarbon pregnone. Cortisones are seen to be in the adrenal fluids [11], Fig 2. Group B: *Pregnenes:* This group of compounds has hormonal activities similar to the female hormone, progesterone. These steroids are found to be more effective than progesterone in specific clinical conditions because of their longer duration of action in the human body [11], Fig 3. Group C: *Androstenes:* These compounds are related to the basic steroid nucleus, androstane. Some of these compounds are the key intermediates for the synthesis of the male hormones testosterones and its esters and methyl testosterones [11], Fig 4. Group D: *19- Nor-steroids:* Esterone obtained from DHA acetate is converted into the basic estrogen used in oral contraceptive formulations [11], Fig 5.

Progress made in different areas in order to convert a *Solanaceae* plant into a potential steroid factory, Fig. 6. Tomatidine and solanidine are also considered in this review, as they are structurally similar to solasodine and could show a possibility of being precursors to steroid production, besides the techniques used in tomato and potato plants could be applicable in *Solanum* plants too. Some microorganisms have the capability of bioconversion of steroidal compounds [12, 13].

Solanaceae on the basis of the presence or the absence of the tubers is divided into two groups; tuberous group represented by approximately 100 species including potato (*S. tuberosum*) and the non-tuberous group. Solanaceae has yielded several types of steroidal bases, and over 200 alkaloids have been isolated from various species of *Solanum* and *Lycopersicon.* All these alkaloids possess

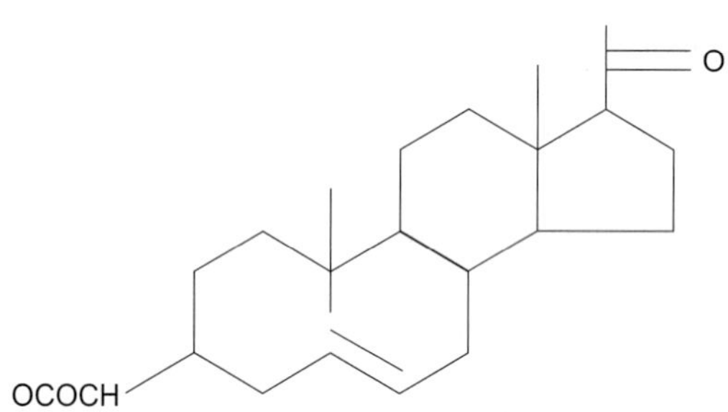

Fig. 1 16-dehydropregnenolone acetate (16-DPA)

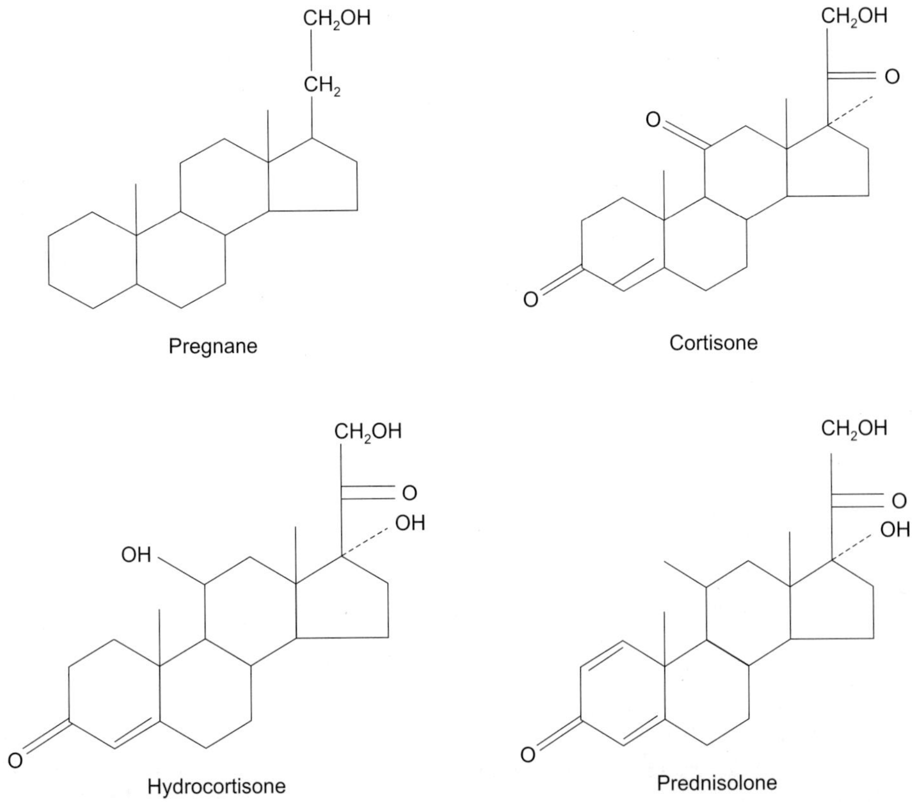

Fig. 2 Strucure of the various corticosteroids manufactured from 16-DPA.

the C_{27} cholestane skeleton.and can be divided into five structural types, solanidine, spirosolane, solacongestidine, solanocapsine and jurbidine [14], Fig 7.

About 50 members of the spirosolane type are found. These alkaloids have a methylpiperidine ring with the J-position joined to C-22 of the steroid moiety to form an oxazaspirane unit. It is known to contain both alkamine glycosides (saponins) and aglycone alkamines (sapogenins) [14]. Spirosolanes are important intermediates in the industrial production of hormonal steroids. Solasodine, belonging to the spirosolane type is present in a number of solanaceous plants is known to be the "diosgenin of the future". Tomatidine belongs to 16-unsubstituted 22, 26-epiminocholestanes and seen to be present in tomato. Solanidine, the alkamine present in potato belongs to the solanidine type, along with galactose, glucose and rhamnose they form the glycoalkaloid solanine. Solanocapsine is the aglycone seen in 3 ß-amino group and jurbidine seen in 3-aminospirostanes has been recently found in *Solanum paniculatum* and *Solanum torvum* [15]. Fig 8 shows the structural similarity between the major aglycones seen in the steroidal glycoalkaloid.

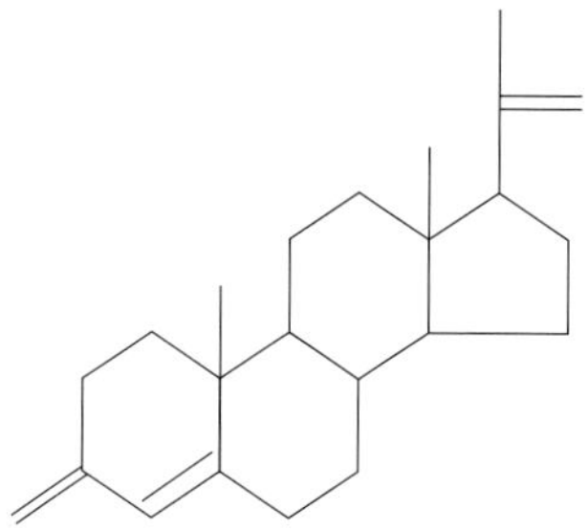

Progesterone

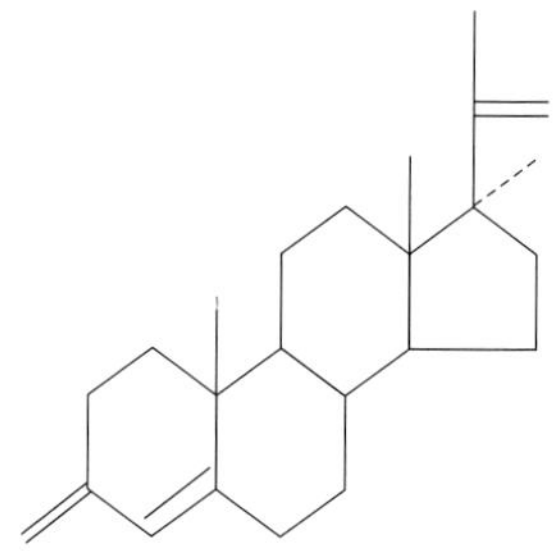

17-β-Hydroxyprogesterone

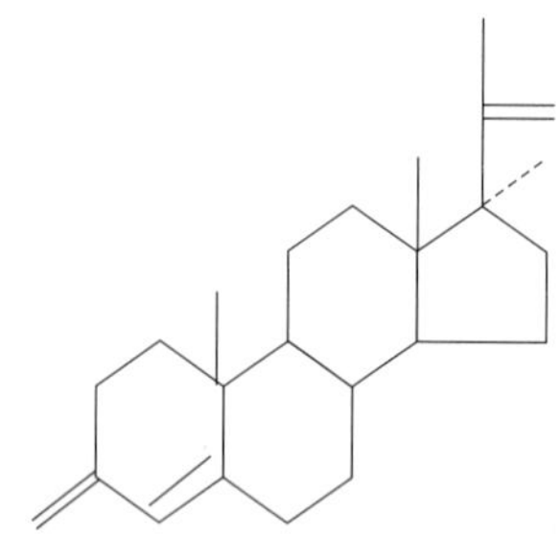

17-β-Acetoxyprogesterone

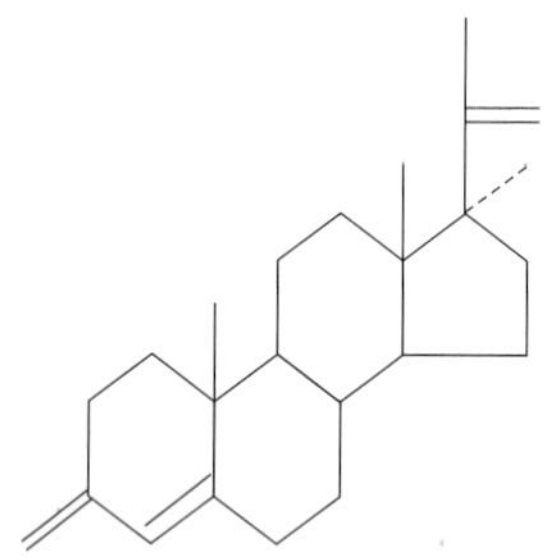

17-β-Caproyloxyprogesterone

Fig. 3 Structure of the various pregnenes manufactured from 16-DPA.

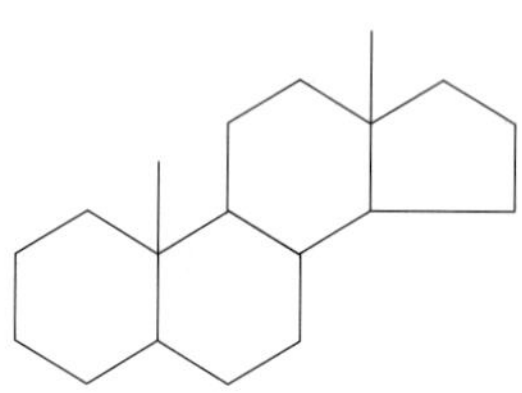

Androstane

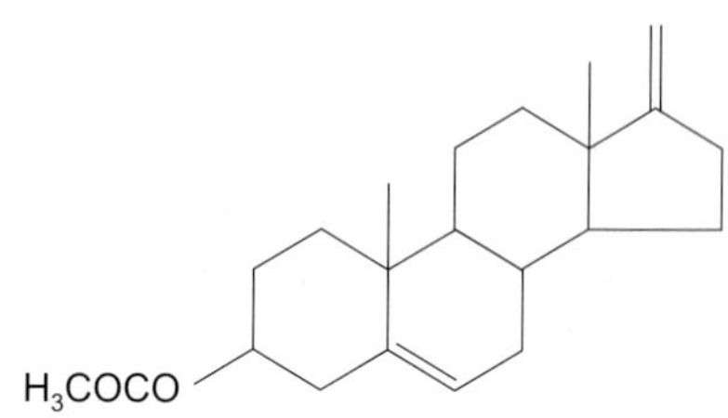

DHA-Acetate

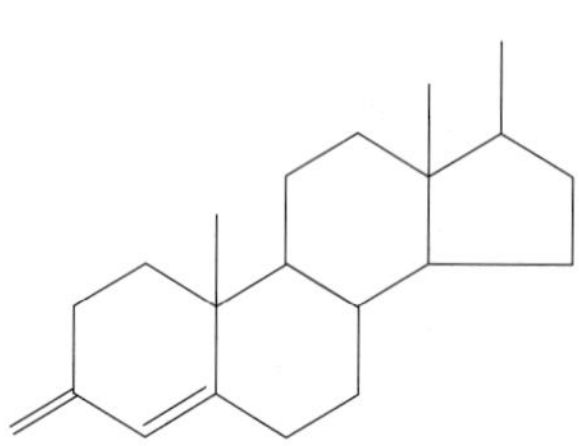

Testosterone

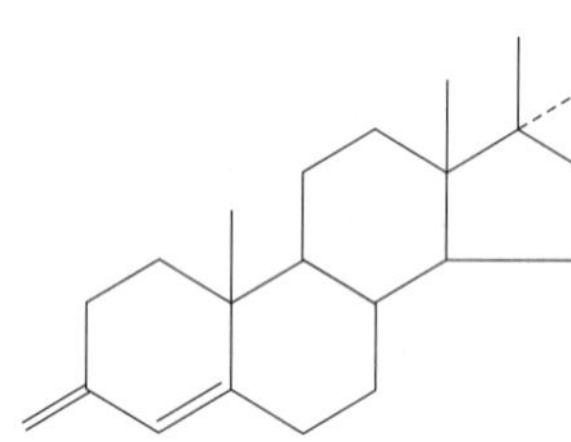

Methyl Testosterone

Fig. 4 Structure of the various androstenes manufactured from 16-DPA.

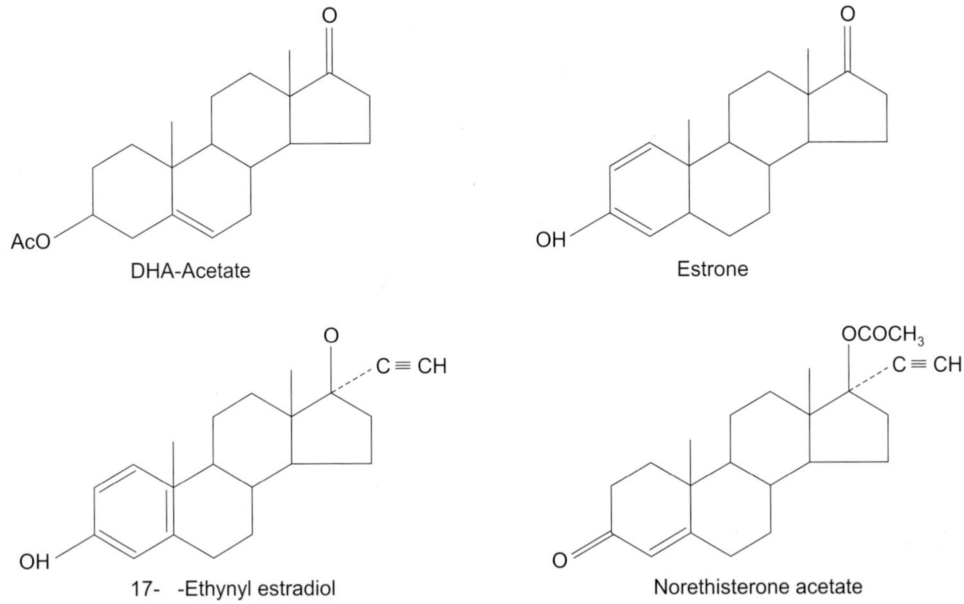

Fig. 5 Strucure of the various steroids used in the preparation of contraceptives.

OCCURRENCE OF *SOLANUM* GLYCOALKALOIDS IN SOLANACEAE

The most widely distributed glycoalkaloid is solasonine and it contains as sugar component a trisaccharide, L-rhamnosido-D-galactosido-D-glucose, for which the name solanose (alpha solasonine) has been proposed (Fig 9). ß-solasonine contains D-Glucose and D-galactose only while gamma- solasonine contains D-galactose only [15,16]. It is not known for certain whether the alpha, ß and gamma forms occur in the plants or are formed as a result of partial hydrolysis. The group R represents the alkamine or aglycone component, which is solasodine $C_{27}H_{43}O_2N$. Solasonine as present in the plant may be isolated by alcohol extraction; on acid hydrolysis it yields the sugar moiety and solasodine [14].

Table 1, lists some of the known glycoalkaloids, their corresponding aglycones and their sugar moiety in the Solanum plants [14], while table 2 gives a list of *Solanum* plants with solasodine as the aglycone [15].

Solanum species considered for commercial production are *S. aviculare* Linn (leaves) *S. nigrum* Linn (berries), *S. robustum* E.Wendl (berries), *S. surattense* Burm.f. (berries) [17], *S. eleagnifolium* Cav. (berries), S. *incanum* Linn (berries), S. *khasianum* Clarke (berries), *S. mammosum* Linn (berries), *S. trilobatum* Linn (berries), *S. verbascifolium* auct. non. Linn (berries) [18], *S. khasianum* var Chatterjeeanum (berries), *S. plantinifolium* Sims (berries) [19], *S. lanciniatum* Ait (berries) [20]. The ripe fruits of *S. khasianum* Clarke have been used and its variety chatterjeeanum Sen Gupta has been cultivated in India for the steroidal precursor. *S. elaeagnifolium* Cav., a spineless variety has received more attention as a good source for the collection of fruits and

Table 1 Some of the known glycoalkaloids, their corresponding aglycones and their sugar moiety in the *Solanum* plants [14, 15].

Glycoalkaloid	*Aglycone*	*Sugar*
Solanine (potato)	Solanidine	L-Rhamnosido-D-galactosido-D-Glucose
Chaconine (potato)	Solanidine	L-Rhamnose-L-Rhamnose-D-Glucose
Demissine (*S.demissum*)	Demissidine	D-Xylose-D-galactose-D-glucose-D-Glucose
Solasodamine*	Solasodine	L-Rhamnosido-L-rahamnosido-D-Galactosido-D-glucose
Solasonine*	Solasodine	L-Rhamnosido-D-galactosido-D-glucose
Solamargine*	Solasodine	L-Rhamnosido-L-rhamnosido-D-Glucose
Solanigrine*	Solasodine	Not identified
Solauricine*	Solasodine	L-Rhamnosido-D-galactosido-D-Glucose
Solaradixine	Solasodine	L-Rhamnose-D-Glucose-D-Glucose-D-Galactose
Solaradinine	Solasodine	Not identified
Solavilline	Solasodine	L-Rhamnose-D-Xylose-D-Glucose-D-Galactose
Soladulcine	Soladulcidine	D-Xylose-D-Xylose-D-Glucose-D-Galactose
Solamarine	Tomatidinol	L-Rhamnose-D-Glucose-D-Galactose
Soladulocamarine	Tomatidinol	L-Rhamnose-L-Rhamnose-L-Arabinose-D-Glucose
Solangustine (*S. pulverulentum*)	Solasodine isomere	D-Glucose
Tomatine (tomato)	Tomatidine	D-Xylose-D-Galactose-D-Glucose-D-Glucose

*Occurance listed in Table 2.

Table 2 *Solanum* plants with solasodine as the aglycone [15].

S. abutiloides (Grieseb) BittEt.Lillo	**Solasodine**	*S. americanum* Mill.	**Solasodine, Solamargine, Solanigrine**	*S. armatum* Vahl	**Solamargine**
S. astroites Forst	**Solasonine, Solamargine, Solamargine**	*S. atriplicifolium*	**Solasonine, Solamargine,**	*S. atropurpureum* Gill.ex Nees	**Solanigrine Solasonine Solamargine**
S. auriculatum Ait	**Solasonine,**	*S. aviculare* Forst.	**Solasonine, Solamargine**	*S. boerhaavii* Thell.Barisbanen	**Solasonine, Solamargine Tomatine**
S. boerhaaviaefolium Sendt.	**Solanigrine (solasodine)**	*S. curolinense* L.	**Solasonine, Solamargine (solasodine)**	*S. chlorocarpum* Schur	**Solasonine, Solamargine (solasodine)**
S. cornutum Lam.	**Solasonine, Solamargine Solamargine (solasodine)**	*S. curtipes* Bitt.	**Solasonine, Solanigrine (solasodine)**	*S. depilatum* Kitag.	**Solasonine Solamargine (solasodine)**
S. douglasii Dun.	**Solasonine Solamargine (solasodine)**	*S. dulcamara* L.	**Solasonine Solamargine (solasodine)**	*S. elaeagnifolium* Cav.	**Solasodine**
S. fasciculatum F.v.Muell.	**Solasonine Solamargine (solasodine)**	*S. flavum* Kit.	**Solasonine Solamargine (solasodine)**	*S. furcatum* Dun.	**Solasonine Solamargine (solasodine)**

(Contd.)

(Contd.)

S. giganteum Jacq.	**Solasonine, Solamargine, (solasodine)**	*S. gracile* Otto	**Solamargine Solanigrine (solasodine)**	*S. gasoba*	**Solasonine Solamargine (solasodine)**
S. guineese (L.) Lam.	**Solasonine, Solamargine**	*S. haematocarpum* hort.	**Solasonine Solamargine (solasodine)**	*S. hartwegii* Benth.	**Solasonine Solamargine (solasodine)**
S. heterodoxum Dun.	**Solasonine, Solamargine (solasodine)**	*S. hetrophyllum* Balb.	**Solasonine, Solamargine, (solasodine)**	*S. hirsutissimum* Standl.	**Solasonine Solamargine (solasodine)**
S. humile Bernh.	**Solasonine, Solanigrine (solasodine)**	*S. indicum* L.	**Solasonine (solasodine)**		
S. inopinum Ewan	**Solasodine, Solamargine (solasodine)**	*S. integrifolium* Poir.	**Solasonine, Solamargine, Solanigrine (solasodine)**	*S. interandinum* Bitt.	**Solasonine, Solamargine (solasodine)**
S. jubatum Willd.	**Solasonine, Solamargine (solasodine)**	*S. khasianum* Clarke.	**Solasonine (solasodine)**	*S. lanciniatum* Ait.	**Solasonine, Solamargine Solanigrine**
S. luteum Mill.	**Solasonine, Solamargine (solasodine)**	*S. macaranthum* Dun.	**(solasodine)**	*S. macrocarpum.* L.	**Solasonine Solamargine Solamarine (solasodine)**
S. mammosum L.	**Solasonine, Solamargine (solasodine)**	*S. marginatum* L.	**Solasonine Solamargine (solasodine)**	*S. melongenal* L.	**Solasonine Solanine Solamargine (solasodine)**
S. memphiticum J. F. Gmel.	**Solasonine, Solamargine (solasodine)**	*S. miniatum* Bernh.	**Solasonine Solamargine Solanigrine (solasodine)**	*S. nigrum* L.	**Solasonine Solamargine Solanigrine (solasodine) Solanine**
S. nitidibaccatum Bitt.	**Solasonine Solamargine (solasodine)**	*S. nodiflorum* Jacq.	**Solasonine Solamargine (solasodine)**	*S. ochroleucum* Bast.	**Solasonine Solamargine (solasodine)**
S. ottonis Hyl.	**Solasonine, Solamargine, (solasodine)**	*S. paranense* Dusén	**Solanigrine Solanigrine (solasodine)**	*S.pyracanthon* Jacq.	**Solasonine Solamargine (solasodine)**
S.quinquangulare Willd.	**Solasonine Solamargine (solasodine)**	*S. quitoense* Lam.	**Solasonine Solamargine (solasodine)**	*S. rostrum* Dun.	**Solasonine Solamargine (solasodine)**
S. roxburghii Dun	**Solasonine Solamargine (solasodine)**	*S. saponaceum* Dun.	**Solasonine Solamargine (solasodine)**	*S. sarachoides* Sendt.	**Solasonine Solamargine (solasodine)**
S. schultesii Opiz	**Solasonine Solamargine (solasodine)**	*S. sodomaeum* L.	**Solasonine Solamargine (solasodine)**	*S. tomatillo* Phil. f.	**Solasonine (solasodine)**

(Contd.)

(Contd.)

S. tomentosum L.	**Solasonine Solamargine (solasodine)**	*S. torvum* Sw.	**Solasonine (solasodine)**	*S.trachycyphum* Bitt.	**Sasonine Solamargine (solasodine)**
S. triflorum Nutt.	**Solasonine Solamargine (solasodine)**	*S. umbellatum* Mill.	**Solasonine Solamargine (solasodine)**	*S. verbascifolium* L.	**Solasonine Solamargine (solasodine)**
S. vescum F.v. Muell.	**Solasonine Solamargine (solasodine)**	*S. villosum.* Mill.	**Solasonine Solamargine (solasodine)**	*S. warscewiczii* hort.	**Solasodine**
S. wittoense	**Solasonine Solamargine (solasodine)**	*S. wrightii* Benth.	**Solanidine Solasonine Solamargine (solasodine) Free solasodine**	*S. xanthocarpum* Schrad. et Wendl	**Solasonine (solasodine)**

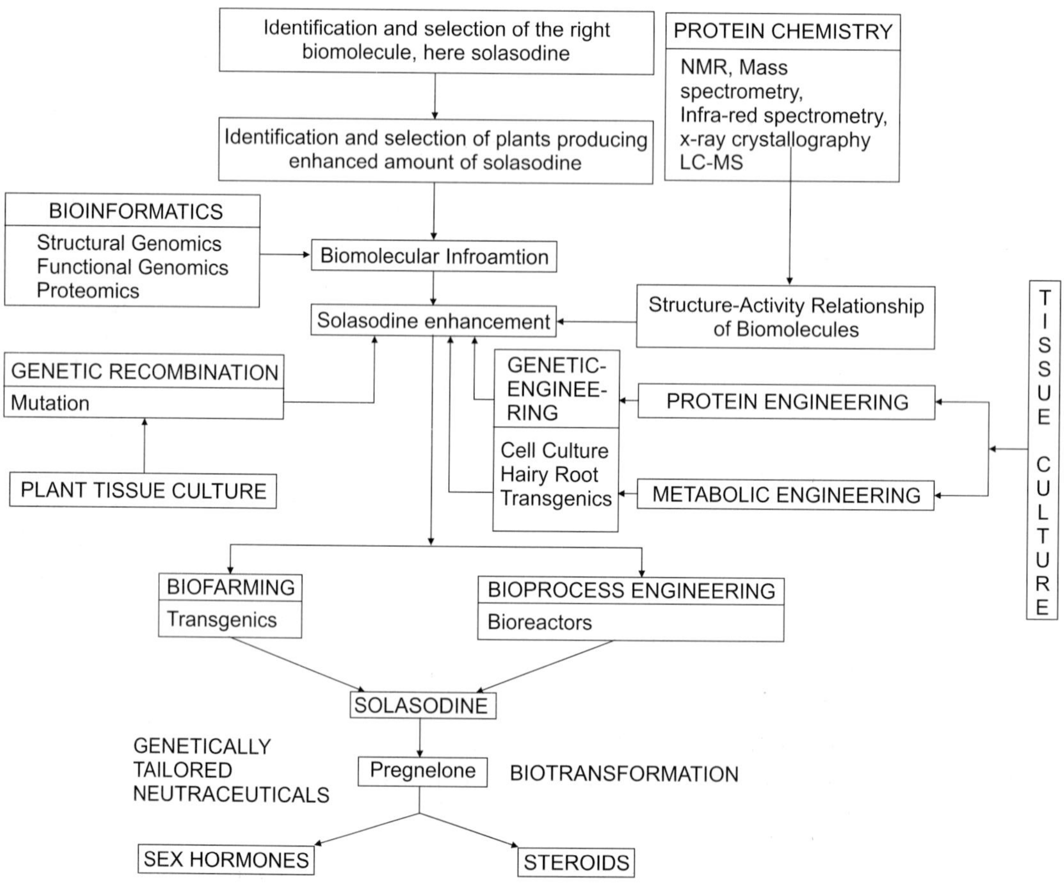

Fig. 6 Biomolecular engineering for maximizing solasodine production, and a subsequent synthesis of the hormones and steroids.

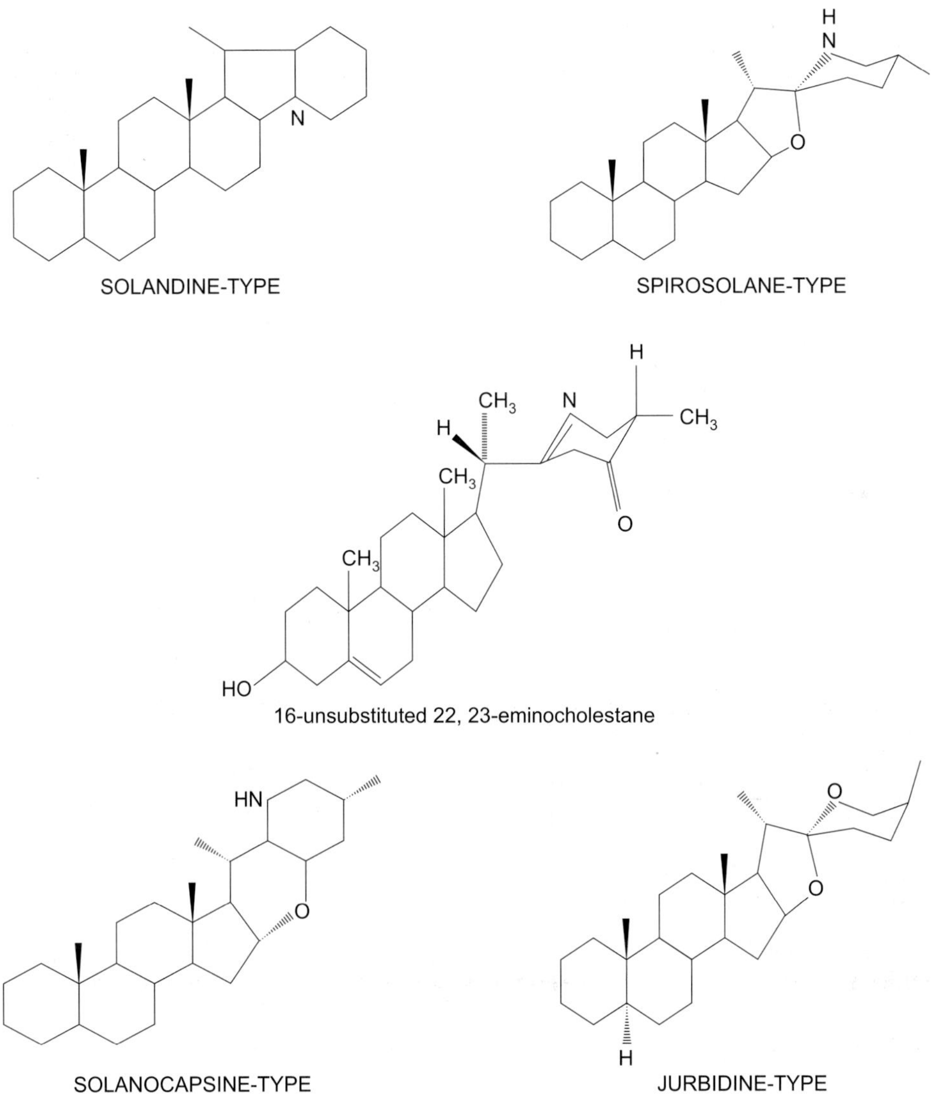

Fig. 7 Different steroidal alkaloids found in *Solanum* species.

extraction of solasodine [1]. *Solanum laciniatum,* Ait have been cultivated in Thailand for this steroidal precursor [12]. Several other workers from different countries are examining the prospects of obtaining increased levels of *Solanum* alkaloids by developing suitable *Solanum* species.

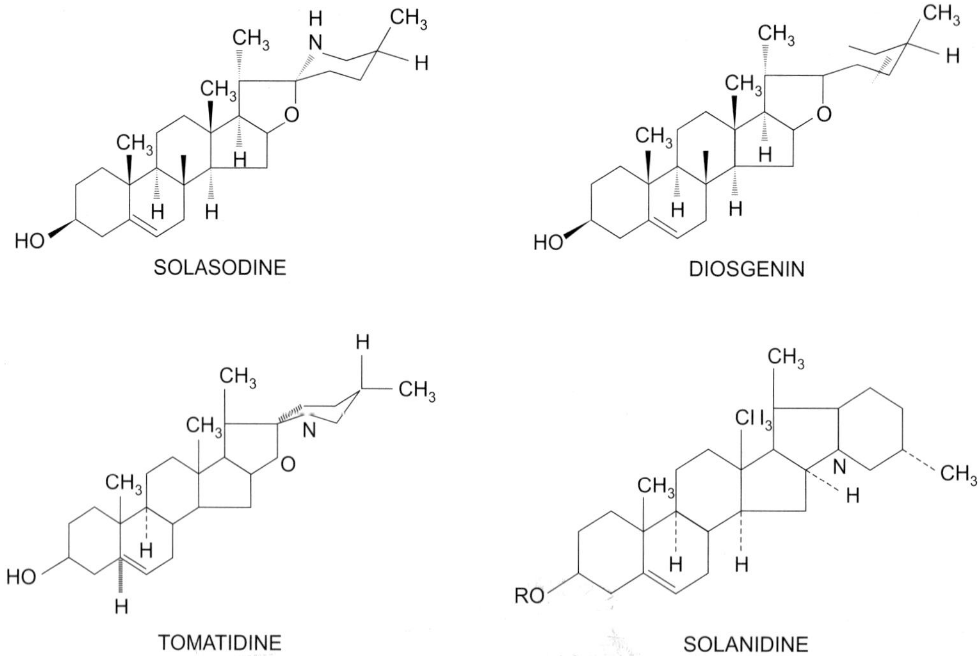

Fig. 8 Chemical structure of the aglycones to show their structural similarity

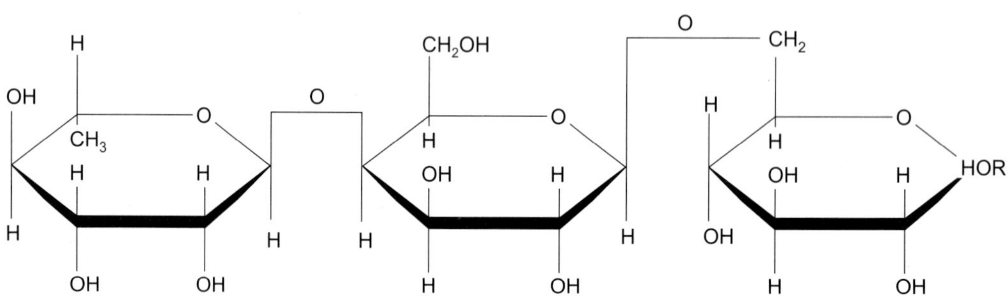

SOLANOSE

Fig. 9 Structure of solanose; the trisaccharide attached to solasodine to form the glycoalkaloid, solasonine [14]

COMMERCIAL PRODUCTION OF SOLASODINE FROM THE SOLANACEAE MEMBERS

In India a number of pharmaceutical companies are involved in the production of solasodine. They extract the alkaloid from berries of *S. khasianum*. High solasodine yielding varieties have been generated through mutation breeding [18]. Improvements over the wild varieties were observed in reducing the spines or prickles to make hand picking of fruits easy, increasing the number of berries per plant, increased solasodine content per gram dry weight and synchronous maturity of fruits for maximum yield. Tetraploids showed great variation in terms of solasodine content and showed delayed flowering and fruiting and the fruit size was smaller thus it is not useful to induce tetraploidy in *Solanum* crop [21]. It was clear that the range of variation in solasodine content increased significantly in M3 after applying a vigorous selection pressure. It was thus possible to shift the mean value of alkaloid content towards a positive direction.

After transplanting, the crop generally takes about six months for harvest. It is essential to dry the berries and leaves as fast as possible to avoid fungal attack as well as metabolic degradation of glycoalkaloid, which is associated with a slow and prolonged drying. Solasodine is prepared from solasonine by hydrolysis of the plant part with 3% hydrochloric acid at 100°C. The cooled reaction mixture deposits the sparingly soluble hydrochloride from which the free base may be regenerated. The technology requires amongst other equipment, glassed steel fermenters, high-pressure stainless steel hydrogenation reactors, stainless steel fermenters, stainless steel solvent distillation units and a range of engineering services including compressed air, cryogenic systems, steam, electricity, tower and auxiliary equipment and installation [22].

USE OF BIOTECHNOLOGY FOR SOLASODINE PRODUCTION

Solasodine Production Using Tissue Culture

Plant tissue culture helps in developing speedy production of a large number of genetically identical plants in a reliable commercial process. This is also the prime requisite for genetic transformation and development of transgenics for increased solasodine content. There have been a number of attempts to produce tissue cultured *Solanum* plants for large-scale clonal multiplication of the species in order to facilitate scale up of solasodine and overcoming problems of conventional breeding. Solasodine can be obtained from the tissue cultures of *Solanum* plants [23, 24, 25, 26, 27]. The two methods used are organogenesis and embryogenesis. The choice of either method for a large-scale cultivation depends on the success rate of the method for producing plants at a realistic cost and the conditions available in the laboratory [28]. Effective methods in direct and indirect organogenesis have been developed for micropropagation of a number of *Solanum* plants using various explants. *S. surattense* [29, 30] *S. khasianum,* [31], *S. paludosum* [32], *S. assuletisimum* [33], *S. nigrum* [27] are some examples of direct organogenesis from various explants such as leaf and stem. The first evidence for the production of solasodine was in the callus culture of *S. surattense* [34]. Callus cultures of a number of other *Solanum* species were raised and used for the solasodine content. *S. lanciniatum* [35, 36], *S. acculeatissimum* [37], *S. jasminoids* [38], *S. verbascifolium* [39], *S. aviculare* [40], *S. nigrum* [41], *S. dulcamara* [42], *S. chacoense*, and *S. tuberosum* [43], *S. khasianum* [23, 24]. *S. surattense* stem calli produced more solasodine than leaf calli [44]. *S. khasianum* seedling callus contained higher amount of solasodine than callus from roots. The seed-callus of *S. khasianum* contained higher amount of solasodine than present in seedling callus. Solasodine content in callus cultures of *S. khasianum* showed to be maximum in the sixth week [45]. Therefore, it is clear that explant source and age affect solasodine production. Anther and embryo cultures have been worked out in *S. khasianum* [46] and *S. surattense* [47] but no reports on solasodine content in them.

High rate of plant regeneration from cultured protoplasts of *S. nigrum, S. bulbocastanum* [48], *S. laciniatum* [49], *S. khasianum* [50], *S. nigrum* [51, 48], *S. brevidens* [52], *S.phureja* [53], *S. tuberosum* [54], *L. esculentum* [55]. This is extremely useful in developing trasgenics using direct gene delivery methods [56]. A number of reports on interspecific protoplast fusion are present. This technology helped in developing disease resistant hybrids [57, 58, 59, 60, 61, 62, 63, 64, 65]. This could also have a marked effect on altering the secondary metabolism of the plant and would make an interesting observation. Regeneration of whole plants from isolated stem parenchyma cells of *S. sisymbriifolium* has been seen [66]. Plant cell suspensions of *S. melongena* have been seen to be extremely effective in large scale cloning [67].

Somatic embryos (SEs) enjoy a number of advantages such as possibility of obtaining transformation in a large number of SEs simultaneously and obtaining clones for solasodine production, their convenience for large scale growth and multiplication in bioreactors [68], sorting, separating and harvesting of them using modern techniques such as digital image analysis [69], production of synthetic seeds in which SEs can be encapsulated in a matrix for its direct and automated seeding [70, 71]. This makes SEs the obvious choice for mass cultivation. Somatic embryogenesis and plant regeneration in *S. melongena* has been seen to be extremely effective for its large scale cloning [69]. SEs of *Solanum tuberosum* has been reported by a number of workers [72, 73, 74]. Recently it has been found that somatic embryo formation in *Arabidopsis* and eggplant is associated with expression of a glycine-rich protein gene (*Atgrp-5*). This gene is expressed in cells that undergo the first anatomical modifications leading to somatic embryos. Therefore the *Atgrp-5* may be useful in distinguishing cell masses competent against a background of non-embryogenic cell population [75].

The tissue culture protocols for the formation of plantlets, somatic embryos and microtubers *in vitro* [76] have significant effects on multiplication rates, tissue quality and acclimatization success.

Efforts have been made for the production of solasodine using cell and lately hairy root cultures of various *Solanum* species. Such a plant free system would enable a scale up in suitable bioreactors with minimum growth phase (when compared to the plant). Other advantages such as seasonal independence, pest/pathogen free fermentation, enhancement of solasodine using optimum physical and nutritional factors, elicitors and precursors followed by subsequent permeabilization and downstreaming processing in solasodine production. The capability of biotransformation of the steroidal alkaloid into the required steroids and sex hormones using various microorganisms [12,13] and cell cultures shall make it possible to shift the present technology of semisynthesis of the steroids and sex hormones to a totally biological one. Cell and hairy root culture form the heart of such fermentation and therefore the work done so far on these two systems shall be dealt with separately.

Solasodine Production Using Cell Suspension Culture

Cell culture is a well worked out systems and can be used as an alternative to whole plant as a biological source for potentially useful compounds. This was demonstrated in *Lithospermum erthrorhyizion,* for the production of shikonin [77] in 750 L bioreactors by Mitsui Petrochemicals, Japan, *Echinacea purpurae* and *Panax ginseng* suspension cultures in 75,000 L bioreactors [78].

The major advantage of *in vitro* culture systems would be: production of high valued fine chemicals under controlled conditions free of contamination, pests or insects, often on synthetic media. Many workers had tried manipulation of media and environment for maximum productivity. The undifferentiated cells provide simple organization; minimum problems of translocation, permeability, segregation of metabolic pools, hence show good precursor incorporation. They also have short cycles and no seasonal variation [79, 80]. Taking explants from parent plants, which accumulate high levels of secondary metabolite [81], can make improvements in the secondary metabolite productivity of plant cell cultures. There is a possible correlation between stress and secondary metabolism in cultured cells [82]. It is plausible that external factors have a high modulating effect on secondary metabolism. Biosynthesis of secondary metabolites sometimes does not take place when there is active growth of cell culture or transformed root culture. This may be because the secondary product synthesizing mechanisms are temporarily inactive and a switching system controls their operation [83]. The latter seems to be maximum when there is an increase in differentiation and decrease in growth [84, 85, 86, 87]. Owing to the inverse relationship between growth and metabolic production, the amount and form in which key nutrients and hormones are supplied greatly influence the switching mechanisms responsible for activating respective pathways [88]. Addition of specific cell-cycle inhibitors such as hydroxyurea to uncouple growth from production has been found to be successful in increasing solasodine productionin in *S. avicurare* [89]. It is understood that hairy root culture shall show increased production of solasodine when compared to the cell culture because of its higher degree of differentiation, but it is well known that cell culture exhibits higher growth rate because of which there shall be increased biomass production for feasibility of growth and solasodine production in a bioreactor. Therefore it is important to study each of these simultaneously.

Nutritional factors especially the carbon, nitrogen and phosphate source has a major effect on the solasodine content. The steroidal alkaloid concentration increased when sucrose concentration was increased from 3% to 4 - 8% in leaf-derived callus of *Solanum laciniatum* [90] and to 4 - 8% in culture of *S. nigrum* [41]. In shoot cultures of *S. laciniatum* the reduction of sucrose concentration and addition of L-arginine, casein hydrolyzate, banana powder in the medium increased the solasodine content [91]. In *S. laciniatum* phosphate depletion resulted in a rise in solasodine content callus cultures. A decrease in the nitrogen content to one-eighth of as used in MS medium increased solasodine production in callus cultures derived from *S. laciniatum* [90]. In suspension cultures of *S. elaeagnifolium* maximum solasodine was obtained with 2:1 ratio of nitrate: ammonia. Further it was seen that NH_4NO_3 was a better nitrogen source than NH_4Cl and $(NH_4)_2SO_4$ [92]. Reduction in the strength of Murashige and Skoog's medium could increase solasodine productivity in the callus cultures of green berry of *S. mauritanum* by half [93, 94]. The addition of growth regulators affects the rate of cell growth, differentiation and product formation [94]. In *S. nigrum* calli, IAA in the medium increased solasodine production [41] whereas 2, 4-D was found to enhance solasodine production in stem and leaf callus cultures of *S. indicum, S. khasianum* and *S. surattense* [44]. Solasodine in callus cultures of *S. laciniatum* was increased with abscisic acid [90].

Solasodine content of cultured cells decreases with subsequent subculturing. Therefore cell cloning and selection of high producing clones is undoubtedly a very useful technique to increase the level of secondary metabolites. However, it is not obvious why cultures contain both high and low yielding cells. Only a few papers concerned with possible mechanisms have appeared. Some indicate that the lack of specific enzyme(s) represents the most important reaction for the inability of plant cell cultures to produce secondary metabolites [95]. Cells with enhanced levels of solasodine were obtained when cell suspension cultures of *S. surattense* were treated with 60 Gy of gamma rays [96].

Light conditions during incubation affect solasodine content of cultured tissues. Solasodine content in *S. elaeagnifolium* calli obtained from seedling was significantly lower in darkness than in calli obtained under light conditions [97]. Contrary results have been observed where dark grown cultures produced more solasodine than light grown ones in *S. laciniatum* [98] and differentiated tissues of *S. nigrum* [41]. *S. tuberosum* showed greening of tubers and sprouting under certain environmental stress increased production of the glycoalkaloids [98]. The chloroplasts isolated from these tubers were able to synthesize the steroidal alkaloid solanidine from CO_2 [98]. Cell suspension cultures of *S. dulcamara* showed maximum solasodine when there was an increase in chlorophyll content on growing the culture in light [99]. These cultures also produced solasodine besides soladulcidine unlike the shoots, which were devoid of solasodine. Subculturing the green cultures in the dark resulted in a concomitant loss of chlorophyll and alkaloid.

Biotransformation is one of the most commercially realistic approaches in plant tissue cultures because of economic reasons. Here inexpensive precursors are converted to expensive end products [100, 101]. Cholesterol was known to be involved in the biosynthesis of solasodine [102]. Manifold increase in solasodine content was reported when cholesterol was included in the medium in suspension cultures established from seedling callus of *S. surattense and S. aviculare* [23].

Cell suspension cultures of *S. surattense* also showed manifold increase in solasodine when supplied with cholesterol. Unfortunately cholesterol also exhibits a feedback inhibition on production of sterols [103]. Therefore, it is important to optimize the concentration of cholesterol supplied in the medium as a precursor. Cholesterol has been found to be an effective precursor for both diosgenin and solasodine [104]. Supplementation of the same in the medium increased solasodine content in callus cultures of *S. nigrum, S. verbascifolium* and *S. khasianum.* Suspension cultures of *S. xanthocarpum* showed enhanced production of solasodine on supplementation of exogenous cholesterol [23].

Cells in suspension culture tend to grow slowly and tend to form large aggregates in suspension cultures. Therefore, immobilization of cell culture was seen to be an advantage [105]. This technology is a rapidly expanding area of biotechnology, especially in the production of secondary metabolism [83, 106, 107, 108, 109, 110]. The technique of immobilization involves the entrapment plant cells in a soft matrix of alginate, agarose, agar or carageenan [105]. Immobilized plant cells are similar in many respects to callus tissues [111] and resemble the situation *in vivo* in that the cells grow together in a particularly organized environment and are maintained in a "stationary" condition with increased stability. A certain amount of aggregation may be important to secondary product formation for many plant species. Immobilized cell cultures facilitate continuous processing, contributing to large increase in productivity and continuous removal of metabolic inhibitors and potentially allow expression of the suppressed pathway. By delinking product formation, an immobilized process permits product optimization without affecting growth. High cell densities, which are formed in immobilized cells, improve accelerated reaction rates as well as volumetric productivity. This greatly increases the potential for the development of commercial plant products. Immobilized cells of *S. aviculare* in polyphenylene oxide beads released steroidal alkaloid into the medium [112]. Similarly immobilized cells of *S. nigrum* in alginate released alkaloid into the medium [107]. *S. surattense* cells have been immobilized in agarose, agar and sodium alginate [113]. Immobilized cells accumulated twofold higher alkaloid content than the free suspended cells. *S. aviculare* cells immobilized within Ca-alginate gel beads enhanced production of scopolin [114]. *Solanum aviculare* cells were immobilized in pectate gel [115]. Immobilized cells of *S. surattense* released increased amount of solasodine into the medium than the cell suspension cultures [116]. Solasodine content in immobilized cells of *S. xanthocarpum* was around 2-fold that of in free cells at the same culture stage. Immobilization in sodium alginate of *S. xanthocarpum* suspension culture of *S. surattense* enhanced solasodine production than the control [115, 116]. Improved production of diosgenin in *Dioscorea delatoidea* was seen in cell cultures immobilized in polyurathene foam [117].

Elicitors are certain signal molecules, which are capable of inducing structural and biochemical responses associated with expression of plant disease resistance. They could be biotic or abiotic. Biotic elicitors are macromolecules originating from host plant (endogenous elicitor) or from the plant pathogen (exogenous elicitor). Many biotic elicitors have been identified as oligosaccharides, glycoproteins, peptides and phospholipids. Elicitor induced responses in plants include rapid ion fluxes which can change the pH of the cell wall, formation of active oxygen species, induction of new gene products and eventual cell death. Recent developments in phytochemical elicitation have shown that simple inorganic and organic molecules can induce product accumulation, if an appropriate elicitor is chosen. Examples of microbial elicitor induction include diosgenin production in the Mexican yam and many others, [118]. Fungal elicitors such as *Alternaria* increased solasodine production in *S. elaeagnifolium* [119]. In *S. surattense* suspension cultures fungal elicitors such as *Aspergillus niger* and *Fusarium udum* enhanced solasodine production [120]. Combining the use of fungal elicitors with immobilization showed stimulation in solasodine production [120].

Plant cell suspension cultures are often treated like microorganisms and can be scaled up in a bioreactor for the production and extraction of secondary metabolites. Cell suspension cultures of *Solanum demissum* has been scaled up in an 800 L bioreactor [121].

Solasodine Production in Root Cultures

Unfortunately cell cultures often differ quantitatively in their spectrum of secondary products from the parent plant and in some cases qualitative changes occur as well. They are also accompanied by an apparent loss of the ability to accumulate secondary compounds due to the lack of organization. The reason may be: 1) the lack of expression in non-specialized cells of genes that control the essential steps in the biosynthetic pathway. 2) The diversion of substrate away from secondary product formation. 3) The non-operation of transport system by which potentially toxic end-products may be removed from the biosynthetic site. 4) The non-availability of storage sites in which secondary metabolites would be normally be sequestered. 5) The unregulated catabolism of synthesized product [122].

The use of organized cultures to study the biosynthesis, transport, storage and turnover of natural products is an attractive approach. Their behaviour is much more predictable when compared to that of cell suspension cultures [123]. Initiation and maintenance of excised root cultures started as early as 1930s but little work was done with regards to secondary metabolite production. Root culture for phytochemicals received a major boost when it was noticed that cell cultures produce only traces of the secondary metabolite but showed a sudden increase in its production when these cultures showed differentiation of roots [124]. Roots are now considered to contain the major sites of synthesis of many secondary metabolites. Roots derived from callus tissue of *S. khasianum* accumulated upto 5.2% steroidal alkaloids including solasonine, solanidine, which were also found in the intact plants [125]. This level is higher than what is in the berries. Steroidal alkaloids and sapogenins from roots of various *Solanum* species have been reported [126]. Root cultures were extremely slow growing and needed an exogenous supply of growth hormones for their growth; therefore hairy root cultures were a welcome replacement.

Solasodine Production Using Hairy Root Cultures

Agrobacterium, the gram negative soil bacteria recognizes some signal molecules such as phenolics or sugars from the wounds of plants and become attached to the cell to cause transformation. Efficiency of transformation is known to differ with different bacterial strains [127, 128, 129]. Virulence genes that form the *vir* region of the Ri plasmid and *chv* genes found on the bacterial chromosome mediate transfer of DNA [130]. It is reported to be the only transkingdom conjugal transfer. *Agrobacterium rhizogenes* contain the Ri plasmid with the T-DNA, while *A. tumefaciens* contains the Ti plasmid. A single copy of T-DNA contains the left T-DNA border (T_L-DNA) and the right T-DNA border (T_R-DNA). The T_R-DNA contains the genes responsible for opine synthesis while the T_L-DNA contains four loci such as rolA, B, C and D, responsible for the induction of hairy roots and its altered hormonal balance [131]. These genes have eukaryotic regulatory sequences enabling their expression in infected plant cells. Hairy roots obtained by infection with different bacterial strains exhibit different morphological properties [132]. The plasmid DNA is stably incorporated into higher plant cells, both by *Agrobacterium tumefaciens,* and *A. rhizogenes* [133, 134, 135]. *Agrobacterium rhizogenes* causes a disease called hairy root syndrome, which is characterized by profuse growth of root which can be cultivated in a simple defined medium without the addition of growth hormones. They are genetically [136] and biochemically stable and thus ensures stable productivity [137, 138]. Hairy roots have been seen to be a boon to harness fine chemicals synthesized in the roots [139, 140, 141, 142] or synthesized in the roots and stored in the aerial parts. *Solanum laciniatum* [141], *S. nigrum* [142, 143], *S. launatum* [144], *S. sysembrifolium* [145], *S. eleagnifolium* [146], *S. mauritianum* [147], *S. aculeatissimum* [148] and *S. aviculare* [149, 150], *S. khasianum* [151], *S. tuberosum* [152, 153], *Lycopersicon esculentum* [154] were genetically transformed by *Agrobacterium rhizogenes.* Transformation was seen to increase solasodine content in the *Solanum* species other than potato. Diosgenin could also be scaled up in hairy root cultures of *Trigonella foenumgraecum* L [155]. Transformed root cultures have been manipulated using different techniques for production of desired compounds. Cultures could be fed with suitable precursors such as cholesterol for the production of solasodine. Light along with temperature can effect the solasodine production in the hairy roots (published elsewhere). Nutrient medium can be manipulated along with the carbon source for increased secondary metabolite production as seen in hairy root cultures of *Solanum aviculare* [156]. Addition of growth hormones to the culture medium induces rapid changes in root morphology [157] and secondary metabolite content [158, 159]. Green roots developed under light conditions could be used for increased secondary metabolite synthesis [160]. Hairy roots of some members of the families such as Cucurbitaceae, Asteraceae besides Solanaceae develop chloroplasts and turn green on exposure to light. The transition to photoautotrophy was accompanied by an increase in CO_2 fixation and a RUBISCO/ PEPCase activity approximating that in the leaves. For example, the total alkaloid content increased several fold in photosynthetic and photoautotrophic hairy root cultures of Datura [160]. These coupled with the perfection of the culture medium, and elicitation would help in increasing the solasodine content. In fact elicitation of root cultures can be done more effectively than that of cell suspensions because of their immobilized nature [161]. Compartmentation plays an important role in the regulation of secondary metabolic pathways [162]. Mevalonate pathway occurs in cytosol [162] and the steroids formed might cause a feedback inhibition on the mevalonate cycle therefore compartmentation of the steroids formed would prevent such a feedback. An extremely useful technique to avoid such a feedback would be *in situ* extraction of solasodine into the culture medium, followed by a removal of the same from the medium. This can be achieved using permeabilizing agents and a continuous mode of fermentation.

Using *A. rhizogenes* mediated transformation any gene placed in binary vector can be transferred to the transformed hairy root clone. The NPT II and GUS gene are example of this, the former is a marker gene and has been incorporated and expressed in *S. nigrum* [163], *S. dulcamara* [164] and *S. tuberosum* [165] while the latter is a reporter gene and integrated and expressed in *S. tuberosum* [165, 166]. The *luc* reporter gene is also expressed in potato [166]. The same authors also transformed potato using particle bombardment; this greatly increases the copy number of the foreign gene. Work is going on in different parts of the world for the use of hairy root culture for enhancing the production of solasodine. For example transgenic roots of *Solanum aviculare* and *Solanum paludosum* containing extra copies of an hmgr gene derived from *Artemisia annua* and driven by CaMV 35S promoter have been obtained via transformation with *Agrobacterium rhizogenes* [150]. 3-hydroxy-3-methylglutaryl-coenzyme A reductase (HMGR) plays a role in the regulation of sterol biosynthesis in plants. Hairy root clones of *Solanum aviculare* which were transgenic for hmgr typically grew faster than those which did not contain extra copies of this gene and also accumulated up to 4.2 times more solasodine when grown under dark, but not light, this could be because CaMV 35S promoter is not light regulated in plants. It is reasonably certain that the steroid biosynthesis in general in both plant and animals takes the same course, starting from acetyl coenzyme A via mevalonic acid. As in the first biosynthetic experiments with radioactive tracers on *Solanaceous* plants [167], the probable path of biogenesis is from cycloartenol via lophenol and cholesterol to the heterocyclic steroids. By administering labeled cholesterol it was shown further that it was a precursor of tomatidine [168]. Mevalonate is formed by the reduction of 3S-3-hydroxy-3-methylglutaryl coenzyme A (HMG-CoA) undergoes reduction and forms mevalonate. This reduction is catalysed by HMG-CoA reductase. Mevalonate is then phosphorylated twice and decarboxylated to form isopentyl pyrophosphate (IPP). IPP and one, two or three molecules of dimethylallyl pyrophosphate (DMAPP) condense together and form farnesyl pyrophosphate (FPP) which are then rearranged, cyclized, hydroxylated and modified by highly specific enzymes to form squalene, which gives rise to sterols such as solanine, tomatine [169] and perhaps solasodine and diosgenin. The formation of squalene from FPP is catalysed by the enzyme squalene synthase [170]. The regulation of this enzyme plays an important role in the direction of intermediates to the sterol pathway. A better understanding of the regulation of squalene synthase at the molecular level was possible through advances in cloning of the squalene synthase gene. It has been cloned in man, yeast, rat, mouse, *Schizosaccharomyces pombe, Arabidopsis thaliana* and *Nicotiana benthamiana.* Squalene epoxidase (SE) and oxidosqualene: lanosterol (OSLC) are two pivotal enzymes in cholesterol biosynthesis [171]. Regulation of the levels of these enzymes *in vivo* modulates cholesterol biosynthesis. The cDNAs encoding several SEs have been cloned and sequenced from man, rat, mouse, pig, yeast and candida. While oxidosqualene: lanosterol cyclase (OSLC) catalyses the conversion of (3S)-2,3-oxidosqualene to lanosterol; the precursor to cholesterol. Several lanosterol synthases have been cloned and sequenced from man, rat, yeast, *Candida albicans, Schizosaccharomyces pombe* and an oxidosqualene-cycloartenol cyclase (OSCC) from *Arabidopsis thaliana* [171]. The availability of both mutations and genes responsible in Arabidopsis sterol biosynthetic pathway raises the possibility of manipulating plant sterol biosynthesis [172]. This biogenesis of solasodine is illustrated in Fig 10. Cholesterol is known to have a feedback inhibition on HMG Co A reductase [103], one of the first enzymes in the sterol biosynthesis and cloned in *S. aviculare* [150]. Therefore it is important to understand the conversion of cholesterol to the next intermediatory compound, which was found to be 26-Hydroxycholesterol as in the production of diosgenin [173] followed by the formation of E and F rings. Unfortunately the regulatory genes that control this conversion is not known so far.

Another approach, which is rather indirect, is to introduce genes that are elicitor responsive and also upregulates the secondary biosynthetic gene. Plants respond to pathogen attack by induction of various defence responses, including the biosynthesis of protective secondary metabolites. In *Catharanthus roseus* there are two genes that are coordinately expressed on fungal attack *viz.* TDC (Tryptophan decarboxylase) gene and STR (Strictoside synthase) gene [174, 175]. These genes when introduced into tobacco plant resulted in the production of strictoside [176]. Similarly genes from the pathogen are also introduced into the plant to constantly activate the secondary metabolite pathway thereby fighting a virtual pathogenic attack. [177], provide a good example of this. Here the gene coding for ODC (ornithine decarboxylase) from yeast (*Sacchromyces cereviseae*) is introduced into *Nicotiana rustica* and a twofold increase in nicotine is seen. A number of genes in potato [178, 179] are isolated that are induced in response to pathogenesis. It would be interesting to see the effect of these genes on the glycoalkaloid production in these plants and also inserting the genes in *Solanum* plants to see its effect on solasodine production.

Transgenics as Bioreactors for Solasodine Production

Foreign DNA sequence are stably integrated and inherited in a mendalian manner and transfered to the progeny [180, 181, 182, 183, 184]. Transformed roots were able to regenerate plants that maintained their genetic stability during further subculturing and plant regeneration in potato [165], tomato [185, 186], *Solanum dulcamara* [164], *S. eleagnifolium* [146], *Solanum khasianum* [151]. Fig. 11 shows the use of genetic transformation for biofarming using transgenics as well as

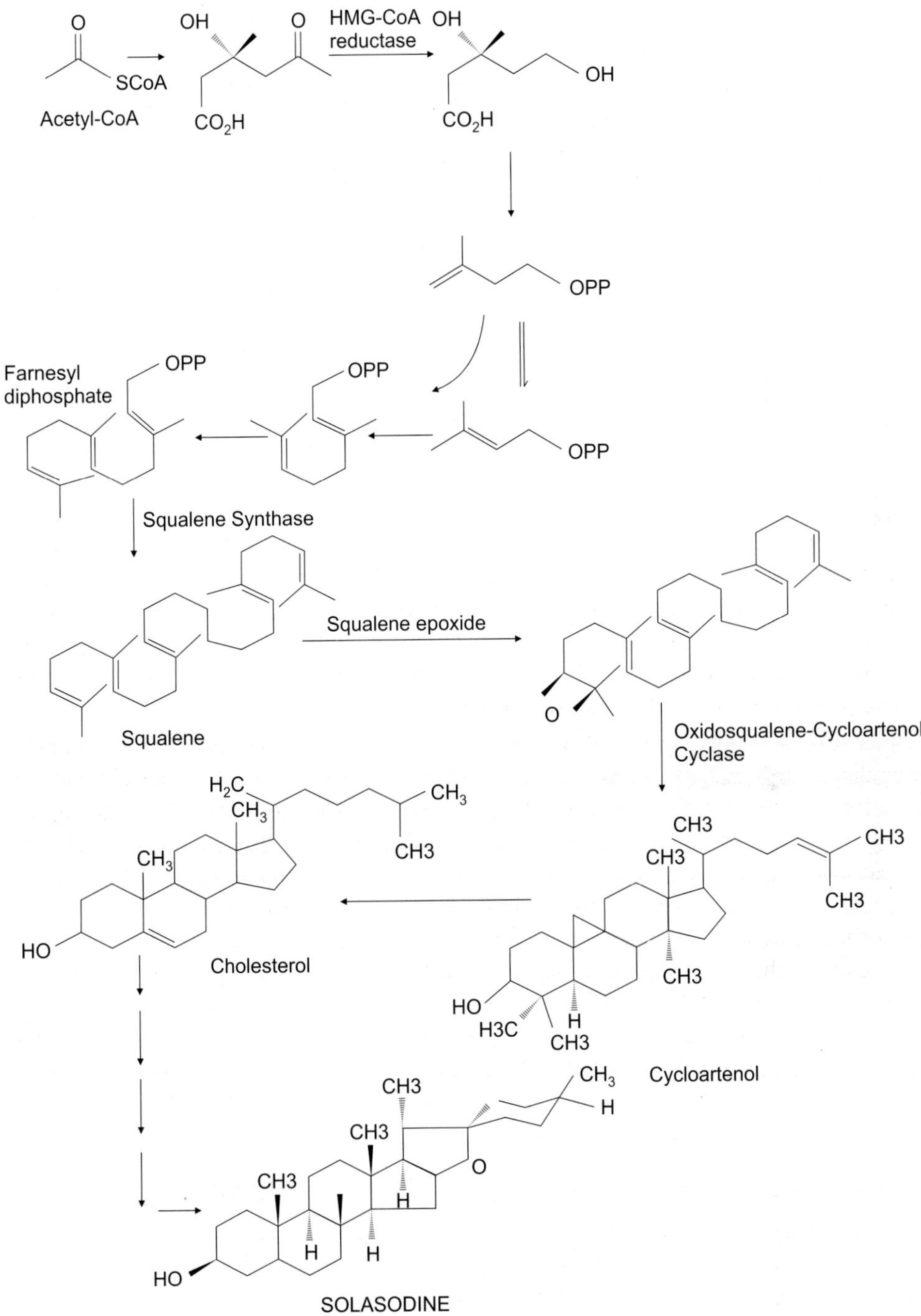

Fig. 10 Biogenesis of solasodine.

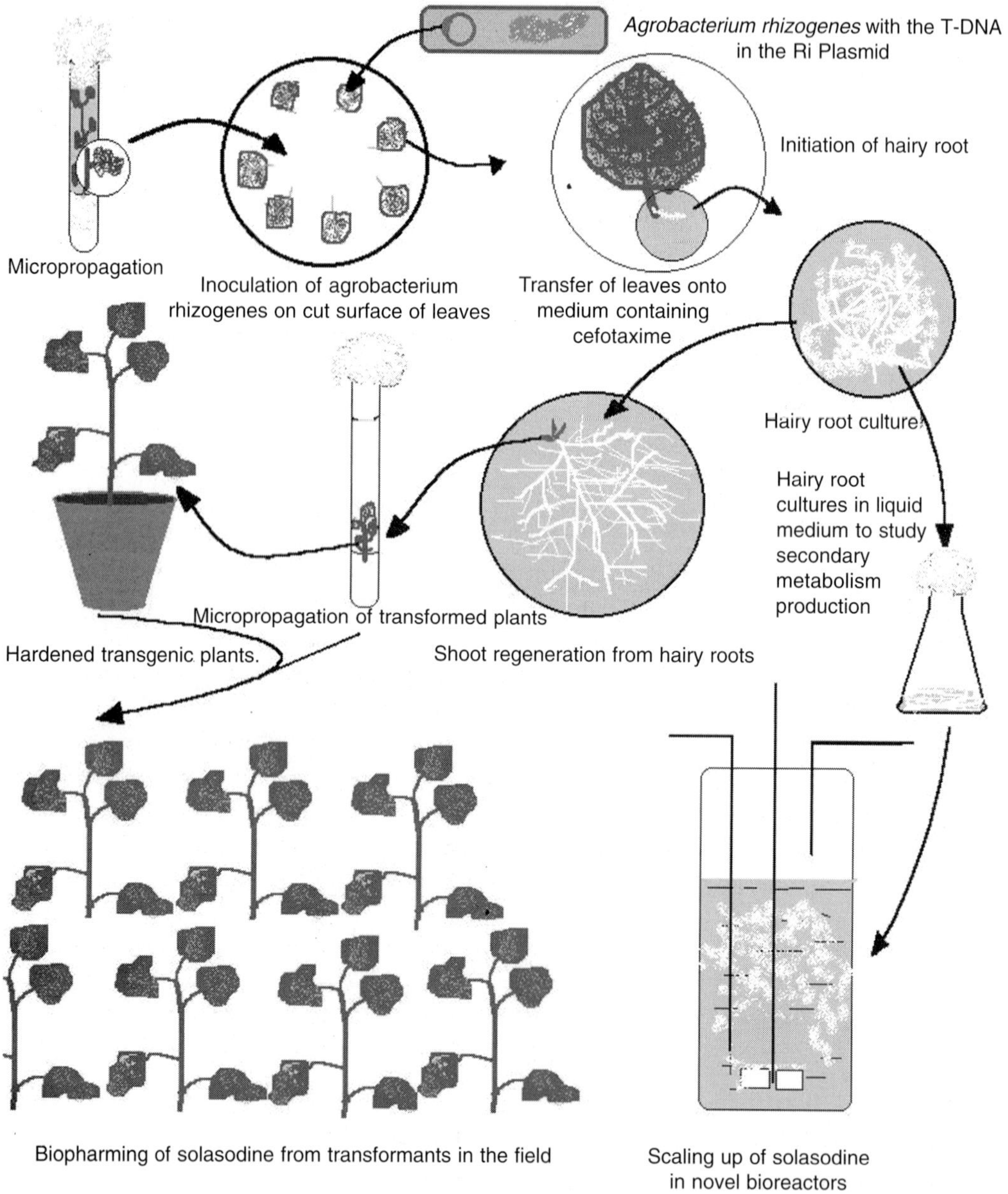

Fig. 11 Schematic diagram of scaling up of solasodine using hairy root cultures in bioreactors and biopharming of the steroidal precursor from transgenics.

bioprocess engineering using hairy root cultures. Having introduced foreign genes responsible for traits such as disease resistance and increased solasodine production in a binary vector, this could be an efficient method of generating plants with elite traits. Transgenic plants can be the commercial alternative to traditional production methods [187, 188, 189, 190, 191] and can be rightly called bioreactors. They can be grown using normal agronomic practices, and a scale up of the steroidal alkaloid would just require increasing the acerage under its cultivation. Unfortunately transgenics also show gene silencing. It is important to look for transgenics that would be stable over the generations and show minimum transgene silencing; both transcriptional gene silencing (TGS) and post-transcriptional gene silencing (PTGS) which arise due to factors such as, a positional effect or homology between the transgenes or between the transgene and endogenous gene or even between the promoters in case of multiple copy number, RNA degradation, methylation etc. [191].

Therefore it is clear that gene silencing disrupts the economic exploitation of these plants as factories [192]. This makes it essential to obtain a large number of transformants and to screen them for the lowest probablity of transgene silencing. A single copy number would greatly reduce the risk of homology and silencing. Careful designing of vector, transformation technique, selection of transformants based on their molecular characteristics and using promoters that are developmentaly regulated and tissue specific should reduce the number of transgenics sensitive to gene silencing [191]. In some instances the cultivation of transformed cell culture could be an alternative as the cells are not interconnected and can be cultivated in a bioreactor. The conditions, both physical and nutritional can be controlled for the expression of the transgene. The example of this silencing was seen in case of hmgr gene in *S. aviculare* [150] where the promoter was light sensitive and showed an increase in solasodine content only when the hairy root cultures were grown in the dark. Plants being photoautotrophic such an expression would not be possible in transgenics.

Agrobacterium changes the genetic make-up of the plant [131] thereby affecting its secondary metabolism [193]. Therefore it is important to screen a large number of transformants for its solasodine content. This is also applicable to cell cultures on treatment of mutagens. In order to do so it is necessary to have rapid screening methods of these cultures for their solasodine content. Radioimmunoassay [194] offers a fast method of screening. On the other hand, total solasodine glycosides were separated and quantified by an immunoaffinity column using a wide-cross reactive anti-solamargine Monoclonal Ab [195], this can also be used to formulate techniques in ELISA and Western Blotting to be able to screen a large number of samples at a time. HPLC [16, 196, 197, 198], and GC by [199] are some of the other methods for accurate quantification of solasodine, IR spectroscopy [200].

Scale up of Hairy Root Cultures in Bioreactors

Use of bioreactor is the answer to any scale up attempt. It involves the usage of various disciplines such as biochemistry, cell and molecular biology, bioorganic and bioinorganic chemistry and has at its core, chemical engineering. Here living organism or a part of it is used for the production of chemicals or biological materials or for the development of new processes. When considering economically feasible production of secondary metabolites by employing hairy root cultures, a process based on continuous operation with release of secondary metabolites into the medium is essential for low priced products. Although with more expensive products, batch operation with standard harvesting of biomass might be sufficient [140]. Paralleling these advances in fermentation, the area of separation and recovery of biological products has seen the development of techniques such as ion exchange, affinity and gel permeation chromatography, electrophoresis, aqueous two-phase separation process and membrane separation process. These are based on biophysical chemistry, colloidal and interfacial phenomena and mass transfer [201]. One of the important limitations for commercial exploitation of hairy roots is the lack of development of technologies for scaling up of root cultures in bioreactors. The meristem dependent growth of root cultures establishes a distinct spatial and temporal order, thereby making the growth of the roots in a bioreactor difficult. These features call for specifically designed novel bioreactors that permit the growth of interconnected, nonhomogenous root tissue, normally unevenly distributed throughout the culture vessel. Thus stable reactor systems are required for better results. A wide variety of bioreactors have been employed for the growth of various hairy root cultures: *N. rustica* grown in air sparged bioreactors [202, 203], *D. stramonium* in stir tank reactors of 12 L capacity [204], *T. patula* and *B. vulgaris* grown in a 6 L bubble column reactor [205, 206], *C. sepium and A. belladonna* grown in a 1 L stir tank reactor with a mesh [207]. Hairy roots of *S. tuberosum* have been scaled up in a 15 L bubble column [208]. In our lab we studied the growth kinetics of hairy root cultures of *S. khasianum* in various bioreactors when compared to the shake flask. It was seen that although bubble column stir tank was the most suitable configuration for the growth of hairy roots of *S. khasianum* when compared to bubble column and mist bioreactor, it still could not achieve the mass transfer efficiency seen in shake flasks. These roots were seen to be extremely sheer sensitive, because of which growth was not as much as what in shake flasks. This calls for an improvement in the design of bioreactors for hairy roots in general and *Solanum* species in particular. A number of Solanaceous members leach out secondary metabolite into the medium. Here the metabolite in concern is usually transported to the aerial parts of the plant and stored there, in case of a root culture or hairy root culture the medium functions as the sink and facilitates leaching of the metabolite into it [209]. This would help us to extract solasodine from the medium and reuse the root biomass Fig 12.

SAFETY OF PLANT BASED STEROIDS WITH SPECIAL REFERENCE TO THOSE OBTAINED THROUGH CELL OR HAIRY ROOT CULTURE

Steroids from *Solanum* were found to be safe and already being manufactured commercially in a number of countries [210]. Fully ripe fruits of *Solanum aviculare* are consumed either raw, boiled or baked by the *Aborgenies* in certain parts of Australia [210] while the leaves of *Solanum nigrum* is eaten in Kenya by the *Piik ap Oom Okiek* tribe [211]. Its a constituent in their daily meal which means that it is safe to be consumed unlike green potatos which are known to be fatal due to the high content of solanine. Solasodine and solasonine are also known for their haemolytic properties. They also possess antifungal and cytostatic properties. Cell cultures were reported to produce solasodine but in an extremely low concentration when compared to the plant. Solasodine from hairy roots or any other part of the transformants are analyzed to be structurally similar (to be published elsewhere by the authors) to that in the plant and need not be tested for toxicity as it is the secondary metabolite

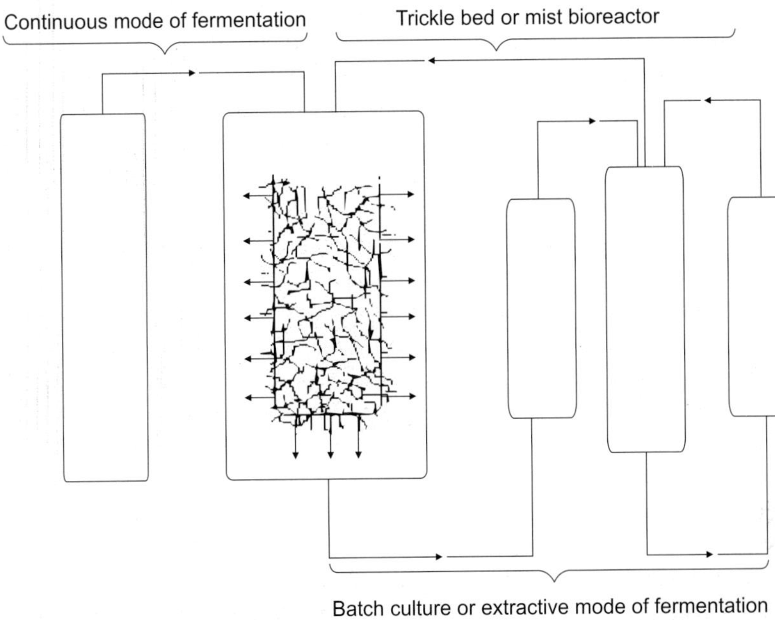

A: Novel bioreactors for hairy root culture with permeabilization or *in vitro* extraction of the glycoalkaloid, **B:** Biotransformation of the glycoalkaloid to its pharmaceutically important aglycone, **C:** Extraction of solasodine, **D:** Concentration and solvent recovery, **R:** reservoir for medium

Fig. 12 Scale up and downstream processing of solasodine from hairy root cultures or for cell cultures too.

from the plant that is being extracted and used as steroid precursor. If the genetically altered plant is to be released to the farmers for commercial cultivation then it needs to be tested for its safety, as it happens to be a Genetically Modified Organism (GMO). Although transgenic plants world over are now being approved for use in agriculture and as food. Never the less there is the responsibility to use them safely for the benefit of humanity and the environment. Genetically engineered medicines and a number of vaccines are already in use [212]. Solasodine too obtained through hairy root cultures or transgenics should be seen as an easy and fast mode of increased production of phytosteroids. Plants are an obvious choice to hairy roots, cell culture or even microorganisms as it does not require expensive fermentation procedures, and therefore enjoys a low cost of production and rapid scale up. Expression of vaccines, therapeutic proteins, nutraceuticals, industrial enzymes and other biopolymers has been achieved in different plants. Some of which are for healthcare and is to be consumed as nutraceuticals or pharmaceuticals. These are being subjected to clinical trials, and are already on the verge of commercialization [191].

This new wave of genetically altered plants and root culture could turn the field and fermentors into production lines for better pharmaceuticals.

FEASIBILITY OF BIOTECHNOLOGICAL PRODUCTION WHEN COMPARED TO THE CONVENTIONAL PRODUCTION

In practice there is considerable variation in the management of individual farms and that is reflected in the cost of cultivation and the profit. The seeds are known to lose their viability within two months of harvest, besides this they are also known to have dormancy. The steroidal alkaloid obtained does not show consistency owing to cross-pollination and the presence of chromosome B that leads to genetic instability. Newer pests (fruit borer, root eating grubs, leaf eating caterpillars and leaf hoppers) and diseases (leaf spot disease, wilting caused by *Fusarium oxysporum*) have been reported on this crop [1]. Drying and attack of berries by fungus causes loss of crop. Lack of information on several aspects of agronomy like optimum plant population, levels of nutrients, and effect of fertilizers on solasodine content etc. Synchronous maturity of fruits for maximum solasodine production is far from reality. These factors make it extremely difficult to cultivate *Solanum* plants for solasodine production thus making its biotechnological production a lucrative alternative. However a strategic analysis of several plant-based compounds show that there is the need to increase growth rate of cultured cells and hairy roots, in order to make it economically productive. Furthermore the steroid market being very huge requires large volumes to provide enough quantity of product to meet the global demand.

FUTURE OF SOLANUM STEROIDS

It is understood that the future of the steroid industry, world over shall be towards *Solanum* alkaloids. Understanding the steroid metabolic pathway, and using "metanomics" or "metabolic engineering" in which the metabolic pathway of the plant is directed for producing increased amount of solasodine, shall be an area of interest. Metabolically engineered plants with higher solasodine content, coupled with disease and stress resistance will be grown. If the regulatory gene that controls the conversion of cholesterol to the solasodine is isolated then the feedback inhibition of cholesterol can be overcome resulting in an increased production of solasodine. It is now well known that solasodine is developmentally regulated in plants, that is to say that the genes responsible for solasodine production are expressed in certain tissue more and at a specific time of development. Solasodine in *Solanum* plants are found to be more in the berries [1], solasodine also seems to be developmentally regulated as the steroidal glycoalkaloid was seen to decrease and finally disappear on ripening of berries [213] solanine content in potatos is also seen to be more in green tubers exposed to light than the ones grown under the soil [98] tomatine too shows a similar trend in being more in green berries than the ripe ones [214]. Therefore, it would be possible to isolate the genes responsible for solasodine production by differential cloning and these could be sequenced to alter regulatory code for tissue specificity and developmental regulations and reintroduced into cultured cells or whole plants. Thus these genes could be overexpressed. These alkaloids are related to chloroplast as was seen in isolated chloroplast of potato that were found to contain solanine [98]. Increasing the chlorophyll content could also result in an increase in the alkaloid content. Using a promoter that is light regulated would probably further increase solasodine content in transgenics by combining the beneficial effects of both hmgr gene and light.

Inserting defence related genes in *Solanum* plants and looking for an increase in solasodine or inserting genes such as ODC from the pathogen for an increase in the steroidal alkaloid is yet to be explored.

Genetic engineering along with large-scale cultivation methods for plant cells and transformed roots could lead to the commercial production of solasodine, devoid of plants. Optimizing the culture conditions, precursor feeding and using both abiotic and biotic elicitors could spell the success of its cultivation in a bioreactor. Perfection of the bioreactor suited for *Solanum* hairy root cultures with online monitors are required. Microbial biotransformation of solasonine to solasodine, and further to the steroids; or identifying *Solanum* species that would produce free solasodine without the glycosides, and in substantially high quantity or engineering the plant not undergo glycosylation would be seen as a major achievement.

Many higher plants have the ability to degrade sterols to pregnenolone and to convert the latter to progesterone [215]. Cell suspension cultures of *Nicotiana tabacum* or *Sophora augustifolia*, were also seen to show interconversion of pregnenolone to progesterone [216], and of androstenedione to testosterone [217]. Biotransformation of solasodine, tomatadine and diosgenin to estrone using either a single microorganism or a series of them; as in the case of diosgenin [13] is a possibility. If the genes responsible for these could be isolated from these plants and microorganisms and reintroduced into tomato or a *Solanum* plant, such as eggplant, could give birth to a new era of 'nutraceuticals' for natural supplements of progesterone for older women, estrone as a natural contraceptive and testosterone, the male hormone. With the advances in molecular techniques, expressed sequence tags (EST), genome sequencing programs and bioinformatics this shall soon be achieved. Biosynthetic and regulatory genes shall be sequenced and cloned in the plant to suit our need. The need for fermentors or elaborate extractive procedures shall probably be a thing of the past with the advent of tailor made *Solanum* plants. It is clear that these shall be some of the areas of research for plant based steroid production in the years to come.

REFERENCES

1. Asolkar L.V. & Chadha Y.R. (1979) Diosgenin and other steroidal drug precursors. Publications and information Directorate. New Delhi: *CSIR*. pp. 60-92.
2. Habermehl G. (1967) The steroid alkaloids: The salamandra group. In: Manske RHF, (Ed.) *The Alkaloids*. Vol 9. New York: Academic Press. pp. 427-440.
3. Khuong-Huu F. & Goutarel R. (1973) Steroidal alkaloids of Apocynaceae and Buxaceae. In: *The Alkaloids*. Vol 3. London: *The Chemical Society*. pp. 258-278.
4. Khuong-Huu-Lainé F., Magdaleine M., Santamaria J. & Goutarel R. (1973) Alcaloides Steroidiques Du Malouetia Brachyloba et du Heudelotii M. *Phytochemistry*. 12: 1813-1816.
5. Sato Y., Miller H.K. & Mossetting E. (1959) Improvement in the preparation of 3¼-acetoxy-pregn-16-en-20-one and 3¼-acetoxypregna-5, 16-diene-20-one from the steroidal alkaloids tomatidine and solasodine. *J. Org. Chem.* 24: 893-894.
6. Fakhil H.I. & Hamied Y.K. (1964) Preparation of pseudosapogenins and 16-dehydropregnenolone acetate. *Ind. J. Chem.* 2: 508-509.
7. Sree A., Rao Y.R., Baujii M., Rao K.M. & Mahaptra S.N. (1988) An improved process for the conversion of solasodine to 16-dehydropregnenolone acetate. Indian Pat No. IN 162, 446.

8. Marker R.E. & Rohrmann E. (1962) JACS 1939; 61: 3592, JACS 1940; 62: 518, 521, 898. cf. Applezweig N. Classification of active compounds. In: *Steroid Drugs*. New York, Toronto, London: Mc Graw-Hill Book Company, Inc. pp. 295-310.
9. Glanes I.T., Webb D.T. & Rosario O. (1998) Steroid production by callus and cell suspension cultures of *S. aviculare. Forst J. Nat. Prod.* 47: 373-376. cf. Kittipongpatana N, Hock RS and Porter JR. Production of solasodine by hairy root, callus and cell suspension cultures of *S. aviculare* Forst. *Plant Cell Tissue Organ Culture.* 52: 133-143.
10. Macek T.E. (1989) *Solanum aviculare* Forst, *Solanum laciniatum* Ait (proporo): *in vitro* culture and production of solasodine. In: Bajaj YPS, (Ed.) Vol 7. *Biotechnology in Agriculture and Forestry.* Berlin: Springer. pp. 443-467.
11. Applezweig N. (1962) Classification of active compounds. In: *Steroid Drugs.* New York, Toronto, London: Mc Graw-Hill Book Company, Inc. pp. 295-310.
12. Manosroi J., Sripalakit P. & Manosroi A. (1998) Bioconversion of hydrocortisone to prednisolone by immobilized bacterial cells in a two - liquid - phase system. *The J. Chem. Tech. Biotech.* 73.
13. Iizuka H. & Naito A. (1967) Microbial transformation of sterols. In: *Microbial Transformation of Steroids and Alkaloids.* Tokyo: University of Tokyo Press. pp. 215-229.
14. Prelog V. & Jeger O. (1960) Steroidal Alkaloids: The Solanum group. In: Manske RHF, (Ed.). *The Alkaloids: Chemistry Chemistry and Physiology.* Vol 7. New York: Academic press. pp. 343-361.
15. Schreiber K. (1968) Steroid alkaloids: The solanum Group. In: Manske RHF, (Ed.). *The Alkaloids: Chemistry Chemistry and Physiology.* Vol 10. New York: Academic press. pp. 1-192.
16. Crabbe P.G. & Fryer C. (1980) Rapid quantitative analysis of solasodine, solasodine glycosides and solasodiene by high-pressure liquid chromatography. *J. Chromatogr.* 187: 87-100.
17. Gadwal V.R. (1977) Commercial cultivation of Solanum crop (Glaxo strain). In: Atal CK, Kapur BM, (Ed.). *Cultivation and Utilization of Medicinal and Aromatic Plants.* Jammu-Tawi: Regional Research Laboratory C.S.I.R. pp. 31-33.
18. Kaul B.L. & Zutshi U. (1982) Cultivation of *Solanum khasianum* Clarke for steroids: problems and promises. In: Atal CK, Kapur BM, (Ed.). *Cultivation and Utilization of Medicinal and Aromatic Plants.* Jammu-Tawi: Regional Research Laboratory C.S.I.R. pp. 23-31.
19. Puri R.K. & Bhatnagar J.K. (1974) Studies on genus Solanum—A review. *Pharmacos.* 19: 7-24.
20. Bordoli D.N., Rabha L.C. & Ganguly D. (1976) Some observations on growth and yield of *Solanum lanciniatum* Ait under conditions of Jorhat. *Indian Drugs* 14: 5-11.
21. Janki, Bhatt. cf Kunjithapadam S. (1982) Utilization of raw materials for the production of steroids. In: Atal CK, Kapur BM, (Ed.). *Cultivation and Utilization of Medicinal Plants.* Jammu-Tawi: Regional Research Laboratory C.S.I.R. pp. 51-63.
22. Kunjithapadam S. (1982) Utilization of raw materials for the production of steroids. In: Atal C.K. and Kapur B.M., (Ed.). *Cultivation and Utilization of Medicinal Plants.* Jammu-Tawi: Regional Research Laboratory C.S.I.R. pp. 51-63.
23. Khanna P., Uddin A., Sharma G.L., Manot S.K. & Rathore A.K. (1976) Isolation and characterization of sapogenin and solasodine from *in vitro* tissue culture of some *Solanaceae* plants. *Indian J. Exp. Biol.* 14: 694-696.
24. Kotate C.K. & Radhawan S.S. (1979) Enrichment of *Solanum khasianum* callus generating rootlets with steroidal alkaloids. *Z Naturforsch.* 34: 634-636.
25. Chaturvedi H.C., Chawdhury A.R. & Uddin A. (1979) Solasodine biosynthesis in seed and seedling callus of *S. khasianum* Clark grown *in vitro*. *Indian J. Exp. Biol.* 17(1): 107-109.

26. Chaturvedi H.C. & Sinha M. (1979) Mass clonal propagation of *Solanum khasianum* through tissue cultures. *Ind. J. Exp. Biol.* 17(1): 107-109.

27. Salih S.M., Al M.M.K. (2000) Plant regeneration from *in vitro* leaf and stem tissues of *Solanum nigrum. Dirasat. Agri. Sci.* 27(1): 64-71.

28. Christie J.A., Kozai T., Takayama S. (1995) Automation in plant tissue culture: general introduction and overview. In: Christie J.A., Kozai T., Smith M.A.L., (Ed.). *Automation and Environmental Control in Plant Tissue Culture.* Dordrecht: Kluwer. pp. 1- 18.

29. Kunisaki J.T. *Hort. Sci.* 12: 141. cf. Prasad P.N. & Chaturvedi H.C. (1977) *In vitro* induction of shoots and formation of plantlets of leaf, stem, and root of *Solanum xanthocarpum* Schrad and Wendl. *Indian J. Exp. Biol.* 16(10): 1121-1122.

30. Prasad P.N. & Chaturvedi H.C. (1978) *In vitro* induction of shoots and formation of plantlets of leaf, stem and root of *Solanum xanthocarpum* Schrad and Wendl. *Indian J. Exp. Biol.* 16(100): 1121-1122.

31. Bhalsing S.R. & Maheshwari V.L. (1997) *In vitro* culture and regeneration of *Solanum khasianum* and extraction of solasodine. *J. Plant Biochem. Biotech.* 6: 39-40.

32. El Badaoui H., Morard P. & Henry M. (1996) Stimulation of growth and solamargine production by *Solanum paludosum* multiple shoot cultures using a new culture medium. *Plant Cell Tissue Organ Culture.* 45: 153-158.

33. Nabeta K. (1993) *Solanum aculeatissimum* Jacq: *In vitro* culture and the production of secondary metabolites. In: Bajaj Y.P.S., editor. Vol 24. *Biotechnology in Agriculture and Forestry.* Berlin: Springer-Verlag. pp. 329-341.

34. Heble M.R., Narayanswami S. & Chadha M.S. (1968) Diosgenin and ¼-sitosterol: Isolation from *Solanum xanthocarpum* tissue cultures. *Science.* 161: 1145.

35. Zenk M.H. (1978) The impact of plant cell culture on industry. In: Thorpe TA, editor. *Frontiers of Plant Tissue Culture.* Calgary, Alberta: University of Calgary, Offset Printing Service. 1-13.

36. Hosoda N. & Yatazawa M. (1979) Sterols, steroidal sapogenin and steroidal alkaloid in callus culture of *Solanum lanciniatum. Ait. Agric. Biol. Chem.* 43: 821-825.

37. Kadkade P.G. & Madrid T.R. (1977) Glycoalkaloids in tissue cultures of *S. acculeatissimum. Naturwissenscaften.* 64: 147.

38. Jain S.C., Khanna P. & Sahoo S.L. (1981) *Solanum jasminoides* Paxt. Tissue cultures I. Production of steroidal sapogenins and glycoalkalooids. *J. Nat. Products.* 44: 125-128.

39. Jain S.C. & Sahoo S.L. (1981) Isolation and characterization of steroidal sapogenins and glycoalkaloids from tissue cultures of *Solanum verbascifolium* Linn. *Chem. Pharm. Bull.* 29: 1765-1767.

40. Kittipongpatana N., Hock R.S. & Porter J.R. (1998) Production of solasodine by hairy root, callus, and cell suspension cultures of *Solanum aviculare* Forst. *Plant Cell Tissue Organ Culture.* 52(3): 133-143.

41. Bhatt P.N., Bhatt D.P. & Sussex I. (1983) Studies on some factors affecting solasodine contents in tissue cultures of *Solanum nigrum. Physiol. Plant.* 57: 159-162.

42. Ehmke A. & Eilert U. (1995) Steroidal glycoalkaloid in cell and shoot teratoma culture of *S. dulcamara. Plant Cell Tissue Organ Culture.* 43: 191-197.

43. Zacharius R.M. & Osman S.F. (1977) Glycoalkaloids in tissue cultures of *Solanum* species. *Plant Sci. Lett.* 10: 283-287.

44. Barnabas N.J., Karmalawala R. & David S.B. (1989) Stimulation of solasodine production by growth regulators in callus cultures of *Solanum* species. *Indian J. Exp. Biol.* 27: 664-665.
45. Udin A. & Chaturvedi H.C. (1979) Solasodine in somatic tissue cultures of *S. khasianum*. *Planta Medica.* 37: 90.
46. Chinchanikar G.S. (1983) Studies in tissue cultures from floral organs of some *Solanaceae* plants. Ph. D Thesis submitted to Poona University, Pune, India.
47. Sinha S., Roy R.P. & Jha K.K. (1979) Callus formation and shoot bud differentiation in anther culture of *Solanum surattense*. *Can. J. Bot.* 57: 2524-2527.
48. Szczerbakowa A.M. & Borkowska. (2000) Plant regeneration from the protoplasts of *Solanum tuberosum, S. nigrum and S. bulbocastanum*. *Acta. Physiol. Plant.* 22(1): 3-10.
49. Serraf I.D. & Sihachakr. (1988) High rate of plant regeneration from cultured protoplasts of two medicinal plants: *Solanum laciniatum* Ait. and *Solanum khasianum* Clarke. *J. Plant Physiol.* 133(4): 498-501.
50. Kowalczyk T.P., Mackenzie I.A. & Cocking E.C. (1983) Plant regeneration from organ explants and protoplasts of the medicinal plant *S. khasianum* Clark var chatterjeeanum sengupta (Syn. Solanum virum Dunal). *Z Pflanzenphysiol.* 3: 55-68.
51. Nehls R. (1978) Isolation and regeneration of protoplasts from *S. nigrum* L. *Plant Sci. Lett.* 12: 183-187.
52. Nelson R.S., Creissen G.P. & Bright S.W.J. (1983) Plant regeneration from protoplasts of *S. brevidens*. *Plant Sci. Lett.* 30: 355- 362.
53. Schumann U., Koblitz H. & Opatrny Z. (1980) Plant recovery from long-term callus cultures and from suspension cultures derived protoplasts of *S. phureja*. *Biochem. Physiol. Planzen.* 175: 670-675.
54. Shepard J.F. & Totten R.E. (1977) Mesophyll cell protoplasts of potato; isolation and proliferation and plant-regeneration. *Plant Physiol.* 60: 313-316.
55. Derks F.H.M., Zelcer A. & Hooymans C. (1990) Plant regeneration of mesophyll protoplasts from a cytoplasmic albino mutant of *Lycopersicon esculentum* cv. Large Red Cherry. *Plant Cell Tissue Organ Culture.* 21: 211-216.
56. Shillito R.D., Saul M.W., Paszkowski J., Muller M. & Potrykus I. (1985) High efficiency direct gene transfer to plants. *Biotechnol.* 3: 1099-1103.
57. Collonnier C., Mulya K., Fock I., Mariska I., Servaes A., Vedel F., Yakovlev S.S., Souvannavong V., Ducreux G. & Sihachakr D. (2001) Source of resistance against *Restonia solanacearum* in fertile somatic hybrids of eggplant (*S. melongena* L) with *Solanum aethiopicum*. *Plant Sci.* 160: 301-313.
58. Jarl C.I., Riitveld E.M. & Hass J.M. (1999) Transfer of fungal tolerance through interspecific somatic hybridization between *S. melongena* and *S. torvum*. *Plant Cell Rep.* 18: 791-796.
59. Serraf I., Tizrontine S., Chaput M., Allot M., Mussio I., Sihachakr D., Rossignol L. & Ducreux D. (1994) Production and characterization of intergeneric somatic hybrids through protoplast electrofusion between potato (*S. tuberosum*) and *L. pennellii*. *Plant Cell Tissue Organ Culture.* 37: 137-144.
60. Xu Y.S., Clark M.S. & Pehu E. (1993) Use of RAPD markers to screen somatic hybrids between *Solanum tuberosum* and *Solanum brevidens*. *Plant Cell Report.* 12: 107-109.
61. Preiszner J., Feher A., Veisz O., Sutka J. & Dudits D. (1994) Characterization of morphological variation and cold resistance in interspecific somatic hybrids between potato (*Solanum tuberosum*) and *Lycopersicon pennellii*. *Plant Cell Tissue Organ Culture.* 37: 137-144.

62. Sihachakr D., Haicour R., Barrientos E., Ducreux G. & Rossignol L. (1989) Somatic hybrid plants produced by electrofusion between *Solanum melongena* L and *Solanum torvum.* Sw. *Theor Appl Genet.* 77: 1-6.

63. Guri A. & Sink K.C. (1988) Organelle composition in somatic hybrids between an atrazine resistant biotype of *solanum nigrum* and *Solanum melongena. Plant Sci.* 58: 51-58.

64. Sihachakr D., Haicour R., Serraf I., Barrientos E., Herberteau C., Ducreux D., Rossignol L. & Souvanavong V. (1988) Electrofusion for the production of somatic hybrid plants of *Solanum melongena* L and *Solanum khasianum* CB Clark. *Plant Sci.* 57: 215-223.

65. Gleddie S., Keller W.A. & Setterfield G. (1986) Production and characterization of somatic hybrids between *Solanum melongena* L and *Solnum sisymbrifolium Lam. Theor Appl. Genet.* 71: 613-621.

66. Fassuliotis G. (1975) Regulation of whole plants from isolated stem parenchyma cells of *S. sisymbrifolium. J. Amer. Soc. Hort. Sci.* 100: 636-638.

67. Gleddie S., Keller W.A. & Setterfield G. (1983) Somatic embryogenesis and plant regeneration from leaf explants and cell suspensions of *S. melongena* (eggplant). *Can. J. Bot.* 61: 656-666.

68. Nishimura S., Terashima T., Higashi K. & Kamada H. Bioreactor culture of somatic embryos for mass propagation of plants. In: Redengaugh K, (Ed.). *Synseed. Boca Raton:* CRC Press. pp. 175-181.

69. Smith M.A.L. (1995) Machine vision analysis of plant cells and somatic embryos. In: Bajaj Y.P.S., (Ed.). *Biotechnology in Agriculture and Forestry.* Vol 30. *Somatic Embryogenesis and Synthetic Seed.* Berlin: Springer. pp. 87-101.

70. Gray D.J. & Purohit A. (1991) Somatic embryogenesis and development of synthetic seed technology. *Crit. Rev. Plant Sci.* 10: 33-36.

71. Gray D.J., Compton M.E., Harrell R.C. & Cantliffe D.J. (1995) Somatic embryogenesis and the technology of synthetic seed. In: Bajaj Y.P.S., (Ed.). *Biotechnology in Agriculture and Forestry.* Vol 30. *Somatic Embryogenesis and Synthetic Seed.* Berlin: Springer. pp. 126-151.

72. Pretova A. & Dedicova B. (1990) Somatic embryogenesis from unripe *Solanum tuberosum* (cv. Desire) embryos (Abstr B4-97). In: Nijkamp H.J.J., van der Pleas L.H.W., van Aartrijk J., (Ed.). 8th *Int. Cong. Plant Tissue Cell. Cult.* Dordrecht: Kluwer. pp. 265

73. Gracia E. de & Martinez S. (1995) Somatic embryogenesis in *Solanum tuberosum* L. cv. Desiree from stem nodal sections. *J. Plant. Physiol.* 145: 526-530.

74. Seabrook J.E.A. & Douglass L.K. (1990) Embryo culture of potato (*Solanum tuberosum* L) (abstract). *Am. Potato. J.* 67: 575-576.

75. Magioli C., Barroco R.M., Rocha C.A.B., Santiago L.D., Fernandes, Mansur E., Engler G., Pinheiro M. & Martins G.S. (2001) Somatic embryo formation in *Arabidopsis* and eggplant is associated with expression of a glycine rich protein gene (*Atgrp-5*). *Plant Sci.* 161: 559-567.

76. Akita M. & Takayama S. (1988) Mass propagation of potato tubers using jar fermentor techniques. *Acta Hort.* 230: 55-56.

77. Curtin M.E. (1983) Harvesting profitable products from plant tissue culture. *Biotechnology.* 1: 649-57.

78. Riiterhem E., Ulrich J. & Westphal K. (1990) Large scale cultivation of plant cell cultures, IAPTC. *News Lett.* 61: 2-10.

79. Overton K.H. (1977) Biosynthesis of mevalonoid derived compounds in cell cultures. In: Barz W., Reinhard E., Zenk M.H., (Ed.). *Plant Tissue Culture and Its Biotechnological Application.* Berlin, New York: Springer-Verlag. pp. 66-75.

80. Tabata M. (1978) Recent advances in the production of medicinal substances by plant cell cultures. In: Thorpe T.A., (Ed.). *Frontiers of Plant Tissue Cultures.* New York: *International association of Plant tissue Culture,* Calgary Press. pp. 3-16.
81. Mantell S.H. & Smith H. (1983) Cultural factors that influence secondary metabolite accumulation in plant cells and tissue cultures. In: Mantell S.H., Smith H., (Ed.). *Plant Biotechnology.* London, New York: Cambridge University Press, Cambridge. pp. 75-108.
82. Di Cosmo F. & Towers G.H.N. (1984) Stress and secondary metabolism in cultured plant cells. In: Timmermann B.N., Steelink C., Louwus F.A., editors. *Recent Advances in Phytochemistry.* Vol 18. New York: Plenum Press. pp. 97-175.
83. Yeoman M.M., Miedzybrodzka M.B., Lindsey K. & Mclauchlan W.R. (1980) The synthetic potential of cultured plant cell cultures. In: Sala F., Parasi B., Cella R., Cifferi O., (Ed.). *Plant Cell Cultures - Results and Perspective.* Holland: Elsevier. pp. 327-343.
84. Zenk M.H., Shagi H., Arens H., Stockgit J., Weiler E.W. & Deus B. (1977) Formation of indole alkaloids, serpentine and ajmalicine in cell suspension cultures of *Catharanthus roseus.* In: Barz W, Reinhard E, (Ed.). *Plant Tissue Culture and its Biotechnological Applications.* Berlin: Springer-Verlag. pp. 27-43.
85. Nickell L.G. (1980) Products. In: Staba, E.J., Raton B., editor. *Plant Tissue Culture as a source of biochemicals.* Florida: C.R.C Press. pp. 256-269.
86. Fowler M.W. (1981) Plant cell biotechnology to produce desirable substances. *Chemistry and Industry.* 229-233.
87. Yeoman M.M., Lindsey K., Miedzbrodzaka M.B. & Mclauchlan W.R. (1982) Accumulation of secondary products as a facet of differentiation in plant cell and tissue cultures. In: Yeoman M.M., Truman D.E.S., (Ed.). Vol 4. Differentiation *in vitro. British Society for Cell Biology Symposium.* Cambridge: Cambridge University Press. pp. 65-82.
88. Kurz W.G.W. (1986) Biological and environmental factors of product synthesis accumulation and biotransformation by plant cell cultures. *New Zealand J Technol.* 2: 77-81.
89. Mak Y. & Doran P.M. (1993) Effect of cell-cycle inhibition on synthesis of steroid alalkloids by *S. aviculare* plant cells. *Biotechnol. Lett.* 15(10): 1031-1034.
90. Chandler S. & Dodds J. (1983) Solasodine production in proliferating tissues of *S. lanciniatum. Plant Cell Rep.* 2: 205-208.
91. Indrayanto G., Erawati T. & Santosa M.H. (1995) Effect of L-arginine, casein hydrolysate, banana powder and sucrose on growth and solasodine production in shoot cultures of *Solanum laciniatum. Plant Cell Tissue Organ Culture.* 43(3): 237-240.
92. Nigra H.M., Alavarez M.A. & Giulietti A.M. (1990) Effect of carbon and nitrogen source on growth and solasodine production in batch suspension cultures of *Solanum eleangnifolium.* Cav. *Plant Celll Tissue Organ Culture.* 21: 55-60.
93. Drews F.E. & van Staden J. (1995) Attempts to produce solasodine in callus and suspension cultures of *Solanum maturitianum* Scop. *Plant Growth Regul.* 17: 21-25.
94. Staba E.J. (1982) Production of useful compounds from plant tissue cultures. In: *Plant tissue Culture.* Proc. 5th International Conference of Plant Tissue and Cell Culture. 25-26.
95. Bohm H. (1978) Regulation of alkaloid production in plant cell culture In: Thorpe TA, editor. *Frontiers of Plant Tissue Culture.* New York: IAPT. pp. 201-210.

96. Malpathak N.P. & David S.B. (1994) Effect of gamma irradiation on growth and solasodine levels in cell cultures of *Solanum Surattense* Burm. *Indian Drugs*. 31(2): 73-74.
97. Nigra H.M., Alavarez M.A. & Giulietti A.M. (1989) The influence of auxins, light and cell differentiation on solasodine production by *Solanum eleagnifolium* Cav. Calli. *Plant Cell Rep.* 8: 230.
98. Ramaswamy N.K., Behere A.G., Nair P.M. (1976) A novel pathway for the synthesis of solanidine in the isolated chloroplast from greening potatoes. *Eur. J. Biochem.* 67: 275-282.
99. Emke A. & Eilert U. (1986) Steroidal alkaloids in tissue cultures and regenerated plants of *Solanum dulcamara. Plant Cell Report.* 5: 350-351.
100. Misawa M. (1985) Production of useful plant metabolites In: Fietcher A, (Ed.). Advances in Biochemical Engineering/ Biotechnology. *Plant Cell Culture.* Vol 31. Berlin: Springer Verlag. pp. 59-88.
101. Whitaker R.J. & Hashimoto T. (1985) Production of secondary metabolites. In: Evans D.A., Sharp W.R., Ammirato P.V., Yamada Y., (Ed.). *Handbook of Plant Cell Cultures,* Vol 4. *Techniques for Propogation and Breeding.* New York: Macmillan Publ. Co. pp. 264-266.
102. Heftman E. (1983) Biogenesis of steroids of Solanaceae. *Phytochem.* 22(9): 18.
103. Bhatt P.N. & Bhatt D.P. (1984) Regulation of sterol biosynthesis in solanum species. *J. Exp. Bot.* 35(155): 890-896.
104. Jaggi R.K., Kapoor V.K. (1993) Tissue culture production of Diosgenin. *Indian Drugs.* 30: 416.
105. Schuler M.L., Sahai O.P. & Hallsby G.A. (1986) Entrapped plant cell tissue cultures. *Ann. NY Acad. Sci.* 413: 373-382.
106. Bordelius P., Deus B., Mosbach K. & Zenk M.H. (1979) Immobilized plant cells for the production of transformation of natural products. *FEBS Lett.* 122(2): 312-316.
107. Lindsey K. & Yeoman M.M. (1983) Novel experimental system for the production of secondary metabolites by plant tissue culture. In: Mantell S.H., Smith H. (Ed.). *Plant Biotechnology.* Cambridge University Press. pp. 39-66.
108. Lindsey K. & Yeoman M.M. (1987) Immobilized plant cells. In: Yeoman M.M., (Ed.). *Plant cell culture technology.* New Delhi: Narosa Publishing House. pp. 228-267.
109. Majerius F. & Pareilleux A. (1986) Alkaloid accumulation of Ca-alginate entrapped cells of *Catharanthus roseus* using limiting growth medium. *Plant Cell Rep.* 5: 446-447.
110. Malpathak N.P. & David S.B. (1992) Stimulation of solasodine production by combining fungal elicitors and immobilized cell suspension cultures of *Solanum surattense* Burm. *Biotech Lett.* 14(9): 965-968.
111. Dainty A.L., Goulding K.H., Robinson P.K., Simpkins I. & Trevan M.D. (1985) Effect of immobilization on plant cell physiology realor imaginary. *Trends in Biotechnol.* 3(3): 59-60.
112. Jirku V., Macek T., Vanek T., Krumphanzi V. & Kubanek V. (1981) Continuous production of steroid glycoalkaloid by immobilized plant cells. *Biotech Lett.* 3(8): 447-450.
113. Subharamani J., Phale P., Bhatt P.N. & Mehta A.R. Immobilization of *Solanum* cells for production of steroidal alkaloid. In: Kureja, (Ed.) *Tissue Culture and Biotechnology of Medicinal and Aromatic Plants.* Lucknow: CIMAP 1998, pp. 19-26.
114. Roisin C. & Gillet M.F. (1997) Enhanced production of scopolin by Solanum aviculare cells immobilised within Ca-alginate gel beads. *Plant Cell Reports.* 16(5): 349-353.

115. Vanek T. & Macek T. (1989) Plant cells immobilized in pectate gel: Biotransformation of verbenol isomers by *Solanum aviculare* free and immobilized cells. *Biotech Techniques.* 3(6): 411-414.

116. Barnabas N.J. & David S.B. (1988) Solasodine production by immobilized cell and suspension cultures of *Solanum surattense. Biotechnol Lett.* 10: 593-596.

117. Ishida B.K. (1988) Improved diosgenin production in *Dioscorea delatoidea* cell cultures by immobilization in polyurathene foam. *Plant Cell Reports.* 7: 270-273.

118. Rokem J.S., Schwarzberg J. & Goldberg I. (1984) Autoclaved fungal mycelia increase diosgenin production in cell suspension cultures of *Dioscorea deltoidea. Plant Cell Rep.* 3: 159-160.

119. Paniego N.B., Rodrigues J., Talon J., Giulitte A.M. (1995) Influence of fungal elicitors on solasodine production by suspension cultures of *Solanum eleangifolium. Fitoterapia.* 66(2): 162-166.

120. Malpathak N.P. & David S.B. (1992) Stimulation of solasodine production by combining fungal elicitors and immobilized cell suspension cultures of *S. surattense* Burm. *Biotech. Lett.* 14(9): 965-968.

121. Scragg A.H. (1990) Fermentation systems for plant cells. In: Charlwood B.V., Rhodes M.J.C., (Eds.). *Secondary Products from Plant Tissue Culture.* Oxford: Clarendon Press. pp. 243-263.

122. Charlwood B.V., Charlwood K.A. & Molina-Torres J. (1990) Accumulation of secondary compounds by organized plant cultures. In: Charlwood B.V., Rhodes M.J.C., (Eds.). *Secondary Products from Plant Tissue Culture.* Oxford: Clarendon Press. pp. 167-200.

123. Paar A.J. (1989) The production of secondary metabolites by plant cell cultures. *J. Biotechnol.* 10: 1-26.

124. Flores H.E. (1987) Use of plant cell and organ culture in the production of biological chemicals. In: Lebaron H., Mumma R.O., Honeycutt R.C., Duesing J.H., (Eds.). Vol 334. *Application of Biotechnology to Agricultural Chemistry.* Washington DC: American Chemical Society. pp. 66-68.

125. Kokate C.K. & Radwan S.S. (1979) Enrichment of *Solanum khasianum* callus generating rootlets with steroidal glycoalkaloids. *Z Naturforch C Biosci.* 34: 634-636.

126. Ripperger H. (1995) Steroidal alkaloids and sapogenins from roots of some *Solanum* species. *Planta Medica.* 61: 292.

127. Hu Z.B. & Alfermann A.W. (1993) Diterpenoid production in hairy root cultures of *Salvia miltiorrhiza. Phytochemistry.* 32: 699-703.

128. Vanahala L., Hiltunen R. & Oksman-Caldentey K.M. (1995) Virulence of different *Agrobacterium* strains on hairy root formation of *Hyoscyamus muticus. Plant Cell Reports.* 14: 236-240.

129. Giri A., Ravindra S.T., Dhingra V. & Narasu S.T. (2001) Influence of different strains of *Agrobacterium rhizogenes* on iduction of hairy roots and artemisinin production in *Artemisia annua. Curr. Sci.* 81(4): 378-3823.

130. Yan-Nong S., Shibuya M., Ebizuka Y. & Sankawa U. (1990) Hydroxyacetosyringone is one of the major virulence gene activating factor in belladonna hairy root cultures, and inositol enhances its activity. *Chem Pharm Bull.* 38: 2063-2065.

131. Spena A., Schmulling T., Koncs C. and Schell J. (1987) Independent and synergistic activity of rol A, B and C loci in stimulating abnormal growth in plants. *EMBO J.* 6: 3891-3899.

132. Nguyen C., Bourgaud F., Forlot P. & Guckert (1992) A. Establishment of hairy roots of *Psoralea* species. *Plant Cell Reports*. 11: 424-427.

133. Chilton M.D., Drummond M.H., Sciaky D., Gordon M.P. & Nester E.W. (1977) Stable incorporation of plasmid DNA into higher plant cells. The molecular basis of crown gall tumorigenesis. *Cell.* 11: 263-271.

134. Chilton M.D., Petit A., David C., Casse-Delbart T., Tempé J. (1977) *Agrobacterium rhizogenes* inserts T-DNA into the genome of the host plant root cells. *Nature*. 295: 432-434.

135. Tepfer D. (1983) The biology of genetic transformation of higher plants by *Agrobacterium rhizogenes*. In: Puhler A (Ed.) Molecular genetics of the bacteria—Plant Interections. Berlin: Springer Verlag. pp. 248-258.

136. Aird E.L.F., Hamill J.D. & Rhodes M.J.C. (1988) Cytogenetic analysis of hairy root cultures from a number of plant species transformed with *Agrobacterium rhizogenes*. *Plant Cell Tissue Organ Culture*. 15: 47-57.

137. Flores H.E. & Filner P. (1985) Metabolic relationships of putrescine GABA and alkaloids in cell and root cultures of Solanaceae. In: Neumann K, Barz W, Reinhard E. *Primary and Secondary Metabolism of Plant Cell Cultures*. Berlin: SpringerVerlag. 174- 185.

138. Kamada H., Okamura N., Satake M., Harada H. & Shimomura S. (1986) Alkaloid production by hairy root cultures of *Atropa belladonna*. *Plant Cell Rep*. 5: 239-242.

139. Shimomura K., Satake M. & Kamada H. (1986). Production of useful secondary metabolites by hairy roots transformed with Ri plasmid. In: Somers D, Genebach BG, Biesboer DD, Hackett WP, Green CE, (Eds.). *Plant Tissue and Cell Culture*. New York: *Liss*. pp. 250.

140. Hamill J.D., Parr A.J., Rhodes M.J.C. & Walton N.J. (1987) New routes to plant secondary products. *Bio Tech*. 5: 800-804.

141. Rhodes M.J.C., Robins R.J., Hamil J.D., Parr A.J. & Waltson N.J. (1987) Secondary product formation using *Agrobacterium rhizogenes* transformed hairy root cultures. IAPTC *Newsl*. 53: 2-15.

142. Signs M.W. & Flores H.E. (1990) The biosynthetic potential of plant roots. *Bioassay*. 12: 7-13.

143. Wei Z.M., Kamada H. & Harada H. (1986) Transformation of *Solanum nigrum* L protoplasts by *Agrobacterium rhizogenes*. *Plant Cell Reports*. 5: 93-96.

144. Chumsri P. & Uboonkhiew U. (1992) Variations of growth and solasodine content in hairy roots of *S. nigrum* L. *Micro Utl Renewable Resour*. 8: 363-374.

145. Brillanceau M.H. (1989) "Etude Chimique des Alcaloides de Deux Espices du Genre Guettarda. Culture *in vitro* de Racines Transformees par *Agrobacterium rhizogenes*", Ph.D Thesis. Université de Paris-Sud, Orsay, France 1986. cf. Tempe J., Delbart F.C. Plant gene vector and genetic transformation: agrobacterium Ri plasmids. In: Schell J, Vasil IK, (Eds.). *Cell Culture and Somatic cell Genetics of Plants*. Vol 6. *Molecular Biology of Plant Nuclear Genes*. San Diego: Academic Press. pp. 26-43.

146. Alvarez M.A., Talou J.R., Paniego N.B. & Giulietti A.M. (1994) Solasodine production in transformed organ cultures (roots and shoots) of *Solanum eleagnifolium* Cav. *Biotechnol Lett*. 16: 393-396.

147. Drews F.E. & van Staden J. (1995) Initiation of and solasodine production in hairy root cultures of *Solanum maturitianum* Scop. *Plant Growth Regul*. 17: 27-31.

148. Ikenaga T., Oyama T. and Muranaka T. (1995) Growth and steroidal saponin production in hairy root cultures of Solanum aculeatissimum. *Plant Cell Rep*. 14: 413-417.

149. Kittipongpatana N., Hock R.S. & Porter J.R. (1998) Production of solasodine by hairy root, callus, and cell suspension cultures of *Solanum aviculare* Forst. *Plant Cell Tissue Organ Culture*. 52(3): 133-143.

150. Argolo A.C.C., Charlwood B.V. & Pletsch M. (2000) The regulation of solasodine production by Agrobacterium rhizogenes-transformed roots of *Solanum aviculare*. *Planta Medica*. 66: 448-451.

151. Malpathak N.P. Solasodine production in multiple shoot cultures derived from hairy roots of *Solanum khasianum* Clark. In: Trivedi P.C. (Eds.). *Plant Biotechnology: Recent Advances.* New Delhi: Parima 2000.
152. Ooms G., Karp A., Burell M.M., Twell D. & Roberts J. (1985) Genetic modification of potato development using Ri T-DNA. *Ther Appl Genet.* 70: 440-446.
153. Ooms G., Twell D., Bossen M.E., Harry C., Hoge C. & Burrell MM. (1986) Developmental regulation of Ri. TL- DNA expression in roots, shoots and tubers of transformed potato (*Solanum tuberosum* cv. Desiree). *Plant Mol Biol.* 6: 321-330.
154. Morgan A.J., Cox P.N., Turner D.A., Pell E., Davey M.R., Gartland K.M.A. & Mulligan B.J. (1987) Transformation of tomato using an Ri plasmid vector. *Plant Sci.* 49: 37-49.
155. Merkli A., Christen P. & Kapetanidis I. (1997) Production of diosgenin by hairy root cultures of *Trigonella foenumgraecum L. Plant Cell Reports.* 16: 632-6.
156. Yu S., Kwok K.H. & Doran P.M. (1996) Effect of sucrose, exogenous product concentration and other culture conditions on growth and steroidal alkaloid production by *Solanum aviculare* hairy roots. *Enzyme Microbial Technol.* 18: 238-243.
157. Lindsey K. & Yeoman M.M. (1983) The relationship between growth rate, differentiation and alkaloid accumulation in cell cultures. *J. Exp. Bot.* 34 (145): 1055-1065.
158. Subroto M.A. & Doran P.M. (1994) Production of steroidal alkaloids by hairy roots of *S. aviculare* and the effect of Gibrillic acid. *Plant Cell Tissue Organ Culture.* 38: 93-102.
159. Rhodes M.J.C., Parr A.J., Ceielutt A. & Aird E.L.H. (1994) Influence of exogenous hormone on growth and secondary metabolite formation in transformed root cultures. *Plant Cell Tissue Organ Culture.* 38:143-151.
160. Flores H.E., Doi Y., Cuello J.L., Maldonado-Mendoza I.E. & Loyola-Vergas V.M. (1993) Green roots: Photosynthesis and photoautotrophy in an underground plant organ. *Plant Physiol.* 101: 363-371.
161. Singh G. (1998) Application of fungal elicitation for enhancing production of secondary metabolites by hairy roots cultured in large-scale bioreactors. In: Doran PM (Ed.). Hairy Roots and its Application. Hardwood Academic Publishers. pp. 209-218.
162. Verpoorte R., van der Heijden R. & Moreno PR. (1997) In: Cordel GA, (Ed.) The Alkaloids. Vol. 49. Academic Press, San Diego. pp. 221-229.
163. Davey M.R., Mulligan B.J., Gartland K.M.A., Peel E., Sargent A.W. & Morgan AJ. (1987) Transformation of *Solanum* and *Nicotianum* species using an Ri plasmid vector. *J. Exp. Bot.* 38: 1507-1516.
164. Mc Innes E., Morgan A.J., Mulligan B.J. & Davey M.R. (1991) Phenotypic effects of isolated pRi A4 Tl-DNA rol genes in the presence of intact TR-DNA in transgenic plants of *Solanum dulcamara* L. *J. Exp. Bot.* 42: 1279-86.
165. Visser R.G.F., Jacobson E., Witholt B. & Feenstra W.J. (1989) Efficient transformation of potato (*Solanum tuberosum* L.) using a binary vector in *Agrobacterium rhizogenes. Theor Appl Genet.* 78: 594-600.
166. Romano A., Racemakers K., Visser R., Mooibroek H. (2001) Transformation of potato (*S. tuberosum*) using particle bombardment. *Plant Cell Culture.* 20: 198-204.
167. Sander H. & Grisebach H. (1958) Gewinnung von ^{14}C-markiertem Tomatin. Z. *Naturforsch.* 13b: 755-776.

168. Tascheshe R. and Hulpke H. (1966) Biogenese von Tomatidin aus Cholesterin. Z. *Naturforsch*. 21b: 893-994.

169. Paiva N.L. (2000) An introduction to the biosynthesis of chemicals used in plant- microbe communication. *J. Plant Growth Regul*. 19: 131-143.

170. Shecter I., Gann G. & Boettcher B.R. (1999) Squalen synthase. In: Barton D, Nakanishi K, Meth-Cohn O, (Eds.). *Comprehensive Natural Products Chemistry*. Vol 2. *Isoprenoids Including Carotenoids and Steroid*. Amsterdam: Elsevier. pp. 245-266.

171. Abe I. & Prestwich G. (1999) Squalene epoxidase and oxidosqualene: Lanosterol Cyclase-Key enzymes in cholesterol biosynthesis. In: Barton D, Nakanishi K, Meth-Cohn O (Eds.). *Comprehensive Natural Products Chemistry*. Vol 2. *Isoprenoids Including Carotenoids and Steroid*. Amsterdam: Elsevier. pp. 267-298.

172. Diener A.C., Li H., Zhou W., Whoriskey W.J., Nes W.D. & Fink G.R. (2000) Sterol Methyltransferase 1 controls the level of cholesterol in plants. *The Plant Cell*. 12: 853-870.

173. Heftman E. (1960) Biogenesis of steroids in Solanaceae. *Phytochemistry*. 22: 1843-1860.

174. Pasquali G., Goddijn O.J.M., de Waal A., Verpoorte R., Schilperoort R.A., Hoge J.H.C. & Memelink J. (1992) Coordinated regulation of two indole alkaloid biosynthetic genes from *Catharathus roseus* by auxin or elicitors. *Plant Mol. Biol*. 18: 1121-1131.

175. Menke F.L.H., Parchmann S., Muller M.J., Kijne J.W. & Memelink J. (1999) Involvement of octadecanoid pathway and protein phosphorylation in fungal elicitor-induced expression of terpenoid indole alkaloid biosynthetic genes in *Catharanthus roseus*. *Plant Physiol*. 119: 1289-1296.

176. Verpoorete R., van der Heijden R., ten Hoopen H.J.G. & Memlink J. (1999) Metabolic engineering of plant secondary metabolite pathways for the production of fine chemicals. *Biotechnol Lett*. 21: 467-479.

177. Hamill J.D., Robins R.J., Parr A.J., Evans D.M., Furze J.M. & Rhodes M.J.C. (1990) Overexpressing a yeast ODC gene in transgenic roots of *Nicotiana rustica* can lead to enhanced nicotine formation. *Plant Mol Biol*. 15: 27-38.

178. Hahn K. & Strittmatter G. (1994) Pathogen-defence gene prp 1-1 from potato encodes an auxin-responsive glutathione-S-transferase. *Eur. J. Biochem*. 226: 619-626.

179. Collinge M. & Boller T. (2001) Differential induction of two potato genes *Stprx Z* and *St NAC*, in response to infection by *Phytopthera infestans* and *to wounding*. 46: 521-529.

180. Costantino P., Spano L., Pomponi M., Benvenuto E. & Ancora G. (1984) The T-DNA of *Agrobacterium rhizogenes* is transmitted through meiosis to the progeny of hairy root plants. *J. Mol. Appl. Genet*. 2: 465-470.

181. De Block M., Herrera-Estrella l., Van Montagu M., schel J. & Zambryski P. (1984) Expression of foreign genes in regenerated plants and their progeny. *EMBO J*. 3: 1681-1689.

182. Peerbolte R., Hoge J.H.C., Wullems G.J. & Schilperoort R.A. (1985) Hairy root: t-DNA, T-DNA transcripts and progeny. In: Magnien E, de Nettancourt D, Junk N, Dordrecht (Eds.). Genetic Engineering of Plant and Microorganisms. Important for Agriculture. The Netherlands: *Nijhoff/Junk*. pp. 76-77.

183. Zambryski P., Tempe J. & Schell J. (1989) Transfer and function of T-DNA genes from *Agrobacterium* Ti and Ri plasmids in plants. *Cell*. 56: 193-201.

184. Hashimoto T., Yun D.J. & Yamada Y. (1993) Production of tropane alkaloids in genetically engineered root cultures. *Phytochemistry*. 32: 713-718.

185. Shahin E.A., Sukhainda K., Simpson R.B. & Spivey R. (1986) Transformation of cultivated tomato by a binary vector in *Agrobacterium rhizogenes:* Transgenic plants with normal phenotypes harbour binary vector T DNA, but no Ri-plasmid T DNA. *Theor. Appl. Genet.* 72: 770-7.

186. Sukhapinda K., Spivey R., Simpson R.B. & Shahin E.A. (1987) Transgenic tomato (*Lycopersicon esculentum* L.) transformed with a binary vector in *Agrobacterium rhizogenes*: non-chimeric origin of callus clone and low copy number of integrated vector T-DNA. *Mol. Gen. Genet.* 206: 491-497.

187. Hood E.E., Witcher D.R., Maddock S., Meyer T., Baszczynski C., Bailey M., Flynn P., Register J., Marshall L., Bond D., Kulisek E., Kusnadi A., Evangelista R., Nicolov Z., Wooge C., Mehigh R.J., Herman R., Kappel W.K., Ritland D., Li C.P. & Horward J. (1997) Commercial production avid from transgenic maize: characterization of transformant, production, processing, extraction and purification. *Mol. Breed.* 21: 17.

188. Tacket C.O., Mason H.S., Losonky G., Cclements J.D., Levine M.M. & Arntzen C.J. (1998) Immunogenicity in humans of a recombinant bacterial antigen delivered in a transgenic potato. *Nature Med.* 4: 607-609.

189. Witcher D.R., Hood E.E., Peterson D., Bailey M., Bond D., Kusnadi A., Evangelista R., Nikolov Z., Wooge C., Mehigh R., Kappel W., Register J. & Howard J.A. (1998) Commercial production of ß-glucuronidase (GUS): a model system for the production of proteins in plants. *Mol. Breed.* 4: 301-312.

190. Wilde C.D., Houdt H.V., Buck S.D., Angenon G., Jager G. & Depicker A. (2000) Plants as bioreactors for protein production: avoiding the problem of transgene silencing. *Plant Mol. Biol.* 43: 347-359.

191. Rishi A.S., Nelson N.D. & Goyal A. (2001) Molecular farming in plants: a current perspective. *J. Plant Biochem. Biotech.* 10: 01-12.

192. Neve M., Buck S., Wilde C., Houdt V.H., Strobbe I., Jacobs A., Van Montagu M. & Depicker A. (1999) Gene silencing results in instability of antibody production in transgenic plants. *Mol. Gen. Genet.* 260: 582-592.

193. Furze J.M., Hamill J.D., Parr A.J., Robus R.J. & Rhodes M.J.C. (1987) Variation in morphology and nicotine alkaloid accumulation in protoplast derived hairy root cultures of *Nicotiana rustica*. *J. Plant Physiol.* 131:237-246.

194. Weiler E.W., Kruger H. & Zenk M.H. (1980) Radioimmunoassay for the determination of steroidal alkaloid solasodine and related compounds in living plants and herbarium specimens. *Planta Med.* 39: 112-124.

195. Shan S.J., Putalun W., Fukuda N., Morinaga O., Tanaka H. and Shoyama Y. (2000) ELISA, western blotting, immunocytolocalization and immunoaffinity column for naturally occurring bioactive compounds using monoclonal antibodies. *J Food Drug Analysis*. 8 (4): 258-269.

196. Hunter I.R., Walden M.K. & Heftmann E. (1980) High performance liquid chromatography of *Solanum* and *Vertrum* alkaloids. *J. Chromatogr.* 198: 363-366.

197. Osman S.F. & Sinden S.L. (1989) High performance liquid chromatographic analysis of *Solanum* steroidal alkaloids. *J. Chromatogr*. 479: 189-193.

198. Kittipongpatana N., Porter J.R. & Hock R.S. (1999) An improvrd high performance liquid chromatographic method for the quantification of solasodine. *Phytochem. Analysis*. 10: 26-31.

199. Weston R.J. (1976) An investigation of methods for isolation of the steroidal alkaloid solasodine from native New Zealand *Solanum Species*. 26: 657-666.

200. Bhattacharya T.K. Chemical examination of certain Indian medicinal plants. Ph. D Thesis submitted to the University of Madras for the Award of Doctorate of Philosophy, Madras, India. 1978.
201. Blanch H.W. & Clark D.S. (1996) Biochemical engineering. In: New York: *Marcel Dekker*.
202. Rhodes M.J.C., Hilton M.G., Parr A.J., Hamill J.D. & Robins R.J. (1986) Nicotine production by 'hairy root' culture; fermentation and product recovery. *Biotechnol. Lett.* 8: 415-420.
203. Wilson P.D.G., Hilton M.G., Robins R.J. & Rhodes M.J.C. (1987) Fermentation studies of transformed root cultures. In: Moody GW, Baker PB, (Eds.). Bioreactors and Biotransformations. London: Elsevier. pp. 38-51.
204. Hilton M.G. & Rhodes M.J.C. (1990) Growth and hyscyamine production in 'hairy root' cultures of *Datura stramonium* in a modified stirred tank reactor. *Appl. Microbiol. Biotechnol*. 33: 132-138.
205. Buitelaar R.M., Langenhoff A.A.M., Heidatra, R. & Tramper J. (1991) Growth and thiophene production by hairy root cultures of *Tagetes patula* in various two-liquid phase bioreactors. *Enzyme Microb. Biotechnol.* 13, 487-494.
206. Mukundan U., Carvalho E.B. & Curtis W.R. (1998) Growth and pigment production by hairy root cultures of *Beta vulgaris* L. in a bubble column reactor. *Biotechnology Letters* 20 (5): 469-474.
207. Jung G. & Tepfer D. Use of genetic transformation by the Ri T-DNA of the *Agrobacterium rhizogenes* to stimulate biomass and tropane alkaloid production in *Atropa belladonna* and *Calystegia sepium* roots grown *in vitro. Plant Sci.* 1987; 50: 145-151.
208. Tescione L.D., Ramakrishnan D. & Curtis W.R. (1997) The role of liquid mixing and gas-phase dispersion in a submerged sparged root reactor. *Enzyme Microbial Technol*. 20 (3): 207-213.
209. Spielmann A. & Simpson R.B. (1986) T DNA structure in transgenic tobacco plants with multiple independent integration sites. *Mol. Gen. Genet.* 205: 34-41.
210. Brooker S.G. & Cooper R.C. (1961) New Zealand medicinal plants. *Econ. Bot*. 15: 1-10.
211. Marshall F. (2001) Agriculture and use of wild and weedy greens by the *Pik ap Oom Okiek* of Kenya. *Econ. Bot.* 55: 32-46.
212. Jeremy R. & Genesis II. (1999) *Span.* Jan-Feb. pp. 42-50.
213. Khanna K.R. & Murty A.S. (1972) Effect of fruit stage and maturity of the glucoalkaloids content in *S. khasianum. Planta Medica.* 30: 153-157.
214. Hand S. & Bin A. (1960) (1982) Tagunsber. Dtsch. Akad. Landwirtschaftswiss. 1961. Berlin 27, 163. cf. Heftman E. Biogenesis of steroids in Solanaceae. *Phytochemistry*. 22: 1843.
215. Heftmann E. *Lloydia*. 38: 195.
216. Furuya T., Hirotani M., Kanaguchi K. Biotransformation of progesterone and pregnenolone by plant suspension cultures. *Phytochemistry.* 1971; 10: 1013-1017.
217. Hirotani M. and Furuya T. (1974) Biotransformation of testosterone and other androgens by suspension cultures of *Nicotiana tabacum* "Bright Yellow". *Phytochemistry*. 13: 2135-2142.

CHAPTER

22

Genetic Transformation of Grain Legumes

Rajani Jaiswal[1], P. S. Kumar[1], M. Z. Abdin[2] and P. Ananda Kumar[1*]

ABSTRACT

Grain legumes are the most important crops yielding nutritious protein rich seeds for human consumption and animal feed. Legumes are also important for improving physical, chemical and biological properties of soil by nitrogen fixation. Improvement of legumes by genetic engineering requires procedures to regenerate them *in vitro*. Being highly recalcitrant to regeneration among all the plant species tropical grain legumes are more difficult to transform. However, a wide range of other legume species have been genetically transformed. Many transgenic legume species have been transformed for herbicide resistance, insect resistance and improved nutritional quality. Among all grain legume species soybean is the most favoured species for genetic transformation. There is a requirement to develop simple, reproducible and efficient transformation protocols for tropical grain legumes. In this review we discuss various aspects of legume transformation viz., different methods used for gene transfer, parameters such as specific bacterial strains, selectable markers and types of explants affecting the transformation efficiency.

INTRODUCTION

Grain legumes occupy an important place in human nutrition as one of the most valuable sources of vegetable protein and vitamins. In many parts of the world legumes are the most important crops where these are grown and consumed along with cereals. Their consumption has increased in recent years due to preference towards vegetarian diet in recent years. High protein content and the presence of fat with low degree of saturation have contributed significantly to the popularity of large seeded grain legumes. There are about 60 domesticated grain legume species throughout the world, the major ones being summarized in Table 1. Of those grain legume species that have been

*Address for correspondence: [1]National Research Centre for Plant Biotechnology, Indian Agricultural Research Institute, New Delhi 110012

[2]Centre for Biotechnology, Hamdard University, Hamdard Nagar, New Delhi

E-mail: polumetla@hotmail.com

domesticated, only few find wider application in human food production. Many efforts have been made to improve different traits in legumes through biotechnological manipulation. Modern biotechnology focuses on the identification and expression of foreign genes to enhance yields or quality of the product, introduce resistance to pests and diseases etc., Some of the most important proteins in the grain legumes are deficient in specific essential amino acids such as methionine and cysteine. Synthesis of these amino acids in grain legumes cannot be achieved by conventional breeding but is possible through genetic engineering. Progress in genetic transformation of grain legumes has been hampered by their recalcitrant nature in tissue culture and lack of compatible gene transfer methods. In this context efficient regeneration and transformation protocols are required for the application of biotechnological tools for legume improvement. A wide range of legumes has been genetically transformed showing some of the genetically determine traits like insecticidal activity, protection against viral infection, resistance to herbicides, tolerance of environmental stresses, improved nutritional quality of seed proteins. There are number of effective

Table 1 Common Grain Legumes

Species	*Common name*
Glycine max (L.) Merrill	Soybean
Arachis hypogea (L.)	Peanut, groundnut
Cicer arietinum (L.)	Chickpea, Bengal gram
Pisum sativum (L.)	Pea, garden pea, field bean
Cajanus cajan (L.) Millsp	Pigeon pea, red gram
Phaseolus vulgaris (L.)	Common bean, French bean, Kidney bean, dry bean, azuki bean, navy bean
Phaseolus aconitifolius Jacq. *(Vigna aconitifolia (L.))*	Moth bean, tapery bean
Phaseolus coccineus	Runner bean
Phaseolus lunatus	Lima bean, butter bean
Vicia faba (L.)	Broad bean, field bean, faba bean,
Vicia narbonensis	Vetch, narban bean
Vigna radiata (L.) Wilczek (*P.aureus* Roxb)	Mungbean, green gram
Vigna mungo (L.) (*P. mungo* Roxb)	Blackgram
Vigna unguiculata (L.) *Walp* (*V. sinesis* [L.] Save ex Hassk)	Cowpea, southern pea, black pea
Vigna sesquipedalis	Asparagus bean
Vigna trilobiata	–
Lens culinaris (L.) Medic *(Lens esculantus* Moench.)	Lentil
Lupinus mutabilis (L.)	–
Lupinus lyteus (L.)	Yellow lupin
Lupinus albus	White lupin
Lupinus angustifolius (L.)	Sweet lupin, blue lupin
Lathyrus sativus	Grass bean, chickling pea
Cyamopsis tetragonoloba	Guar
Canavalia ensiformis (L.)	Jack bean, sword bean
Dolichus biflorus (L.)	Horse gram
Psophocarpus tetragonolobus	Winged bean

DNA delivery systems that work in variety of plants. The most commonly used method is *Agrobacterium*-mediated transformation. This method is a nondirect method of gene transfer because the gene must first be transferred to *Agrobacterium*. *Agrobacterium*-mediated transformation has provided a reliable means of creating transformants in a wide variety of species that are amenable to tissue culture and regeneration. *Agrobacterium*-mediated gene transfer system is effective in several species but not all the species can be transformed because of the host range of the *Agrobacterium* although there are reports of successful transformation in legumes and cereal crops. Other methods that have been used include chemical treatment by calcium phosphate and polyethylene glycol (PEG), electroporation, particle bombardment or biolistic transformation, microinjection, macroinjection, liposome fusion, silicon carbide fibre-mediated, biotin-mediated gene transfer, pollen tube pathway method, sonication, sonication-assisted *Agrobacterium* transformation (SAAT), *in planta* transformation, vacuum infiltration and imbibition of dry seeds or embryos with DNA. All of these techniques have certain limitations. The reliability of various gene transfer methods is extensively reviewed by Potrycus (1991). The method of transformation selected depends on the plant species and the type of the tissue. The successful production of transgenic plants depends on various factors like the method used for the transformation, stable integration of foreign DNA in the plant genome, followed by regeneration of the transformed cells to produce intact plants and the subsequent expression of the introduced gene(s) at the correct time and place in the transgenic plant. The objective of this review is to review the literature dealing with grain legume transformation and suggest future directions.

In past few years grain legumes have been transformed using various marker and reporter genes, agronomically useful genes imparting resistance to herbicides, insects, viruses and genes responsible for nutritional quality (high fatty acids, lipids and oils, carbohydrates and proteins). Progress achieved in legume transformation is summarised in Table 2.

GENETIC TRANSFORMATION

Agrobacterium-Mediated Gene Transfer

Several gene transfer system have been tested in grain legume species. The majority of the techniques need *in vitro* culture and a reliable regeneration protocol. *Agrobacterium (A. tumefaciens* and *A. rhizogenes)*, a gram-negative plant pathogen is a natural genetic engineer, which is a very effective vehicle of introducing foreign DNA into plant species. Upon infection at plant wound site *Agrobacterium* transfers its segment of oncogenic DNA in the plant genome by illegitimate recombination. The T- DNA is flanked by 26-bp direct repeats, called border repeats and the DNA that lies between these border repeats is transferred and integrated in the plant genome. A foreign gene inserted in the T-DNA of the bacterial plasmid (Ti or Ri) is thus integrated in the plant genome.

Direct Gene Transfer

It is an alternative method of gene transfer by which direct uptake of naked DNA by plant cell occurs. It does not require any biological vector system and since it is a physical process there is no problem of host range specificity. Insertion of DNA fragments into the cells requires proper

Table 2

Legume Crops	*Explant*	*Transformation method*	*Plasmid*	*Marker gene*	*Other gene*	*Result/ Transformant*	*Reference*
Glycine max	Cotyledon	A. t.		NptII, GUS	EPSP	Plant	Hinchee *et al.*, 1988
	Embryonic axis	PB				Plant	McCabe *et al.*, 1988
	Cotyledon	A.t.	pGA482	NptII	-	Plant	Chee *et al.*, 1989
	Cotyledon	A348, LBA4404	pTiA6, pTiB0542, pBin6	Chek in ref	-	-	Owens & Smigocki, 1988
	Cotyledon	A.t	pZac(l)	Gus	Maize controlling elements	Callus	Zhou & Atherly, 1989, 1990
	Seedling	A281, GV3834	pTiB0542, pGV2100	-	-	tumour	Austin & Cress, 1994
	Cotyledon	A281	pZa7, pTiB0542	Gus, nptII	-	-	Luo *et al.*, 1994
					Bt	Plant	Parrott *et al.*, 1994
	Cotyledonary node	A.t.	-	-	BPMV (Bean Pod Mottle corno Virus) coat protein precursor gene	-	Di *et al.*, 1996
	Hypocotyl, cotyledonary node	A.t.	pKT54B7C5	NptII	Bt gene	-	Xu *et al.*, 1997
		CP4	-	-	EPSPS	Plant	Padgette *et al.*, 1995
	Cotyledon	A.t.	-	-	Barnase Barstar	Plant	Guo & Lin, 1997
	Somatic embryo	PB		hptII	Bt cry1Ac	Plant	Stewart *et al.*, 1996
	Suspension culture	PB		hptII , GUS	-		Hadi *et al.*, 1996
	Hypocotyls	PB		-	Antisense DNA from Glym Bd 30 K	-	Tokano *et al.*, 1996

(Contd.)

(Contd.)

	Hypocotyl of cotyledonary stage somatic embryo	PB		GUS, nptII	Rat liver stearyl co A delta-9 desaturase (D9DS)	Plant	Li *et al.*, 1996
		PB		GUS, nptII	-	Plant	Christou, 1990; Yang & Christou, 1990
		PB		PAT, bar		Plant	Christou & Swain, 1990
	Protoplast from immature embryo	Electroporation		NptII	-	Callus	Kuchuk *et al.*, 1990
	Protoplast from immature cotyledon	Electroporation		GUS, hptII	-	Plant	Dhir *et al.*, 1991
	Ovary	Microinjection		-	Atrazine resistant gene	Plant	Yue *et al.*, 1996
	Ovary	Microinjection		-	PSbA gene of *Solanum nigrum*-atrazine resistant gene	Plant	Liu *et al.*, 1990
	Callus	Biotin mediated		-	Macromoleccule: bovine, insulin, bovin ribonuclease, human haemoglobin bovin serum albumin	Callus	Horn *et al.*, 1990
	Suspension culture	SAAT		HptII	-	Callus	Trick & Finner, 1998
	Immature cotyledon	SAAT		GUS	-	Transient expression	Santarem *et al.*, 1998
	Immature seed	A.t. EHA101, LBA4404	pSAOR1221, pTRA415	GUS	PRI	Plant	
	Immature embryo	PB		hph			Santarem *et al.*, 1998
	Cotyledonary node	A.t. EHA 101, EHA105		Bar			Zhang *et al.*, 1999

(Contd.)

(Contd.)

	Immature cotyledon	A.t. EHA 105		Hpt			Yan *et al.*, 2000
	Embryonic axis	PB		ahas			Aragao *et al.*, 2000
	Cotyledonary node	A.t.		Hph			Olhoft *et al.*, 2003
		Pollen tube pathway				early ripening plant	Li *et al.*, 1994
		Pollen tube pathway	Total DNA from wild soybean *G. soja*			Plant with high protein content	Li *et al.*, 1994
		Pollen tube pathway	DNA from soybean			Early maturing plant	Li *et al.*, 1996
		Starch gel electrophoresis	Total DNA from sunflower			Plant	Li *et al.*, 1996
Phaseolus vulgaris	Epicotyl	A.r.	-	-		Root	Barross *et al.*, 1997
	Shoot tip	A.t. A281	pGV3850, pKIWI105	GUS		-	Lewis & Bliss, 1994
	Seed meristem	PB		PWRG2204-pUC19GUS, bar	BGMV coat protein gene	Plant	Russell *et al.*, 1993
	Apex of embryonic axes	PB		PSOG-1016, GUS	-	Palnt	Kim & Minamikawa, 1996
	Shoot apex	PB		pSAG-734, GUSA	-	Plant	Kim & Minamikawa, 1997
	Protoplast cell suspension culture	Electroporation		GUS, luciferase	-	Plant	Leon *et al.*, 1991
	Intact cell	Electroporation		pDPG165, bar	-	Plant	Sakar & Kuhne, 1998
	Embryonic axis	PB		Bar			Aragao *et al.*, 2002
Vigna angularis Wild.	Elongated epicotyls	A.t. EHA 105		Ualics			Yamada *et al.*, 2001
Phaseolus acutifolius	Germinating embryo	PG		GUS, nptII	-	-	Bhargava & Smigoeki, 1994
	Callus	A.t. C58C1		GUS		Plant	Dillen *et al.*, 1997

(Contd.)

	Bud explant	A.t. C58C1RifR		NptII			De Clercq *et al.*, 2002
Cicer arientum	Embryo axes	A.t . LBA4404	PBI121	nptII, GUS		Plant	Fonatana *et al.*, 1993; Kar *et.al.*, 1996; Ramana *et al.*, 1996
	Embryo axes	A.t., GV2260, EHA101	p35SGU-SNT	NptII, GUS		Plant	Krishnamurthy *et al.*, 2000
	Cotyledonary node	PG		NptII	Bt Cry1Ac	Plant	Kar *et al.*, 1997
Vigna mungo	Germinated seeds, cut stem	A.t.	-	DNA-A & DNA-B from MYMV (Mungbean yellow mosaic virus)		Plant	Mandal *et al.*, 1997
	Primary leaves	A. t. LBA4404-EHA105	pGA472	NptII		Callus	Karthikeyan *et al.*, 1996
	Germinating embryo	PG		GUS, nptII	-	-	Bhargava & Smigoki, 1994
Vigna rediata	Cotyledon	A.t.	pWC12GOS	GUS, nptII		Plant	
	Hypocotyl	A.r.	-	NptII, GUS		Plant & root	Jaiwal *et al.*, 2001
	Germinating embryo	PG		Gus nptII	-	-	Bhargava & Smigoki, 1994
Vigna unigiculata	Immature cotyledon	PG	DP532	GUS, bar	-	Plant	Koler *et al.*, 1997
	Leaf disc	A.t.			m-RNA of CPMV	callus	Garcia *et al.*, 1986, 1987
	Immature seeds	PG	pBI121, pBC1	GUS	-	Callus	Matsuoka *et al.*, 1997
	Embryo from mature seed, nodal explant	Electroporation		GUS		Plant	Penza *et al.*, 1992; Akella & Lurquin, 1993; Chowrira *et al.*, 1995
	De-embryonated cotyledon	A.t.		HptII, GUS		Plant	Gnanam *et al.*, 1995; Muthukumar *et al.*, 1996

(Contd.)

(Contd.)

Pisum sativum	Epicotyl and nodal explant	A.t.GV2260, GV 3850	p35SGUSINT	GUS, nptII, hpt		Plant	De Kathen & Jacobson, 1990
		A.t. EHA101	pU53	nos		Plant	Jorden & Hobbs, 1993,
						Plant	Shade *et al.*, 1994
		A.t AGL1	pLN27	nptII	Barnase	Plant	Grant *et al.*, 1995
		A.t.			αAi		Schroeder *et al.*, 1993, 95
							Ishimoto *et al.*, 1996
	Cotyledonary meristem	A.t. EHA105	SLJI561	Bar		Plant	Bean *et al.*, 1997
	Immature cotyledon	A.t		Bar	Na-PI	Plant	
	Immature cotyledon	PG		GUS	-	-	Ozcan, 1995
	Chloroplast	Direct uptake		R46, sulfonamide resistance gene	-	Plant	
	Protoplast	Electroporation		pKS, nptII	DS element of maize Ac-DS transformatio system	Callus	Puonti-Kaerlas *et al.*, 1992
	Protoplast	Electroporation		GUS	-	Callus	
	Immature cotyledon	At (Ag1)		NptII			Grant *et al.*, 1998
	Cotyledons	At-EHA105, C58C1, LBA4404		NptII, bar			Nadolska-Orczyk & Orczyk, 2000
	Embryonic axis	At (EHA 105)		Bar; nptII, als			Polowick *et al.*, 2000
Arachis hypogea		PG		GUS	-	Plant	Brar *et al.*, 1994
					Bt cry1Ac	Plant	Singsit *et al.*, 1997
	Embryo	PG		pDO432, pHygr & pGNI, hptII	-	-	Livingstone & Birch, 1999

(Contd.)

(Contd.)

	Embryonic culture	PG		GUS, hptII -	Plant	-	Wang *et al.*, 1998
Cajanus cajan	Embryonic axis	At (GV2260)			Coepea protease inhibitor	Plant	Lawrence & Koundal, 2000, 2001
	Embryonic axis and cotyledonary nodes	At (EHA105)		NptII,	Hamagglutinin protein of Rinderpest virus	Plant	Satyavathi *et al.*, 2003
	Shoot apex, cotyledonary node	A. t.-LBA4404		NptII, GUS		Plant	Geetha *et al.*, 1999
	Petiole	PG		NptII, GUS		Plant	Dayal *et al.*, 2003
	Cotylednar node	A. T.		NptII, GUS		Plant	Thu *et al.*, 2003
Vicia faba	Epicotyle	At (EHA101 and 105)		NptII			Böttinger *et al.*, 2001
Vicia narbonensis	Epicotyle and shoot tip	At (EHA 101)		NaptII			Czihal *et al.*, 1999
Psophocarpus tetragonoloba	Protoplast	PG	pABDI, pHP23	NptII		Callus	Gill, 1990
Lens culinaris	Shoot apex, other seedling explants	A. t. (GV2260)	p35SGUSINT	GUS	Transient expression		Warkentin & MeHughen, 1992
	Leaf prtoplast	Liposome mediated	pCaMVCN, pBI121, pBI221	NptII, GUS, cat	Transient expression		Maccarone *et al.*, 1992
	Cotyledonary node	PG		GUS		Transient expression	Öktem *et al.*, 1999
	Cotyledonary node	A. t. (GV2260)-vacuum infiltration	pGUSINT	GUS		Transient expression	
	Cotyledonary node	Particle bombardment			Acetolactate synthase	Plant	Gulati *et al.*, 2002
Cymopsis tetragonoloba	Cotyledons	A. t. (LBA 4404)	pBKL4	NptII, GUS			Joerbo *et al.*, 1999
		A. t.			Alpha galactosidase gene from *Senna* seeds	Plant	Joersbo *et al.*, 2001

(Contd.)

(Contd.)

Lupinus luteus	Meristem	A. t.	pPZBNIa	bar		Plant	Li *et al.*, 2002
Lupinus angustifolius	Shoot apices, embryonic axis	A. t.		Bar		Plant	Pigeaire *et al.*, 1997
		A. t.			Sunflower seed albumin gene	Plant	Molvig *et al.*, 1997
Lupinus mutabilis	Shoot apical *mutabilis*	A.t. LBA4404	PTOK47	NptII, GUS		Plant	Babaoglu *et al.*, 2000

instrumentation, a high degree of dexterity, and an element of luck. The method selected must have a reasonable expectation of success and must inflict minimal damage on the cell. Direct DNA transfer by chemical method needs the removal of cell wall, which is achieved by enzymatic degradation. In chemical method uptake of naked DNA in plant protoplasts is facilitated by calcium phosphate or polyethylene glycol. In electroporation, plant protoplast or nodal meristems or ovaries (Chowrira *et al.*, 1995; Yue *et al.*, 1996) subjected to high voltage electrical charge for short period that leads in the formation of the pores in membrane through which DNA moves inside the protoplast and get incorporated in the plant genome. The mechanism by which integration occurs is not fully understood but presumably by illegitimate recombination. In Particle bombardment, acceleration of microprojectiles (1-4 mm) coated with DNA is carried out at high speed (300-600 m/s) that is sufficient for the penetration of plant cell walls and membranes. Propelling force is generated either by gunpowder, helium gas, nitrogen gas, compressed air or electric discharge. As compared to the other techniques of gene transfer this method is simple and rapid as it depends only on physical parameters and has the potential to transform any type of cell. Transformation by this method is genotype independent. The most important parameter in this method is the acceleration of the microprojectiles so that they penetrate the cells without causing excessive damage to the cell. One widely used procedure for DNA insertion is microinjection. A microscopic needle penetrates the cell wall and the plasma membrane and injects DNA into the nucleus. DNA can be delivered precisely into small cells such as microspores, and few celled proembryos. Despite its straightforward approach, microinjection is a laborious method with a high failure rate. The disadvantage is that only one cell can be injected at a time, which requires more skill and 10,000 or more injections may be required before there is a successful DNA transfer to the nucleus. Macroinjection employs needles with diameters greater than cell diameter. In rye (*Secale cereale*) a marker gene was macroinjected into the stem below the immature floral meristem, so as to reach the sporogenous tissues. Another DNA insertion method uses needle-like fibers, or "whiskers" of silicon carbide. When combined with cells and fragments of DNA, the fibers punch innumerable holes in the target cells. The hole permits passage of DNA fragment into the cells (Kaeppler *et al.*, 1990). Enhancement in transformation frequency was observed when cells were mechanically injured by sonication prior to *Agrobacterium* infection (Trick & Finer, 1997). In liposome mediated transformation method introduction of naked DNA into cells is carried out via liposomes. It is one of the first methods of DNA deliver into plant protoplast. Higher frequency transformation with PEG and electroporation is reported. *In planta* transformation is very successful in Arabidopsis. Thousands of transformed lines were produced in a few years by applying *Agrobacterium* culture to *Arabidopsis* seeds (Feldmann and Marks, 1987; Feldmann, 1992). Chang *et al.* (1994) used another approach called as "clip'n Squirt" method for *in planta* transformation for *Arabidopsis*. In this method *Agrobacterium* was applied to the centre of plant rosette after clipping off the reproductive inflorescence. *Agrobacterium* was reapplied after cutting of the new inflorescence formed. Plants were then allowed to grow and set seeds. Beachtold *et al.* (1993) again modified the *in planta* transformation of *Arabidopsis*. Transformed plants were recovered by "vacuum infiltration" method. Plants at early stages of flowering were uprooted and submerged in a fresh culture of *Agrobacterium* and the bacteria are introduced into plants by vacuum infiltration. Treated plants were then grown and selfed and allowed to set seeds. Successful *in planta transformation* with high transformation rate (2.9% to 76%) in forage legume *Medicago truncatula* by *Agrobacterium* vacuum infiltration has been reported (Trieu *et al.*, 2000). Transgenic plants were obtained only when seedlings were subjected to 4°C/14 d vernelization treatment. This approach is very useful in plants like legumes where regeneration protocols are either not available or regenerating parts of plants are not accessible to *Agrobacterium*-mediated transformation. The generation of transgenic plants in the absence of tissue culture reduces the risk of somaclonal variation. In addition to being useful for transformation it can also be used to create the large number of independent transgenic. In pollen transformation DNA can be taken up into germinating pollen and can either integrate into the sperm genome or reach the zygote with the pollen tube. In this transformation method pollen grains soaked in DNA solution were used for pollination. The method is highly attractive in view of its simplicity and general applicability. But the presence of cell wall, external and internal nucleases may cause the problem for the efficient transformation of the foreign DNA. In pollen tube pathway method stigma of a flower is cut off after pollination, and DNA solution is applied onto the cut surface. This method is easy and simple, provided consistent results and stable transformations are achieved.

Reporter Genes

Detection of foreign DNA that has been integrated in plant genome is essential so that cells that are transformed can be identified. Sometimes quantification of the level of expression of foreign gene with easily identified product is needed. Reporter genes can be used to select the transformed cell and to quantify their expression. A number of prokaryotic and eukaryotic genes have been tested for plant transformation. Many studies on plant promoters and on the regulation of the gene expression have been made possible by the use of the reporter gene. The activity of some reporter gene products like β-D-*glucuronidase* and *luciferase* can be detected in intact plant cells. Reporter genes that allow the visualization of the expression in whole tissue includes *luciferase (luc)*, β-D-*glucuronidase* (*uid or gus*), aequorein and green fluorescent protein (*GFP*). The most widely used reporter gene in plant molecular biology is the β-D-*glucuronidase* (*uid or gus*) gene from *E. coli.* (Jefferson *et al.*, 1986). It encodes a stable enzyme resistant to thermal denaturation, detergents and alkalis.The *gus* activity in transformed cells can localized by observing the blue colour that is formed after the hydrolysis of the substrate 5-bromo-4-chloro-3 indolyl β-D-glucuronic acid to produce indoxyl derivative which subsequently undergoes an oxidative dimerization to form an insoluble blue colour dye. This blue colour can be easily visualized and provides precise marker of the cell type in which the gene is expressed. Activity of *gus* gene in plant extract can be quantitatively assayed by a flurometric analysis in which hydrolysis of the substrate 4-methylumbelliferyl-β-D-glucuronide to a fluorescent product occurs. The *luciferase* gene (DeWet *et al.,* 1985) from firefly (*Photinus pyralis*) encodes an enzyme that catalyses the oxidation of luciferin in the presence of ATP and O_2 to generate oxyluciferin and yellow green light. A high level of *luciferase* activity was detected in plants transformed with 35S construct (Ow *et al.*, 1986). Due to inefficient transport of the substrate through the plant vascular tissue may sometimes result in the low light production though the level of the expression of the gene is high. The use of *luciferase* as a reporter gene is limited because the enzyme is labile and difficult to assay accurately. Histochemical assay is impossible due to the nature of the end product that is visible light. A green fluorescent protein (GFP) from jellyfish (*Aequoria victoria*) emits bright green fluorescence upon excitation with ultraviolet (395 nm) or blue (475 nm) light (Heim *et al.*, 1994). GFP finds immense application in transgenic plants. GFP fluorescence occurs with UV/blue light and oxygen without any externally added substrate. Synthetic versions of GFP were made to avoid cryptic intron and to change the codon preference to that of flowering plant genes. These changes dramatically increased the intensity of fluorescence in transgenic *Arabidobsis* and tobacco plants and full length GFP was translated. Davis and Viestra (1998) carried out two-site directed mutagenesis experiments to generate special variants. A soluble-modified red shift version of GFP (smRS GFP) and a soluble modified blue fluorescent version of GFP (smBFP) were generated. Plants transformed with GFP appears green in contrast to reddish wild type plants. The fluorescent properties associated with GFP make it an ideal marker for plant transformation, however poor expression efficiency and low quantum yield of wild type protein have slowed its widespread application on plants. Successful use of GFP is reported in tobacco (Chiu *et al.*, 1996; Pang *et al.*, 1996; Leffel *et al.*, 1997; Moliner *et al.*, 2000), Arabidopsis (Pang *et al.*, 1996; Haseloff *et al.*, 1997), corn and wheat (Pang *et al.*, 1996; Jordenm, 2000), barley (Ahlandsberg *et al.*, 1999; Fang *et al.*, 2002), *Medicago truncatula* (Kamaté *et al.*, 2000), oat (Kaeppler *et al.*, 2000), *Astragalus sinicus* (Cho and Widholm, 2002). Chloramphenicol acetyl transferase (*cat*) from transposon T9 of *E. coli* (Alton *et al.*, 1979) has been widely used as a reporter gene in mammalian cells and to a lesser extent in plants. It modifies the antibiotic compound chloramphenicol. The expression can be assayed with cellular extract and chloramphenicol by thin layer chromatography of the reaction products. Endogenus *cat* activity has been observed in *Brassica* species (Balazs and Bonneville, 1987). The lac Z gene encoding b-galactosidase (β-*GAL*) has been expressed in tobacco crown gall tissues (Matsumato *et al.*, 1988).

Selectable Marker Genes

A selectable marker gene is used to recover transformants after a gene transfer experiments. It encodes a protein that confers on transformed cells the ability to grow on media containing a compound toxic for untransformed cells. The introduced gene encodes for the enzyme insensitive to inhibition by the selective agent. The commonly used selectable markers are based on antibiotic resistance genes such as *nptII* gene also known as *aphA2* or *aph(3')II* gene encoding neomycin phosphotransferase from transposon Tn5 from *E. coli.* was the first selectable marker used for transformation (Herrera-Estrella *et al.*, 1983; Bevan *et al.*, 1983). This gene detoxifies a number of aminoglycoside antibiotics including neomycin, kanamycin and genticin. Another gene encoding for a detoxifying enzyme used for plant transformation is the *hptII (hph or aph (3') IV)* gene encoding hygromycin phosphotransferase from *E. coli.* This enzyme detoxifies the antibiotic hygromycin B. Several crop species have been transformed with *hph* gene. The streptomycin phosphotransferase gene (*spt*) from transposon Tn5 was first developed as a selectable marker for plant transformation (Jones *et al.*, 1987). Herbicide resistance gene *pat* from *S. viridiochromogenes* encoding phosphinothricin acetyltransferase that allows selection with the herbicide Basta or closely related chemicals is widely used. The Bialaphos resistance gene (*bar*) from *Streptomyces hygroscopicus* encodes an enzyme phosphinothricin acetyltransferase that is able to detoxify the herbicide phosphinothricin (Thompson *et al.*, 1987).

Promoters/Enhancers

Foreign genes must have an appropriate promoter, 5' leader and 3' terminator sequence for efficient expression in plant cells. A variety of promoter sequences have been used to drive genes in plant cells. Widely used promoters include *nos* (nopaline synthase), *ocs* (octopine synthase), *mas* (manopine synthase) promoters from *Agrobacterium* and 35S RNA gene promoter from Cauliflower Mosaic Virus (CaMV). 35S promoter is the most commonly used constitutive promoter both in dicots and in monocots. It is 10-40 fold more efficient than the *nos* promoter. This promoter achieves a high level of transcripts in nearly all plant tissues. Induced modifications of some of the promoters, for example, 35S RNA promoter have resulted in increased promoter efficiency. The 35S promoter consists of several domains, which confer developmental, and tissue specific expression patterns. The 35S promoter possesses a transcriptional enhancer located upstream of the TATA box. The duplication of certain enhancer sequence present within the 35S promoter has increased promoter activity by several folds (Kay *et al.*, 1987). Though it is presumed to be a constitutive promoter, some reports suggest that it is not expressed in all cell types. Custers *et al.* (1999) reported that in *Brassica napus* the first histochemical detection of gus activity was at the mid-cotyledon stage in developing embryos and not in the very young embryos. During embryo development in cotton the activity of the 35S promoter using GFP as a reporter gene was investigated and it was found that during early stage of embryogenesis the GFP expression was not detected. GFP activity was not detected in globular, heart, and torpedo stage embryos. First detectable fluorescence was observed in the embryos at the stage when the cotyledon just begins to expand. As the embryo development advanced the expression became stronger. GFP expression was evident in the remaining stages of embryo development until maturity. It proved that the expression of 35S promoter was developmentally regulated (Sunilkumar *et al.*, 2002). The maize alcohol dehydrogenase 1 (Adh1) promoter shows anaerobic induction/expression in roots. The activity of Adh1 promoter is either comparable or higher than that of 35S promoter which is not an efficient promoter for the monocot species. Saalbach *et al.,* (1994) found the high expression of methionine rich 2S albumin gene of Brazil nut under the control of seed specific legumin B4 promoter in transgenic seeds of *Vicia narbonensis.* Similarly high activity of yeast invertase gene under the control of legumin B promoter was observed in *V. narbonensis* seed cotyledons (Weber *et al.*, 1998).

Factors Affecting Transformation Efficiency

Successful genetic transformation depends on various factors like specific bacteria-plant interactions, genotype, chemical compounds, pre-culture and co-cultivation conditions and selection schemes, type, morphology and age of the explant, transformation competency of the target tissue and plant growth regulators. Some regulatory sequences like promoters and enhancers also play a major role in transformation and in the expression of the transgene. When cotyledon explants of soybean were inoculated with mixed strains of *Agrobacterium tumefaciens* with super virulent L, L-succinamopine type strain A281 (pTiBo542) and strain LBA 4404 carrying an octopine type virulence region and a binary vector (pBin6) with a chimeric gene for kanamycin resistance gave rise to tumours of which 25% were both kanamycin resistant and capable of hormone independent growth. On the other hand single strain inoculation with LBA4404 (pBin6) failed to give rise to kanamycin resistant callus (Owens and Smigoki, 1988). In Chinese soybean cultivars tumour formation frequency was found higher with *A. tumefaciens* Chry5 strain as compared to highly virulent strain A281 and T37 (Li *et al.,* 1994). The susceptibility of chickpea to different strains of *A. tumefaciens* and *A. rhizogenes* was evaluated. The study revealed specific strain/genotype interactions (Riazuddin and Husnain, 1993; Islam *et al.*, 1994; Siefkes *et al.*, 1995; Altinkut *et al.,* 1996; Husnain *et al.,* 1997; Srinivasan *et al.*, 1998). It was reported that successful transformation in chickpea was dependent on bacterial strain-plant genotype interactions. Agropine strain A281 was the most effective for tumour formation in peanut and chickpea (Lacorte *et al.,* 1991; Islam *et al.,* 1994). *Agrobacterium* strain EHA 101 (or its derivative EHA 105, or its oncogenic strain A281) is generally considered highly virulent strain for leguminous species such as soybean (Meurer *et al.*, 1998; Donaldson and Simmonds, 2000), pea (Nadolskar-Orczyk and Orczyk, 2000) and peanut (Egnin *et al*, 1998), and in other crops such as apple (De Bondt *et al.*, 1994), cabbage (Takasaki *et al.*, 1997), blueberry (Cao *et al.*, 1998) wheat and barley (Guo *et al.*, 1998). However, this strain is not effective in common bean (Zhang *et al.*, 1997), cotton (Sunilkumar and Rathore, 2001), rice (Hiei *et al.*, 1994,1997), *Kalanchoe* (Jia *et al.*, 1989), black locust (Igasaki *et al.*, 2000) and *Hibiscus* (Srivatnakul *et al.*, 2001). This study suggests that testing of different strains for various species and genotype is needed for successful transformation. Specific strain-genotype interaction was observed in faba bean (*Vicia faba*) and chickpea when different genotypes were inoculated with eight *Agrobacterium* root inducing strain. Hairy root size and morphology differed substantially between faba bean and chickpea (Siefkes *et al.,* 1995). Transformation with *A. tumefaciens* and *A. rhizogenes* wild type strains revealed contrasting differences between pea genotypes in tumour and root formation respectively. Specificity of the four wild *Agrobacterium tumefaciens* to five genotypes of peanut was studied and it was found that strain A281 was the most effective for tumour induction (Lacorte *et al.*, 1991).In pigeonpea,. susceptibility of *Agrobacterium* has been reported with wild strains of *Agrobacterium tumefaciens* A281, A6 and T37.Strain A281 induced tumour formation in 5 cultivars, A6 in 4 cultivars and T37 in 3 cultivars. The tumours were largest in plants infected with Strain T37. The genotype Bahar and Pant A106 gave the highest response followed by H82-1, UPAS 120, and ICPL 151 (Rathore and Chand, 1997). Different genotypes of common bean were tested for susceptibility to *A. tumefaciens* and *A. rhizogenes.* Tumour formation was observed when several cultivars of bean were inoculated with wild *A. tumefaciens* strain. Genotype specific response in terms of frequency and size of tumours were observed (Genga *et al.*, 1990). Transformation of cotyledons and stem tissues by *A. rhizogenes* was carried out to study the infectivity of the bacterium to the *V. faba* plant. The response to infection depended on tissue type and plant genotype. The change in the ploidy level up to the octoploidy and structural rearrangement of the chromosome of the transformed root clones were recorded (Ramsay and Kumar, 1990). Susceptibility of lentil to *Agrobacterium* was studied using different strains and explants (Warkentin and Mc Hughen 1991, Lurquin *et al.*, 1998; Khawar and Özcan, 2002). Four trains of *Agrobacterium* C58, Ach5, GV3111 and A281 were tested for tumour formation. GV3111 and Ach5 produced the largest and heaviest tumours on inoculated stem of lentil *in vivo*, while A281 produced the heaviest tumour *in vitro* (Warkentin and Mc Hughen, 1991). Pea lines with a typical morphogenetic responses to transformation were observed. The existing differences between lines for root formation and tumour formation allowed genetic analysis of these traits. F1 hybrids were uniform with segregation in the F2. The capacity for root formation was inherited as a dominant trait, but tumour formation was a recessive trait. Using exogenous phytohormones, phenocopies of the typical morphogenetic response to transformation were obtained. The results suggest an interdependence of these morphogenetic processes with endogenous phytohormone balance as determined by plant genotype (Lutova *et al.*, 1996). When embryogenic cultures of soybean bombarded, it was found that the cell were not competent for transformation until cultures were at least six months old. Histological analysis of cell lines differing in transformation ability showed that cytoplasmic rich cells in the outermost layer of the tissues are more competent for transformation, while the cells that are less transformable contained prominent vacuoles. When the cultures transferred to fresh medium it shows the high mitotic activity, which lasts from the second to sixth day following subculture, and peaked on the fourth day. Tissues at this stage produced more transient expression of the reporter gene. This indicates that any treatments that increase the mitotic index may increase the transformation frequency (Hazel *et al.*, 1998). In an experiment conducted with immature zygotic cotyledons from six different soybean varieties transformation was performed by particle bombardment. Transient expression was checked for *gus* activity. Expression of *gus* varied among genotypes independently of the pressure of bombardment. It was observed that the ability to express the reporter gene depended on the developmental stage of the seed. *Gus* expression was least in the younger stage of the cotyledon (Moore *et al.*, 1994). In common bean it was reported that seedlings germinated for 1-3 days are more susceptible to *Agrobacterium* infection than seedlings germinated for 5-7 days. The tumours that formed on the 1-3 days old seedlings were significantly larger (McLean *et al.*, 1991; Lewis and Bliss, 1994). The best transformation condition for common bean is the use of susceptible genotype and explants derived from mature seeds, preconditioning of explants in a medium containing 20 mM BA in darkness or on a filter paper, dipping explants in high concentration of *Agrobacterium* cell suspension (O.D. 650 1 0.8-1.0) followed by a long term (6 days) co-cultivation period on a semisolid agar and in the presence of cytokinin or 3- day co-cultivation on a moistened filter paper and the use of lethal levels of selective agents (Zhang *et al.*, 1997). Morphology of the explant is an important factor in biolistic transformation. In some cultivars of common bean the apical meristematic region of shoot apex partially covered by the primordial leaves hindered the biolistic transformation (Arago and Rech, 1997). Transformation frequency also depends on the type of the plasmid used for transformation. When soybean protoplasts were electroporated with the plasmid containing number of foreign genes either linked on a single plasmid or as a separate plasmid transformation frequency was different. In the case of linked gene 50% transformation frequency was observed in transgenic cell lines while transformation frequency was 18-27% in the case where genes were on the separate plasmid (Christou and Swain, 1990). Effect of lipoxygenase activity present in soybean embryo on *Agrobacterium*-mediated transformation was investigated. The mutant line lacking lipoxygenase 3 showed significantly greater frequency of *Agrobacterium* mediated transformation than other lines. Stages of embryo development which showed the greatest differences in lipoxygenase 3 activity between mutant and wild type also showed the greatest differences in transformation frequency with the absence of lipoxygenase 3 was only seen when both lipoxygenase 1 and 2 were present (Bhatt *et al.,* 1991). *Agrobacterium* transformation was quite effective when syringaldehyde and acetosyringone, a *vir* gene activator compound was used during transformation with excised cotyledons of soybeans by the weakly virulent octopine strain A348 (pTiA6) (Owens and Smigoki, 1988). Successful transformation via particle bombardment depends on various factors like osmotic treatment to the explants, gas and vacuum pressure, distance between the explants, velocity of the microprojectiles and amount of DNA. Preincubation of peanut embryos and leaflets on osmoticum resulted in a 10-fold and 4-fold increase in the transient expression (Livingstone and Birch, 1995). Embryos and embryonic leaflets bombarded with higher helium pressure and hence higher microprojectile velocities showed increased transient expression rates (Livingstone and Birch, 1995). Reduced transient *gus* expression was observed in bombarded peanut seed when the precipitation mixture contains higher amounts of DNA (Lacorte *et al.,* 1997). Electroporation was used to evaluate parameters affecting transient gene expression in soybean protoplasts. Protoplast viability and reporter enzyme activity of *cat* and *gus* gene depended on the field strength employed. Transient gene expression increased with increasing plasmid DNA concentration and with the time after electroporation. Addition of polyethylene glycol at 5.6% and heat shock (5 minutes at 45°C) given to the protoplasts before adding DNA further enhanced the transformation efficiency in soybean (Dhir *et al.,* 1991). The effect of auxin on the expression of isopentyl transferase gene (*ipt*) was studied in common bean. The *ipt* transcript was not detected in tumour single cell clone cultured on picloram supplemented medium, but it was detected in tumour single-cell clones cultured on hormone-free medium. However, the nopaline synthase gene (*nos*) transcript was detected in the tumour single cell clones from both culture conditions. So it was concluded that picloram regulates the *ipt* mRNA steady state level, either at the transcriptional level or by affecting *ipt* mRNA stability (Young *et al.,* 1995). Incubation of tissues on kinetin enhanced the expression of *gus* gene in *Vigna* species (Bhargava and Smigoeki 1994). In chickpea, the expression of marker gene was 20-40% higher using a CaMV 35S promoter in comparison to the Win 6 and Actin promoters when zygotic embryos of chickpea were bombarded with tungsten particles coated with plasmid carrying *gus* and *nptII* gene (Hunsain *et al.,* 1997). The most successful transformation system in grain legumes exploited embryonic axes (Schroeder *et al.*, 1993, 1995), stem nodal segments, cotyledon hypocotyl sections (De Kathen and Jacobsen 1990; Nauerby *et al.*, 1991; Davis *et al.*, 1993), and apical explants (Pickardt *et al*; 1991; Russel *et al.*, 1993). Such explants all have terminal or axillary meristems having high shoot regeneration capacity. In pea, meristematic cells have been considered not to be competent or to have low competence to *Agrobacterium* mediated transformation (De Kathen and Jacobsen, 1995). Transgenic plants have been obtained from meristematic cells in pea (Hussey *et al.*, 1989; Bean *et al.*, 1997), chickpea (Fontana *et al.*, 1993; Kar *et al.*, 1996, 1997; Krishnamurthy *et al.*, 2000), common bean (Lewis and Bliss, 1994), sweet lupin (Babaoglu *et al.*, 2000), and blue lupin (Pigeaire *et al.*, 1997). The transformation procedure involves excision of the apical meristematic layers or injuring the apical meristems prior to inoculation with *Agrobacterium.* Pigeaire *et al.* (1997) transformed blue lupin with herbicide resistance gene by stabbing the apical dome and primordial of the third pair of leaves of shoot explants with *A. tumefaciens* carrying *bar* gene. Babaoglu *et.al.* (2000) could induce multiple shoots only after removal of the cell layers of the apices of the sweet lupin to a maximum depth of 300mm below the apical dome. Transgenic shoots of lettuce were obtained using such explants (Curtis *et al.*, 1994).

TRANSGENIC LEGUME CROPS EXPRESSING VARIOUS TRAITS

Legume Species with Only Selectable Marker and Reporter Genes

There are numerous examples of successful transformation of leguminous crop species using selectable marker and reporter genes. Soybean is a model system and it is the most extensively studied legume for transformation. It was one of the first legumes to be genetically transformed. Worldwide, soybean is the only transgenic grain legume being cultivated in nearly 63% of the total area under transgenics (James, 2003). Chimeric transgenic plant of soybean were obtained from immature zygotic cotyledon explants by *Agrobacterium*-mediated transformation. Transformed progeny could not be obtained due to the chimeric nature of the T_0 plants (Parrott *et al.*, 1989). Chee *et al.* (1989) reported a non-tissue culture method for *Agrobacterium*-mediated transformation of soybean. The plumule, cotyledonary node and the adjacent cotyledon tissues of germinating seeds were inoculated with *A. tumefaciens* containing the gene for kanamycin resistance. The transformation frequency was about 0.7%. The integration of the kanamycin resistance gene in 3 T_1 plants was confirmed by genomic blot hybridisation. The maize controlling element Ac was introduced in the soybean cells by *Agrobacterium*-mediated transformation. Calli transformed with intact *gus* gene showed even staining of the tissues, staining was rare in the callus transformed with *gus* including Ac transposable element in the untranslated leader region and no staining was observed in the callus transformed with *gus* including Ds element in the untranslated leader region. Southern hybridisation and *in situ* hybridisation showed that the Ac is active in callus. But no transgenic plant from transformed callus was recovered (Zhou and Atherly, 1989,1990). Transgenic soybean plants expressing the *nptII* and *GUS* reporter gene were produced from the meristem of embryonic axes by electric discharge particle bombardment (McCabe *et al.*, 1988; Christou, 1990). To study the co-transformation, 12 different plasmids were transformed in embryogenic suspension culture of soybean tissue, via particle bombardment. Ten different plasmids containing RFLP markers for maize and 2 plasmids separately having hygromycin resistance and beta glucuronidase genes were used for transformation. All of the co-transforming plasmids were present in most of the transgenic soybean clones and there was no preferential uptake and integration of any of the plasmid (Hadi *et al.,* 1996). Transformation of embryonic suspension culture of soybean using the sonication-assisted *Agrobacterium* transformation (SAAT) technique has been reported in order to achieve high frequency transformation. It was found that this technique markedly improved the efficiency of agroinfection by introducing number of microwounds into the target tissue. Hygromycine-resistant clones were obtained from suspension culture (Trick and Finer, 1998). Transient expression of *gus* gene was optimised by SAAT. Immature cotyledons of soybean were used as explants to explore the gus activity. The extent of tissue disruption caused by sonication determined. The highest gus expression was obtained when immature cotyledon were sonicated for 2 seconds in presence of *Agrobacterium* followed by co-cultivation for 3 days at 27°C. The addition of acetosyringone to the co-culture medium enhanced transient expression of *gus* gene. Cotyledons sonicated for 2 seconds had moderate tissue disruption while longer treatments resulted in more extensive damage (Santarem *et al.,* 1998). Transformation frequency of the soybean plant has been studied with using different selectable marker gene and thiol compounds. Enhancement in the transformation frequency was reported in these experiments (Olhoft and Somers, 2001; Yan *et al.*, 2000). In soybean, hygromycin as selection marker greatly increased the transformation frequency and reduced the number of escapes and culture period (Olhoft *et al.*, 2003). It was found that the frequency of transformed cells was increased only when L-cysteine was present during co-cultivation period (Olhoft *et al.*, 2001). Transformed soybean clones containing *nptII* gene were obtained when protoplasts from immature embryos were transformed by electroporation with the plasmid carrying a gene for neomycin phosphotransferase (*nptII*). But the transgenic plants could not be recovered from the transformed clones (Kuchuk *et al.,* 1990). Stable transformation of soybean protoplasts isolated from immature cotyledons was achieved following electroporation with the plasmid DNA carrying *gus* and *nptII* genes. Transformed colonies were selected on hygromycin. Shoots were obtained from transformed callus (Dhir *et al.,* 1991). Cell type specific expression of *gus* gene along with 35S CaMV promoter was studied through direct DNA transfer via electric discharge mediated particle acceleration. It was found that pericycle cells in the root, parenchyma cells in the xylem and phloem tissues of the stem and leaf had high expression level of *gus* gene. Procambial, phloem and cortex cells in the leaf midrib expressed low or no gus activity, while in leaf mesophyll cells, stem, leaf midrib, trichomes and epidermal guard cells, the GUS activity was at intermediate level (Yang and Christou, 1990). A novel system to select transgenic meristematic cells after the physical introduction of foreign genes has been developed with using the selectable marker gene, ahas, from *A. thaliana* and imazapyr, a herbicidal molecule that concentrates in the apical meristematic region of the plant. Multiple shoots were obtained from the meristematic cells. Production at high frequency of fertile transgenic soybean plants was genotype-independent (Arago *et al.,* 2000). Transgenic chickpea plants carrying *nptII* and *GUS* gene from different explants have been reported (Fontana *et al.,* 1993; Kar *et al.,* 1996; Raman *et al.,* 1996; Krishnamurthy *et al.,* 2000). Transgenic calli were obtained in common bean when leaf discs were cocultivated with engineered *A. tumefaciens.* The transformed nature of the callus was confirmed by *Npt*II assays (Genga *et al.,* 1990). Kanamycin resistant callus was produced from leaf disc and hypocotyls explants of common bean (*P. vulgaris*) when cultured on a defined medium containing kanamycin after 4 days of co-cultivation with *A. tumefaciens* strain EHA 101 and EHA 105 carrying the plasmid pKYLX71*GUS* and CHS15-*GUS* respectively. The construct CHS-*GUS* contains the bean chalcone synthase (CHS15) promoter-*GUS* fusion in the binary vector pBI101.1. ELISA confirmed the presence of *Npt*II in crude cellular extract from kanamycin resistant callus. The expression of GUS was confirmed by histochemical and flurometric analysis. Southern analysis confirmed the integration of the *uid*A gene in the bean genome (Franklin *et al.,* 1993). Stable transformation and regeneration was developed for a grain legume, azuki bean (*Vigna angularis*). Two constructs containing *nptII* and *uid*A gene or the modified green fluorescent protein gene were introduced independently via *A. tumefaciens*-mediated transformation. Molecular analysis confirmed the stable integration and expression of foreign genes (Yamada *et al.*, 2001). Significant differences for tumour frequency and size was observed when shoot tips of germinating seeds were stabbed with *A. tumefaciens* strain C28. Histochemical GUS assay of inoculated shoots showed transient expression of GUS (Lewis and Bliss, 1994). Germinating seeds were infected with nopaline, octopine, and agropine biotypes of *A. tumefaciens*. It was reported that seedlings germinated for 5-7 days and the galls that are formed on the 1-3 days seedlings were significantly larger. Transformed callus tissue and roots carrying *nptII* gene were obtained when cotyledonary nodes and hypocotyls of bean were inoculated with disarmed *A. tumefaciens* strain C58Z707 and *A. rhizogenes* A4RS (pRiB278b). Integration of the gene was confirmed by genomic blot hybridisation (McLean *et al.,* 1991). When apical meristem of common bean was infected with *A. tumefaciens* after injuring with tungsten microprojectile bombardment the explant showed the high frequency of tumour formation. Similarly, when bombarded meristems were inoculated with an *A. tumefaciens* disarmed strain, 44% of the explants showed substantial sectors of GUS activity (Brasileiro *et al.,* 1996). Activity of GUS was detected in 30 % of the regenerated shoots when cotyledonary explants were cocultivated with *A. tumefaciens*. Transgenic roots were obtained when epicotyls were inoculated with *A. rhizogenes* (Barros *et al.,* 1997). Transient expression of luciferase and *uid*A gene was observed in the protoplasts of *P. vulgaris* transformed by electroporation and PEG treatment (Leon *et al.,* 1991). Transient expression of *uid*A gene was observed when embryonic axes of *P. vulgaris* were electroporated with plasmid pDE4, containing the *uid*A gene under the control of CaMV35S promoter (Dillen *et al.,* 1995). Stable transformation of *uid*A gene was achieved when shoot apices of embryonic axes were bombarded with plasmid pSOG-1016 carrying a chimeric gene encoding GUS under the control of concanavalin A (con A) promoter. PCR and Southern blotting confirmed the integration of the gene (Kim and Minamikawa, 1996; Aragao *et al.,* 1996). The organ and maturation stage specific expression of the *uid*A gene was observed histochemically in seeds of transgenic plants of *P. vulgaris* when shoot apices were bombarded with gold particles coated with plasmid pSAG-734 carrying *uid*A gene under the control of canavalin gene promoter, a seed specific promoter (Kim and Minamikawa, 1997). Transgenic pigeonpea plants with *nptII* and *uid*A gene were obtained via *Agrobacterium*–mediated transformation as well as through particle bombardment method (Geetha *et al.,* 1999; Thu *et al.,* 2003; Dayal *et al.,* 2003). Cotyledonary node explants were used for most of the transformation studies in pigeonpea. In the above reports, transgenic pigeonpea plants were obtained either through multiple shoot induction or callus formation from cotyledonary node explants when infected with *A. tumefaciens*. In pea, high frequency of transient GUS expression was observed on regeneration of the cotyledons 2 days after the bombardment immature cotyledon explants (Ozcan, 1995). Transgenic pea plants have been recovered by *Agrobacterium*-mediated transformation of thin slices from developing embryo axes. All vectors contained the *uid*A gene but with different chemical selection genes. In total, 323 transgenic plants were recovered from 39 independent transformation events. Transgenic plants were recovered from each genotype and each selection system, but not from all combinations. The transformation frequency was 0.6%. The inserted genes were functional and inherited in a Mendelian fashion. Although more plants were recovered by selection on chlorsulfuron, GUS activity was generally greater in plants selected on kanamycin (Polowick *et al,* 2000). The first successful transformed plants and seeds of *P. acutifolius* were obtained from callus by *A. tumefaciens* (strain C58C1) transformation (Dillen *et al.,* 1997). Transient GUS expression was observed on all the parts of germinating embryos after particle bombardment of tissues of *P. aconitifolia* (Bhargava and Smigoeki, 1994). To improve *Agrobacterium*-mediated transformation of *P. acutifolius*, the effect of different factors on T-DNA transfer were examined by measuring transient expression levels of an intron-containing *uidA* gene. Improved transformation frequencies were obtained with an *A. tumefaciens* strain carrying nopaline-type virulence genes and when calli were infected with *Agrobacterium* cells in the early-log growth phase. Optimized co-cultivation was performed at 22°C under a 16/8-h (day/night) photoperiod in an acidic medium (pH 5.5) in the presence of 200 μ*M* acetosyringone. By combining the best treatments, an efficient and reproducible transformation procedure was established for the *P. acutifolius* genotype NI576. Southern and immunoblot analyses confirmed the stable integration and expression of the transgene in the primary transgenic plants and their progeny (De Clercq *et al.*, 2002). Direct gene transfer of plasmid pABD1, pHP23 to epicotyl-derived protoplasts, obtained transformed calluses of *Psophocarpus tetragonolobus*. Transformed calluses expressed *Npt*II activity and also exhibited the presence of the plasmid gene integrated into the genome. But transformed callus was unable to regenerate into the shoot (Gill, 1990). Transgenic cowpea plants carrying *uid*A gene were obtained through electroporation using embryo from mature seeds and intact nodal meristem *in planta* (Penza *et al.*, 1992; Akella and Lurquin, 1993; Chowrira *et al.*, 1996). Transgenic shoots of cowpea carrying *hpt*II *and uid*A gene were obtained through *Agrobacterium*-mediated transformation using de-embryonated cotyledons (Gnanam *et al.*, 1995; Muthukumar *et al.*, 1996). In an experiment of gene transfer to cowpea by using particle gun, immature seeds of 5 cultivars were used to induce callus with B5 medium supplemented with BA and NAA. Efficiency of gene transfer by transient GUS expression was checked by bombarding embryos with gold microprojectiles coated with *uid*A reporter gene constructs. Maximum GUS gene expression occurred when the culture was initiated one day before the bombardment (Matsuoka *et al.,* 1997). Chimeric transgenic plant of the *V.* sesquipedlis subspecies of *V. unguiculata* containing *npt*II, *uid*A and *pat* genes was obtained through *Agrobacterium*-mediated transformation (Ignacimuthu, 2000). Transient GUS expression were observed on all the parts of germinating embryos after particle bombardment of tissues of *V. radiata* and *V. mungo* (Bhargava and Smigoeki, 1994). Transgenic plants of mungbean containing *uid*A and *npt*II gene were obtained through *Agrobacterium*-mediated transformation from cotyledonary node explant. Ten fertile transgenic shoots were recovered (Jaiwal *et al.*, 2001). Transformed calli were obtained through *Agrobacterium*-mediated transformation of primary leaves of *V. mungo* (Karthikeyan *et al.*, 1996). Transformed peanut plants have been recovered from primary leaf, zygotic embryos, leaf discs of peanut plant when co-cultivated with *A. tumefaciens* carrying the *uid*A and *npt*II genes. Stable integration and expression of the transgene were confirmed by *Npt*II assay, Southern blot analysis and GUS assay (Eapen and George, 1994, McKently *et al.*, 1995; Cheng *et al.,* 1996, 1997). By using microprojectile bombardment transgenic peanut plants with *uid*A gene were recovered (Brar *et al.,* 1994). But the approach they have used was labour intensive for screening. Embryogenic cultures from peanut when bombarded with two plasmid constructs containing *uid*A gene controlled by either CaMV 35S promoter or soybean vegetative seed promoter (vspB) over 200 transgenic plants were regenerated (Wang *et al.,* 1998). An efficient transformation and regeneration system was developed by particle bombardment into embryonic leaflet and embryogenic callus from mature seeds. Peanut cv. Gajah and NC-7 were used for callus production through embryo (Livingstone and Birch, 1995, 1999). *Vicia faba* seedlings were inoculated with *A. rhizogenes* resulted in the formation of roots at the inoculation sites. Transformation was monitored by detection of GUS activity (Schiemann and Eisenreich, 1989). Initial studies in genetic transformation of lupin reported the production of kanamycin resistant hairy roots and kanamycin resistant plants carrying *npt*II gene in *L. mutabilis* and *L. angustifolius* when intact hypocotyl and epicotyl, shoot apices were infected with *A. rhizogenes* and *A. tumefaciens* (Babaoglu *et al.*, 2000). Response of different explants of lentil for *Agrobacterium*-mediated transformation was studied. It was found that shoot apex; epicotyl and root explants are capable of expressing intron containing the *uid*A gene when infected with *A. tumifaciens* strain GV2260 (Warkentin and McHugen, 1992). Transient expression of GUS in cotyledonary node of lentil was shown to be negative (Warkentin and McHugen, 1993). But in other experiment with same *A. tumefaciens* strain GV2260 and binary vector causing a much higher transient expression of *uid*A gene was observed in lentil cotyledonary node explants probably because of the use of higher concentration of asetosyringione (200mM) and the vacuum infiltration of the explants (Mahmoudian *et al.*, 2002). Successful transfection of lentil protoplasts and characterization of the DNA/liposome complexes was reported. Transient activity of CAT and GUS was found in the cell lysate after 24 hours from the incubation of protoplast with the lipid/ DNA complex (Maccarone *et al.*, 1992). *In planta* electroporation technique has been tried in lentil (Chowrira *et al.*, 1995, 1996). Transient GUS activity has been detected in cotyledonary nodes of lentil after particle bombardment (Öktem *et al.*, 1999). A rapid and easy protocol for transient expression based on vacuum infiltration of *Agrobacterium* cells into lentil cotyledonary nodes was reported. It has been observed that vacuum infiltrated tissue showed higher transient GUS expression than the infiltrated tissue. Under optimal conditions 95% on the explants exhibited an average of 16 blue colour foci (Mahmoudian *et al.*, 2002). No transgenic lentil plant has been recovered in these transformation experiments. *Agrobacterium*-mediated transformation from cotyledon explants of *Cymopsis tetragonoloba* was reported. Stable transformation of *uid*A and *npt*II genes was observed in T_1, T_2 and T_3 generations with single copy of the gene. Study on transmission of the transgene to offspring was carried out and it was found that major parts of the transgenes when present in multiple copies displayed aberrant transmission at high frequency in the process of going from the primary transformants to the first offspring. Genes segregated in Mendelian pattern (Joersbo *et al.*, 1999). Production of transgenic pea plants by *A. tumefaciens*-mediated gene transfer was reported by culturing explants from axenic shoot cultures and seedling epicotyls (Puonti-Kaerlas *et al.*, 1989, 1990, Puonti Kaerlas, 1993). Transgenic plants of pea were obtained from epicotyl and nodal explant by *A. tumefaciens*. In this study, acetosyringone did not contribute to the transformation rate (De Kathan *et al.*, 1990). A single transgenic plant was obtained from an explant inoculated with strain EHA 101 having two inserts. Expression of *nos* gene was observed in the T_2 generation indicating inheritance and expression of the foreign DNA over at least two generations (Jordan and Hobbs, 1993).

Insect Resistant Legume Crops

Insects have always been a major cause of yield losses in pulses. The most damaging insects are usually lepidopterans followed by coleopteran and dipterans. There are two strategies for incorporating the insecticidal genes in plants. 1) Genes from microorganisms such as a soil bacterium *Bacillus thuringiensis* (Bt) which contains the insecticidal proteins and 2) plant derived genes which encode inhibitors of protein or carbohydrate digestion in insects such as protease inhibitors, alpha-amylase inhibitors and lectins. Insect resistant soybean expressing Bt δ-endotoxin gene was developed (Parrott *et al.*, 1994; Stewart *et al.*, 1996; Xu *et. al.*, 1997). Somatic embryos of soybean were bombarded with a synthetic Bt-*cry1Ac* insecticidal crystal protein gene. The transgenic plants were selected on hygromycin containing medium. The Cry1Ac protein accumulated up to 46-ng/mg extractable proteins. The leaf bioassay showed protection against Corn earworm (*Helicoverpa zea*), Soybean looper (*Pseudoplusia includens)*, Tobacco budworm (*Heliothis virescens*), and Velvetbean caterpillar (*Anticarsia gemmatalis*). All the transgenic plants were fertile (Stewart *et al.,* 1996). The plasmid pKT54B7C5 containing Bt δ-endotoxin gene was used to transform two soybean cultivars Heinong 37 and Heinong 39 by *A. tumefaciens*. The regenerated plants were obtained from hypocotyls and cotyledon nodes. The transformants were selected on kanamycin containg medium. The presence of the gene was confirmed by PCR and dot hybridization (Xu *et al.*, 1997). Two chickpea cultivars, ICCV-1 and ICCV-6 were genetically transformed with Bt-*cry1Ac* gene by particle bombardment. Transgenic kanamycin resistant chickpea plants were obtained through multiple shoot formation. Molecular analysis confirmed the presence of the *cry1Ac* gene in the transgenic plants. Insect bioassay showed that the expression level of the *cry1Ac* gene was inhibitory to the development of the larvae of *Heliothis armigera* (Kar *et al.,* 1997). Transformation of peanut by a synthetic *cry1Ac* gene resulted in various levels of resistance to the lesser cornstalk borer, from complete larval mortality to a 66% reduction in larval weight (Singsit *et al.*, 1997).

Many plants have evolved natural defence mechanisms against herbivorous insects based on the production of proteinaceous protease inhibitors. In contrast to Bt delta-endotoxins these proteins have antimetabolic activity against the wide range of insect species. The protease classes found in the digestive tracts of phytophagous insect pests are either serine proteases, which are the major class of proteases in Lepidoptera, or cysteine proteases, the major protease class found in Coleoptera (Ryan, 1990). Inhibitors of α-amylases are the second type of enzyme inhibitors used to modify crop plants. Seeds of common bean (*Phaseolus vulgaris*) are naturally resistant to storage pest bruchid beetles (*Callosobruchus maculatus*) because of the presence of an α-amylase inhibitor, a seed protein that is toxic to the larvae. A cDNA encoding α-Ai protein found in seeds of *Phaseolus vulgaris* was transferred to pea via *Agrobacterium*. The α-*Ai* gene was stably expressed in transgenic pea seeds. It was reported that α-Ai protein accumulated in seeds up to 3% of the soluble protein (Schroeder *et al.,* 1993,1995). A cDNA clone encoding a multi domain proteinase inhibitor precursor from *Nicotiana alata* (*Na-PI*) was transformed into pea under the control of a promoter from rubisco small subunit gene (Charity *et al.,* 1999). The levels of α-Ai protein in the transgenic pea seeds were found to be as high as in bean seed. The α-*Ai-Pv* gene driven by a strong seed specific promoter was transferred to pea. The transformed peas were resistant to the cowpea and azuki bean weevils (Shade *et al.,* 1994). Ishimoto *et al.* (1996) transformed azuki bean with α-amylase inhibitor gene and tested for protection against bruchid beetles. Transgenic *Phaseolus acutifolius* expressing a gene from *P. vulgaris* encoding Arcelin-5a-protein has been reported via

Agrobacterium mediated transformation (Dillen *et al.*, 1997). High level of Arcelin-5a-protein was found in the transformed seeds. This seed storage protein confers resistance to the insect *Zabrotes subfasciatus*, a major pest of *P. vulgaris*. Cowpea protease inhibitor gene was introduced in pigeonpea by *Agrobacterium* mediated transformation (Lawrence and Koundal, 2000, 2001).

Virus Resistance

Viral diseases of crop plants constitue a major problem reducing the yield. Infection of plant tissue requires damage to the cell wall and/or plasma membrane, which, for insect-born viruses, is achieved by the penetration by the insect stylet during feeding. Although viruses can be removed from the individual plants by meristem culture, this method is not so easy and reliable. Virus coat protein-mediated resistance, based on the "genetically engineered cross protection" and antisense-RNA derived resistance, has been reported in legumes. Accumulation of coat protein confers resistance to infection and/or disease development by the virus from which the CP gene was derived and by related viruses.Virus resistant soybean plants with bean pod mottle virus (BPMV) coat protein precursor (CP-P) gene were obtained via *Agrobacterium*-mediated transformation. The level BPMV CP-P that accumulated in transgenic plants was 1-2 μg/g wet weight of tissue. Symptomatology and ELISA assay showed that homozygous but not hemizygous plants exhibit the resistant phenotype (Di *et al.*, 1996). Soybean was transformed by *A. rhizogenes* carrying the plasmid pRiA4b having the *SMV-CP* gene. Integration of the gene in the soybean genome was confirmed by PCR and Southern analysis (Xu *et al.*, 1996). The nucleotide protein (N) gene of the lettuce isolate of Tomato Spotted Wilt Virus (TSWV) was transferred to peanut via microprojectile bombardment. Expression analysis indicated that the single or low copy number of the transgene resulted in the predictable expression (Yang *et al.*, 1998). An efficient method for transformation of peanut using cotyledon explants by *Agrobacterium*-mediated transformation has been reported (Magbamea *et al.*, 2000). Transgenic plants with *npt*II, *uidA* and coat protein gene of Indian Peanut Clump Virus (IPCVcp) were obtained (Sharma and Anjaiah, 2000). *Agrobacteruim*-mediated transformation of *V. mungo* by the cloned genomic components (DNA-A and DNA-B) of mungbean yellow mosaic virus (MYMV) was reported (Mandal *et al.*, 1997). Transformed calli with kanamycin resistance gene and full-length copy of m-RNA of CPMV was obtained from leaf discs by *Agrobacterium*-mediated transformation (Garcia *et al.*, 1986, 1987). Many independent transgenic lines of common bean containing *uid*A, *bar* and BGMV coat protein genes were produced by electric discharge particle acceleration method. PCR and Southern analysis confirmed the successful transformation. Particle bombardment was used to transform fusion constructs of *uid*A and *npt*II genes with the methionine rich 2s albumin gene isolated from the Brazil nut and the antisence sequence of Ac1, Ac2, Ac3 and Bc1. The genes were inherited in a Mendelian manner in the transgenic bean lines (Aragao *et al.*, 1996; Aragao and Rech, 1997). Transgenic pea plants with partial resistance to the alfalfa mosaic virus (AMV) were developed by *Agrobacterium*-mediated transformation (Grant *et al.*, 1998a,b).

Fungal Resistance

Transgenic peanut plants expressing *uid*A, *npt*II, and tobacco chitinase gene were obtained through tissue culture independent method where embryos in the mature seeds were pricked 4-5 times throughout their axes with a sterilized needle and immersed in *Agrobacteruim* suspension. These embryos were then directly germinated on Soilrite. The transgenic plants expressing the chitinase gene were resistant to attack by the fungal pathogen *Cercospora arachidicola* (Rohini and Rao 2000, 2001).

Herbicide Resistance

The production of herbicide resistant plants is based on the utilization of enzymes which detoxify the herbicide. Resistant plant modifies herbicides to inactivate derivative often by conjugation with other compounds.

Glyphosate tolerance

Glyphosate is an extremely reliable, phloem-mobile, broad spectrum herbicide with little residual soil activity. It acts as an inhibitor of aromatic amino acid synthesis by blocking shikimate biosynthesis, the target enzyme being 5-enolpyruvilshikimate-3-phosphate synthase (EPSP). Glyphosate tolerant plants can be produced by incorporating three types of transgenes, namely, *EPSP* gene from *Petunia* and *Arabidopsis* which results in the overproduction of the target enzyme, mutated gene from *Salmonella typhimurium, E. coli, Agrobacterium* sp. tolerant to glyphosate and a gene encoding glyphosate-oxidoreductase (GOX) from *Achromobacter*, which can metabolise glyphosate. The mutated EPSP gene for glyphosate tolerance from petunia with kanamycin resistance gene and *uid*A has been used for *Agrobacterium*-mediated transformation of cotyledon explants of soybean (Hinchee *et al.*, 1988). About 6% of the shoots regenerated by *de novo* organogenesis from seedling cotyledons were transformed. It was the first report of successful recovery of transgenic soybean plants, though Facciotti *et al.* (1985) first reported the transformation of soybean tissues. Glyphosate-tolerant soybean (GTS) plants carrying the bacterial 5-enolpyruvylshikmate-3-phosphate synthase (EPSP) gene were obtained by *Agrobacterium*-mediated transformation. The molecular characterization showed the single gene insert. Transgene was stable over several years. The composition of the seeds was compared with that of the parental soybean line. The analytical data showed that the GTS lines were equivalent to the parental conventional soybean cultivar (Padgette *et al.*, 1995,1996). Yield evaluation study showed that there was no significant reduction when transgenic plants were treated with glyphosate at various rates at different developmental stages at any locations (Delannay *et al.*, 1995). Production of GTS assures the reduced production cost and reliable weed control system. These plants, which express tolerance to the nonselective herbicide 'Roundup', were introduced in US market in 1996.

Phosphinothricin (PPT) tolerance

Phosphinothricin is an analogue of glutamine and acts as an herbicide by inhibiting glutamine synthesis. Bacterial gene *bar* from *Streptomycis hygroscopicus* has the ability to detoxify PPT by acetylation. In most of the grain legumes, the *bar* gene has been used as a selectable marker. The biolistic process was used to insert *bar* gene in *P. vulgaris* and the integration was confirmed by Southern blot analysis. Two transgenic events, PHV119 and PHV122 were found tolerant to 500 g ha^{-1} of glufosinate ammonium under green house conditions, with no visible symptoms and developmental growth comparable to nontransgenic *P. vulgaris*. Field evaluation carried out with the PHV119 event has shown that the plants tolerated up to 400 g GA ha^{-1}, with no visible symptoms. The transgenes segregated in a Mendelian fashion (Aragao *et al.*, 2002). Shoots of *P. vulgairs* were transformed by electroporation of intact cells with the plasmid pDPG165containing the *bar* gene conferring herbicide resistance to plants. Insertion of the *bar* gene in the genome of herbicide resistant shoots were confirmed by PCR and Southern blotting (Saker and Kuhne, 1998). Another report on pea transformation was reported from immature cotyledons infected with *A. tumefaciens* strain AGL1, harbouring binary vector pLN 27 contained *bar* and *npt*II genes. The *bar* and *npt*II genes were shown to be stably inherited through the first sexual generation of transformed plants (Grant *et al.,* 1995). The lateral cotyledonary meristem of germinating pea cultivar Puget seed used for *Agrobacteruim*-mediated transformation. The hyper-virulent *Agrobacterium* strain EHA105 was used in conjunction with plasmid SLJI561 which contained the *bar* gene fused between the *nos* promoter and the terminator sequences. Transformants were selected on medium containing Phosphinothricin (Bean *et al.*, 1997). Transgenic yellow lupin (*Lupinus luteus* L.) plants have been generated by meristem co-cultivation with *A. tumefaciens* carrying the binary plasmid pPZBNIa containing the *bar* gene under the control of CaMV 35 S promote. The transgenic nature of plants grown to the T_6 generation was confirmed by PCR and Southern analyses (Li *et al.*, 2000). Herbicide resistant transgenic plants of *Lupinus angustifolius* containing *bar* gene were obtained from shoot apices producing multiple shoots via *Agrobacterium*-mediated transformation (Pigeaire *et al.*, 1997).

Atrazine resistance

Atrazine is a herbicide, which inhibits photosynthesis by binding to the chloroplast thylakoid membrane proteins, resulting in failure of electron transport in photosystem II. Atrazine resistance gene from *Solanum nigrum* was transferred to soybean chloroplast genome through direct injection in the ovaries. The plants derived from ovaries injected with DNA when sprayed with atrazine solution showed resistance to atrazine (Liu *et al.,* 1989, 1990; Fu *et al.*, 1993; Yue *et al.,* 1996). Lentil cotyledonary nodes excised from germinated seedlings were bombarded with a plasmid containing a mutant acetolactate synthase gene (ALS) from tobacco conferring resistance to sulfonylurea herbicides. T_0 and selfed progeny plants were screened using metasulfuron herbicide. The non-transformed escapes died and the transformed plants survived the test. The surviving plants were phenotypically normal and produced viable seeds. The presence and stable transmission of the transgene into genomic DNA of screened T_1 transformants was confirmed by PCR and Southern hybridization. This method for producing transformed plants will allow new opportunities for lentil breeding to produce improved cultivars (Gulati *et al.*, 2002).

Improved Quality

In several species of legumes, a number of genotypes have been found to be defective in the expression of specific seed protein genes. Lysine content in the seeds of soybean was increased by transferring the lysine feedback-insensitive bacterial dihydropicolinic acid synthase (DHDPS) and AK enzymes, encoded by the *Corynebacterium* dapA gene and a mutant *E. coli* lysC gene via particle bombardment. Ten to several hundredfold increase in free lysine and up to fivefold increase in total seed lysine content contributing up to 33% of the total seed amino acid content was observed (Falco *et al.*, 1995). In peanut transient expression of *uid*A and *2S albumin* gene from Brazil nut has been reported (Lacorte *et al.*, 1997). Transient expression of seed protein phaseolin gene in the protoplasts derived from cell suspension culture was observed. The expression of the gene resulted in accumulation of the correctly glycosylated and assembled protein. This system when compared to the tobacco protoplasts, largely avoids fragmentation of phaseolin and the presence of contaminant polypeptides in the immunoprecipitates suggesting that the bean protoplasts are a good system to study the expression of wild type a well as *in vitro* modified bean seed protein (Giovinazzo *et al.*, 1997). Böttinger *et al.* (2001) reported the recovery of stably transformed lines of faba bean with an *A. tumefaciens*-mediated gene transfer system. Stem segments from aseptically germinated seedlings were used for transformation. Binary vectors conferring GUS, a mutant lysC gene, coding for a bacterial aspartate kinase insensitive to feedback control by threonine, and the coding sequence for a methionine-rich sunflower 2S albumin, each in combination with *npt*II as selectable marker were used. Kanamycin-resistant calli were obtained on callus initiation medium at a frequency of 10-30%. Molecular analysis confirmed the integration of the transgenes in the plant genome. Inheritance and expression of the foreign genes were demonstrated by Southern blotting, PCR, Western analysis and enzyme activity assays. Methionine rich 2S albumin synthetic gene of the Brazil nut (*Berthollletia excelsa*) was transferred to *Vicia narbonensis* by *Agrobacterium*-mediated transformation (Saalbach *et al.*, 1994). High expression of this gene under the control of 35 S promoter was observed in leaves and roots but the expression was very low in seeds. The expression of the gene was improved by fusing the 2S albumin gene with seed specific legumine B4 promoter from *Vicia faba* (Pickardt *et al.*, 1995; Saalbach *et al.*, 1995a,b). Seed specific gene expression was found in T_0, T_1 and T_2 generations. Some transformants exhibited a threefold increase in the methionine content of the seed. Changes in carbohydrate metabolism were reported in transgenic *V. narbonensis* (Weber *et al.*, 1998). A gene coding for a heat stable amylase was isolated from *Bacillus licheniformis* and introduced in *Vicia narbonensis* by *A. tumefaciens*. The amylase was stably expressed and inherited in plants and able to degrade 40% of the seed starch after heat treatment of the seeds (Saalbach *et al.*, 1997). The transgenic narbon bean plants containing the *Bacillus licheniformis* alpha-amylase gene have been reported. The transgene was controlled by the seed specific USP gene promoter and encoded a heat-stable enzyme. The transgene product accumulated in the storage vacuoles of cotyledon cells and retained its thermostability and starch degradation specificity. The enzyme is able to liquefy the endogenous seed starch reserves of this grain legume (Czihal *et al.*, 1999). Seed specific expression of a sulphur rich sunflower seed albuin was reported in transgenic Lupin. Transformation was carried out by *Agrobacterium*. The transgenic seed contained less sulphate and more total amino acid sulphur than the non-transgenic parent line. Methionine content was increased by 94% and cysteine content was reduced by 12%. No significant change in other amino acids or in the total nitrogen and total sulphur content of the seed was observed (Molvig *et al.*, 1997). To evaluate the potential role of alpha-galactosidase for the control of the final galactose content in guar an alpha galactosidase gene from immature senna seeds was cloned and transformed by *A. tumefaciens* into the related high yield species of guar. The gene was driven by a wheat of high molecular weight glutenin promoter. Reduction in the galactose content of the endosperm with galactomannans was observed in 30% of the transformants. After self-fertilization it was found that this character was inherited stably to the progeny and that it was coupled to the presence of the senna alpha-galactosidase gene suggesting that this gene can be involved in the determination of the final galactose content of endosperm galactomannans. This proves that cell wall polysaccharides can be modified *in vivo* (Joersbo *et al.*, 2001).

Other Traits

To produce new germplasm, the DNA from a maize inbred line and a soybean line was injected in the young pods of another soybean cultivar. Significant change in the agronomic traits such as maturity, seed protein content, plant height, stem diameter, lodging, number of branches, pods and seeds per plant was observed (Zhang *et al.,* 1994). In order to obtain a soybean line lacking Glym bd 30 K, a major allergenic seed protein, the antisense DNA from Glym Bd 30 K was introduced in soybean hypocotyl segments via particle bombardment. No transgenic plants were obtained because all the shoots eventually died (Takano *et al.*, 1996). Early ripening transgenic soybean plants were derived by introducing foreign DNA via pollen tube. RAPD analysis confirmed the presence of the DNA fragment in the genome of the progeny (Li *et al.,* 1994). Transgenic soybean plants producing 11% higher yield than the control plant were obtained containing high protein when total DNA from wild soybean (*Glycine soja*) was introduced in the cultivated *Glycine max* via pollen tube pathway (Lei *et al.,* 1995). Transgenic soybean plants expressing the male sterility and fertility restorer gene were recovered from cotyledon explants by A. *tumefaciens* carrying barnase and barstar genes. Regenerated plants transformed with the barnase gene were male sterile with shrivelled anthers which were devoid of pollen grains. Transgenic plants expressing the barstar gene were restorer for the male fertility (Guo and Lin, 1997). Early maturing lines of soybean cultivar Heining 26 were obtained when foreign DNA of early maturing cultivar Suining 8 was introduced directly through the pollen tube pathway after self-pollination. Three lines maturing 15 days before the recipient were obtained (Lei *et al.,* 1996). Impaired male and female fertility was obtained when total DNA from *Cicer arietinum* was transferred using pollen tube pathway technique. The sterility maintained by self pollination (Zhao *et al.,* 1995). Soybean cultivars were transformed with Sunflower (*Helianthus annus*) DNA by starch gel electrophoresis. The glucose-6-phosphate isomerase (GPI), isocitrate dehydrogenase (IDH) and phosphoglucomutase (PGM) patterns of progeny were analysed by PAGE. In transformed plants, the three enzymes produced clear patterns and there were differences between the progeny, receptors and donor. Some of the progeny had sunflower bands, and some specific bands, which did not occur in either parent. The result indicates that the exogenous DNA had inserted into the genome and was being expressed (Li *et al.,* 1996). Biotin mediated delivery of macromolecules such as bovine insulin (MW about 5700), bovine ribonuclease (MW about 14000), human haemoglobin (MW about 64000) and bovine serum albumin (MW 68000) into soybean cells was reported (Horn *et al.*, 1990). In another recent report transgenic pigeonpea plants expressing haemagglutinin protein of Rinderpest virus have been recovered by *Agrobacterium*-mediated transformation (Satyavathi *et al.*, 2003).

CONCLUSION AND FUTURE PROSPECTS

There is an urgent need to increase the productivity and enhance the nutritional quality of the pulse crops. However, genetic improvement of pulses by molecular approaches need routine procedures of genetic transformation, which are not available for many important tropical legume species. Though various methods of genetic transformation of plants have made it possible to introduce the desired characters in some legume species, several problems hinder routine production of the transgenic crop plants. It has been observed that most of the legume species are difficult to transform due to various reasons. Procedures to transform species such as pigeonpea and chickpea are far from routine. There is a need to develop efficient and reproducible *in vitro* regeneration protocols. Direct organogenesis is the most common regeneration system in grain legumes, and therefore *Agrobacterium*-mediated transformation that has been successful in other crops has not been as efficient in grain legumes. Regeneration via embryogenesis is reported in many legumes such as pigeonpea and chickpea but genetic transformation by this route is not yet reported. The reason may be the poor frequency of regeneration. The rooting of transgenic shoots combined with poor shoot growth is another major limiting factor. The efficient recovery of the transgenic shoots depends not only on the mode of regeneration, transformation method but also on selectable markers. Although kanamycin has been used in most of the transformation protocols it has not proved an efficient selectable marker for grain legumes. Alternate systems such as marker-free transformation systems need worthy of investigation.

REFERENCES

Ahlandsberg S., Satish P., Sun C. & Jansson C. (1999) Green Fluorescent protein as a reporter system in the transformation of barley cultivars. *Physiol Plant.* 107: 194-200.

Akella V. & Lurquin P.F. (1993) Expression in cowpea seedlings of chimeric transgenes after electroporation into seed-derived embryo. *Plant Cell Rep.* 12: 110-117.

Altinkut A., Gozukirmiz N., Bajroric K., Altman A. & Ziv M. (Ed) (1996) High percentage of regeneration and transformation in chickpea; Proceedings of the third international ISHS symposium on *in vitro* culture and horticultural breeding Jarusalem, Israel, 16-21 June: *Acta. Hort.* 447: 319-320.

Alton N.K. & Vapnek D. (1979) Nucleotide sequence analysis of the chloramphenicol resistance transposon tn9. *Nature.* 282: 864-869.

Aragao F.J.L., Barros L.M.G., Brasileiro A.C.M., Ribeiro S.G., Smith F.D., Sanford J.C., Faria J.C. & Rech E.L. (1996) Inheritance of foreign gene in transgenic bean (*Phaseolus vulgaris* L.) co-transferred via particle bombardment. *Theor. Appl. Genet.* 93: 142-150.

Aragao F.J.L. & Rech E.L. (1997) Morphological factors influencing recovery of transgenic bean plants (*Phaseolus vulgaris* L.) of a carioca cultivar. *Int. J. Plant Sci.* 158: 157-163.

Aragao F.J.L., Sarokin L., Vianna G.R. & Rech E.L. (2000) Selection of transgenic meristematic cells utilizing a herbicidal molecule results in the recovery of fertile transgenic soybean [*Glycine max* (L.) Merril] plants at a high frequency. *Theor. Appl. Genet.* 101: 1-6.

Aragao F.J.L., Vianna G.R., Albino M.M.C. & Rech E.L. (2002) Transgenic dry bean tolerant to the herbicide glufosinate ammonium. *Crop Sci.* 42: 1298-1302.

Austin H.A. & Cress W.A. (1994) The detection of Agrobaceterial Ti transformation of soybean: *S Afr J Sci.* 90: 299-302.

Babaoglu M., Davey M.R. & Power J.B. (2000) Genetic transformation of grain legumes: Key transformation events. Ag Biotech Net vol 2 June ABNO50.http://www.agbiotechnet.com/ reviews/june00/html/ Babaoglu.htm

Balazs E. & Bonneville J.M. (1987) Chlorampenicol acetyl transferase activity in *Brassica* spp, *Plant Sci.* 50:65.

Barros L.M.G., Gama M.I.C.S., Goncalves C.H.R., de P. Barreto C.C., Santana E.F. & Carneiw V.T. de C. (1997) Bean tissue culture with a view to introducing foreign genes. Pesquisa-Agropemaria- Brasilaria 32: 67-275.

Bean S.J., Gaoding P.S., Mullineaux P.M. & Davies D.R. (1997) A simple system for pea transformation. *Plant Cell. Rep.* 16: 513-519.

Beachtold N., Ellis J. & Pelletier G. (1993) *In planta Agrobacterium*-mediated gene transfer by infiltration of adult *Arabidopsis thaliana* plants. *C.R. Acad. Sci. Paris Life Sci.* 316: 1194-1199.

Bevan M., Flavell R.B. & Chilton M.D. (1983) A chimeric antibiotic resistance gene as a selectable marker for plant transformation. *Nature* 304: 184-187.

Bhargava S.C. & Smigoeki A.C. (1994) Transformation of tropical grain legume using particle bombardment. *Curr Sci* 66: 439-442.

Bhatt D., Parrott W.A., Collins G.B. & Hilderbrand D.F. (1991) *Agrobacterium* induced gall formation in lipoxigenase mutant isolines of soybeans. *Plant Cell Rep.* 9: 651-654.

Brar G.S., Cohen B.A., Vick C.L. & Johnson G.W. (1994) Recovery of transgenic peanut (*Arachis hypogaea* L.) plants from elite cultivars utilizing ACCGL technology. *Plant J.* 5: 745-753.

Brasileiro A.C.M., Arago F.J.L., Rossi S., Dusi D.N.A., Barras L.M.G. & Rech E.L., (1996). Susceptibility of common and tapery bean to *Agrobacterium* spp strain & improvement of *Agrobacterium* mediated transformation using microprojectile bombardment: *J Am Soc Hortic Sci.* 121: 810-815.

Böttinger P., Steinmetz A., Schieder O. & PicKardt T., (2001). *Agrobacterium* mediated transformation of *Vicia faba. Mol. Breed.* 8: 243-254.

Cao X., Liu Q., Rowland L.T. & Hammerschag F.A. (1998) GUS expression in blueberry (*Vacxcinium* spp.): Factors influencing *Agrobacterium*-mediated gene transfer efficiency. *Plant Cell Rep.* 18: 266-270.

Chang S.S., Park S.K., Kim B.C., Kang B.J., Kim D.U. & Nam H.G. (1994) Stable genetic transformation of *Arabidopsis thaliana* by *Agrobacterium* inoculation *in planta. Plant J.* 5: 551-558.

Charity J.A., Anderson M.A., Bittisnich D.J., Whitecross M. & Higgins T.J.V. (1999) Transgenic tobacco and pea expressing a proteinase inhibitor from *Nicitiana alata* have increased insect resistance. *Mol. Breed.* 5: 357-365.

Chee P.P., Fober K.A. & Slightom J.L. (1989) Transformation of soybean (*Glycine max*) by infecting germinating seeds with *Agrobacteriun tumefaciens. Plant Physiol.* 91: 1212-1218.

Cheng M., Jarret R.L. & Liz Deueski J.W. (1997) Expression and inheritance of foreign gene in transgenic plants generated by *Agrobacterium* mediated transformation. *Plant Cell Rep.* 16: 514-544.

Cheng M., Robert L.J.Z., Li A., Xing J.W. & Dewski. (1996) Production of fertile transgenic peanut (*Arachis hypgaea* L.) plants using *Agrobacterium tumefaciens. Plant Cell Rep.* 15: 653-657.

Cheng M., His D.C.H. & Phillips G.L. (1994) Recovery of transformants of valenncia-type peanut using *Agrobacteroium tumefaciens. Peanut Sci.* 21: 84-88.

Chiu W.L., Niwa Y., Zeng W., Hirano T., Kobayashi H. & Sheen J. (1996) Engineered GFP as vital reporter in plants. *Curr. Biol.* 6: 325-330.

Chowrira M.G., Akella V., & Lurquin P.F. (1995) Electroporation mediated gene transfer into intact nodal meristem *in planta*. Generating transgenic plants without *in vitro* tissue culture. *Mol. Biotechnol.* 3: 17-23.

Chowrira M.G., Akella V., Fuerst P.E. & Lurquin P.F. (1996) Transgenic grain legumes obtained by *in planta* electroporation-mediated gene transfer. *Mol Biotechnol* 5: 85-96.

Cho H.J. & Widholm J.M. (2002) *Agrobacterium* mediated transformation of the legume *Astrgalus sinicus* using Kanamycin resistance selection and green fluorescent protein expression. *Plant Cell Tiss. Org.* 69: 251-258.

Chrisou P. (1990) Soybean transformation by electric discharge particle acceleration: *Physiol Plantarum.* 79: 210-212.

Christou P. & Swain W.F. (1990) Contransformation frequencies of foreign genes in soybean cell cultures. *Theor. Appl. Genet.* 79: 337-341.

Curtis I.S., Power J.B., Blackhall N.W., De Loat A.M.M. & Davey M.R. (1994) Genotype independent transformation of Lettuce using *Agrobacterium tumefaciens*. *J. Exp. Bot.* 45: 1441-1449.

Custers J.B.M., Snepvangers S.C.H.J., Janson H.J., Zhang C. & Van Lookeren Camagne M.M. (1999) The 35-SCaMV promoter is silent during early embryogenesis sporophytic development in microspore culture. *Protoplasma.* 208: 257-264.

Czihal A., Conrad B., Buchner P., Brevis R., Farouk A.A., Manteuffel R., Adler K., Wobus U., Hofemeister J. & Bäumlein H. (1999) Gene farming in plants: expression of heat stable *Bacillus* amylase in transgenic legume seeds. *J. Plant Physiol.* 155: 183-189.

Davis D., Hamilton R.J. & Mullineaux P. (1993) Transformation of peas. *Plant Cell Rep.* 12: 180-183.

Davis S.J. & Viestra R.D. (1998) Soluble, highly fluorescent variant of green fluorescent protein (GFP) for use in higher plants. *Plant Mol. Biol.* 36: 521-528.

Dayal S., Lavanya M., Devi P. & Sharma K.K. (2003) An efficient protocol for shoot regeneration and genetic transformation of Pigeonpea [*Cajanus cajan* (L.) Millsp.] using leaf explants. *Plant Cell Rep.* 21:1072-1079.

De Bondt A., Eggermont K., Druart P., De Vil M., Guderis I., Vanderleyden J. & Broekert W.F., 1994. *Agrobacterium*-mediated transformation of apple (*Malus* × *domestica* Borkh): an assessment of factors affecting gene transfer efficiency during early transformation steps. *Plant Cell Rep.* 13: 587-593.

De Clercq J., Zambre M., Van Montagu M., Dillen W. & Angenon G. (2002) An optimized *Agrobacterium* mediated transformation procedure for *Phaseolus acutifolius* A. Gray. *Plant Cell Rep.* 21: 333-340.

De Kathen A. & Jacobson H.J. (1995) Cell competent for *Agrobacterium*-mediated DNA transfer in *Pisum sativum* L. *Transgenic Res.* 4: 184-191.

De Kathen A. & Jacobson H.J. (1990) *Agrobacterium tumefaciens* mediated transformation of *Pisum sativum* L. using binary and cointegrate vectors. *Plant Cell Rep.* 9: 276-279.

De WetJ.R., Wood K.V., Helinski D.R. & DeLuca M. (1985) Cloning of firefly luciferase cDNA and the expression of active luciferase in *Escherichia coli*. *Proc. Natl. Acad. Sci.* USA 82: 7870-7873.

Delannay X., Bauman T.T., Beighley D.H., Buettner M.J., Coble H.D., DeFelice M.S., Derting C.W., Diedrick T.J., GriffinJ.L., Hagood E.S., Hancock F.G., Hart S.E., LaVallee B.J., Loux M.M., Lueschen C.W., Matson K.W., Moots C.K., Murdock E., Nickell A.D., Owen M.D.K., Paschall E.H.II, Prochasaka L.M., Ramond P.J., Reynolds D.B., Rhodes W.K., Roeth F.W., Sprankle P.L., Tarochione L.J., Tinius C.N., Walker R.H., Wax L.M., Weigelt H.D. & Padgette S.R. (1995) Yield evaluation of a glyphosate- tolerant soybean line after treatment with glyphosate. *Crop Sci.* 35: 1461-1467.

Dhir S.K., Dhir S., Hepburn A. & Widholm J.M. (1991) Factors affecting transient gene expression in electroporated *Glycine max* protoplasts: *Plant Cell Rep.* 10: 106-110.

Dhir S.K., Dhir S., Sturtenant A.P. & Widholm J.M. (1991) Regeneration of transformed shoots from electroporated soybean (*Glycine max* (L.) Merr.): *Plant Cell Rep.* 10: 97-101.

Di Rong Purcell V., Collins G.B. & Ghabrial S.A. (1996) Production of transgenic soybean lines expressing the bean pod mottle virus coat protein precursor gene. *Plant Cell Rep* 15: 746-750.

Dillen W., De Cleug J., Goossens A., Van Montagu M. & Angenson G. (1997) *Agrobacterium*-mediated transformation of *Phasiolus acutifolius* A. Gray. *Theor. Appl. Genet.* 94: 151-158.

Dillen W., Engler G., Mantagu M. van., Angenon G. & Van-Montagu M. (1995) Electroporation-mediated DNA delivery to seedling of *Phaseolus vulgaris* L. (common bean). *Plant Cell Rep.* 15: 119-124.

Donaldson P.A. & Simmonds D.H. (2000) Susceptibility to *Agrobacterium tumefaciens* and cotyledonary node transformation in short-season soybean. *Plant Cell Rep.* 19: 478-484.

Eapen S. & George L. (1994) *Agrobacterium tumefaciens* mediated gene transfer in peanut (*Arachis hpogaea* L.). *Plant Cell Rep.* 13: 582-586.

Egnin M., Mora A. & Prakash C.S. (1998) Factors enhancing *Agrobacterium tumefaciens*-mediated gene transfer in peanut (*Arachis hypogaea* L.). *In Vitro Cell Dev. Biol. Plant.* 34: 310-318.

Facciotti D., O'Neal J.K., Lee S. & Shewmaker C.K. (1985) Light inducible expression of a chimeric gene in soybean tissue transformed with *Agrobacterium*. *BioTechnology* 3: 341-346.

Falco S.C., Guisa T., Locke M., Mauvais J., Sanders C., Ward R.T. & Webber P. (1995) Transgenic canola and soybean seeds with increased lysine. *Biotechnol* (New York) 13: 577-581.

Fang Y.D., Akula C. & Altpeter F. (2002) *Agrobacterium*-mediated barley (*Hordeum vulgare* L.) transformation using green fluorescent protein as a visual marker and sequence analysis of the T-DNA: barley genomic DNA junctions. *J. Plant Physiol.* 159: 1131-1138.

Feldmann K.A. & Marks M.D. (1987) *Agrobacterium*-mediated transformation of germinating seeds of *Arabidopsis thaliana*: a non tissue culture approach. *Mol. Gen. Genet.* 208: 1-9.

Feldmann K.A. (1992) T-DNA insertion mutagenesis in Arabidopsis: seed infection transformation. In Koncz C., Chua N.-H., Shell J., eds, Methods in Arabidopsis Research. World Scientific, Singapore. pp. 274-289.

Fonatana G.S., Santini L., Carepto S., Frugis G. & Mariotti D. (1993) Genetic transformation in the grain legume *Cicer arietinium* L. (Chickpea.) *Plant Cell Rep.* 12: 194-198.

Franklin C.I., Trieu T.N., Casridy B.G., Dixon R.A. & Nelson R.S. (1993) Genetic transformation green bean callus via *Agrobacterium* mediated DNA transfer. *Plant Cell Rep.* 12: 74-79.

Fu J.H., Li L.C., Yuan H.L., Yue S.X. & Zhu L.H. (1993) Expression of the atrazine-resistant transgenic soybean plants in field test. *Acta Agronomica Sinica* 19(6): 497-500.

Garcia J.A., Hille J. & Goldbach R. (1986) Transformation of cowpea *Vigna unguiculata* cells with an antibiotic resistant gene using a Ti-plasmid- derived vector. *Plant Sci.* 44: 37-46.

Garcia J.A., Hille J., Vas P. & Goldbach R. (1987) Transformation of cowpea *Vigna unguiculata* with a full length DNA copy of cowpea mosaic virus m-RNA. *Plant Sci.* 48: 89-98.

Geetha N., Venkatachalam P. & LaxmiSita G., (1999). *Agrobacterium* mediated Genetic Transformation of Pigeonpea (*Cajanus cajan* L) and Development of transgenic plant via Direct organogenesis. *Plant Biotechnol.* 16: 213-286.

Genga A., Allavena A., Ceriotti A. & Billini R. (1990) Towards genetic transformation of bean by *Agrobacterium tumefaciens* (poster) *Acta Horticulturae*. No 280,527-536; Proceedings of the first ISHS symposium on *in vitro* culture and horticultural breeding, Cesena, Italy, 30 May- 3 June 1989.

Gill R. (1990) Direct gene transfer in *Psophocarpus tetragonolobus* resistant to kanamycin. *Ann. Bot.* 66: 31-39.

Giovinazzo G., Greco V., Vitale A. & Bollini R. (1997) Bean (*Phaseolus vulgaris* L.) as a model system to study the expression and stability of recombinant seed proteins. *Plant Cell Rep* 16: 705-709.

Gnanum A., Mutukumar B., Mammen M., Veluthambi K. & Mathis P. (1995) Genetic transformation of cowpea (*Vigna unguiculata* L. Walp) by *A. tumefaciens* using cotyledons as explants. Proceedings of the Xth international Photosynthesis Congress, Monterpellier, France. 707-710.

Grant E.J., Cooper P.A., Alastair E. McAra. & Frew T.J. (1995) Transformation of pea (*Pissum sativum* L.) using immature cotyledons. *Plant Cell Rep.* 15: 254-258.

Grant E.J., Pither-Joyce M., Fifield W., Cooper P.A., Erasmuson S.K. & Timmerman-Vaughan G.M. (1998a) Resistance to alfalfa mosaic virus in transgenic pea (*Pisum sativum* L.). In: abstract IX *International Congress Plant Tissue Cell Culture*, Jerusalem, pp. 147.

Grant E.J, Cooper P.A., Pither-Joyce M., Fifield W., Cooper P.A. & Timmerman-Vaughan G.M. (1998b) In: AEP (Ed) Opportunities for High Quality, Health andAadded-Value Crops to Meet European Demands, Proceedings of 3rd Europian Conference on grain Legume, Valladolid, Spain. Aep, Paris, pp. 372-373.

Gulati A., Schryer P. & McHughen A. (2002) Production of fertile transgenic lentil (*Lens culinaris* Medik) plants using particle bombardment. *In vitro Cell Dev. Biol. Plant.* 38: 316-324.

Guo G., Maiwald F., Lorenzen P. & Steinbiss H.M. (1998) Factors influencing T-DNA transfer into wheat and barley by *Agrobacterium tumefaciens. Cereal Res. Commun.* 26: 15-22.

Guo Z.B. & Lin J.X. (1997) Genetic Engineering of male sterile and fertility restorer in soybean: *Soybean Genet. Newslett.* 24: 50.

Hadi N.Z., McMuller M.D. & Finer J.J. (1996) Transformation of 12 different plasmid into soybean via particle bombardment. *Plant Cell Rep.* 15: 500-505.

Hasseloff J., Siermering K.R., Prascher D.C. & Hodge S. (1997) Removal of cryptic intron and subcellular localization of green fluorescent protein are required to mark transgenic Arabidopsis plant brightly. *Proc. Natl. Acad. Sci.* USA 94: 2122-2127.

Hazel C.B., Klein T.M., Anis M., Wilde H.D. & Parrot W.A. (1998) Growth characteristics and transformability of soybean embryogenic cultures. *Plant Cell Rep* 17: 765-772.

Heim R., Prasher D.C. & Tsien R.Y. (1994) Wavelength mutations and post-translational autoxidation of green fluorescent protein. *Proc. Natl. Acad. Sci.* USA. 91: 12501-12504.

Herrera-Estrella L., De block M., Messens E., Haenalsteens J.P. Van Montagu M. & Schell J. (1983) Chimeric genes as dominant selectable markers in plant cells. EMBOJ 2: 987-995.

Hiei Y., Komari T. & Kubo T. (1997) Transformation of rice mediated by *Agrobacterium tumefaciens. Plant Mol. Biol.* 35: 205-218.

Hiei Y., Ohta S., Komari T. & Lumashiro T. (1994) Efficient transformation of rice (*Oryza sativa* L.) mediated by *Agrobacterium* and sequence analysis of the boundaries of the T-DNA. *Plant J.* 6: 271-282.

Hinchee N.A.W., Connor-ward D.V., Newell C.A., McDonnell R.E., Soto S.J., Gasser C.S., Fiscgoff D.A., Re D.B., Franley R.T. & Horsch R.B. (1988) Production of transgenic soybean plants using *Agrobacterium* mediated DNA transfer. *Bio-technol* (New York) 6: 915-922.

Horn M.A., HeinStein P.E. & Low P.S. (1990) Biotin mediated delivery of exogenous macromolecule onto soybean cells. *Plant Physiol.* 93: 1492-1496.

Hunsain T., Malik T., Raizuddin S. & Gorden M.P. (1997) Studies on the expression of marker gene in chickpea. *Plant Cell Tiss. Org.* 49: 7-16.

Hussey G., Jahnson R.D. & Warren S. (1989) Transformation of meristematic cells in the shoot apex of cultured pea shoots by *Agrobacterium tumefaciens* and *Agrobacterium rhizogenes*. *Protoplasma* 148: 101-105.

Igasaki T., Mohri T., Ichikawa H. & Shinohara K. (2000) *Agrobacterium tumefaciens* mediated transformation of *Robinia pseudoacacia*. *Plant Cell Rep.* 19: 448-453.

Ignacimuthu S. (2000) *Agrobacterium* mediated transformation of *Vigna sesquipedalis* Koern (asparagus bean). *Indian J. Exp. Biol.* 38: 493-498.

Islam R., Malik T., Hunsain T. & Raizuddin S. (1994) Strain and cultivar specificity in the *Agrobacterium* – Chickpea interaction. *Plant Cell Rep.* 13: 561-563.

Ishimoto M., Sato T., Chrispeels M.J. & Kitamura K. (1996) Bruchid resistance of transgenic azuki bean expressing seed amylase inhibitor of common bean. *Entomol. Exp. Appl.* 79: 309-315.

Jaiwal P.K., Kumari R., Ignacimuthu S., Potrykus I. & Saulter C. (2001) *Agrobacterium* tumefaciens-mediated transformation of mungbean (*Vigna radiata* L. Wilczek) a recalcitrant grain legume. *Plant Sci.* 161: 239-247.

James C. (2003) Global status of transgenic crops. ISAAA Brief 26. *International Service for Acquisition of Agri-biotech Appl.* (ISAAA), Ithaca, USA.

Jefferson R.A., Burgess S.M. & Hirsh D. (1986) β-Glucuronidase from *Escherichia coli* as a gene-fusion marker. *Proc. Natl. Acad. Sci.* USA 83: 8447-8451.

Jia S.R., Yang M.Z., Ott R. & Chua N.H. (1989) High frequency transformation of *Kalanchoe laciniata Plant Cell Rep.* 8: 336-340.

Joersbo M., Brunstedt J., Marcussen J. & Okkels F.T. (1999) Transformation of the endospermous legume guar (*Cymopsis tetragonoloba* L.) and analysis of transgene transmission. *Mol. Breed.* 5: 521-529.

Joersbo M., Marcussen J. & Brunstedt J. (2001) *In vivo* modification of the cell wall polysaccharide galactomannan of guar transformed with an alphagalactosidase gene cloned from senna. *Mol. Breed.* 7: 211-219.

Jones J.D.G., Svab Z., Harper E.C., Hurwitz C.D. & Maliga P. (1987) A dominant marker for plant cell transformation. *Mol. Gene. Genet.* 210: 86-91.

Jordenm M.C. (2000) Green fluorescent protein as a visual marker for wheat transformation. *Plant Cell Rep.* 19: 1069-1075.

Jorden M.C. & Hobbs S.A. (1993) Evolution of a cotyledonary node regeneration system for *Agrobacterium* mediated transformation of pea (*Pisum sativum* L.). *In vitro Cell Dev. Biol. Plant.* 29: 77-82.

Kaeppler H.F., Menono G.K., Skadsen R.W., Nuutila A.M. & Carlson A.R. (2000) Transgenic oat plants via visual selection of cells expressing green fluorescent protein. *Plant Cell Rep.* 19: 661-666.

Kaeppler H.F., Weining G., Somers D.A., Rines H.W. & Cockburn A.F. (1990) Silicon carbide fiber- mediated DNA delivery into plant cells. *Plant Cell Rep.* 9: 415-418.

Kamaté K., Rodriguez-Llorente I.D., Scholte M., Durand P., Ratet P., Konodorosi E., Kondorosi A. & Trinh T.H. (2000) Transformation of floral organs with GFP in *Medicago truncatula*. *Plant Cell Rep.* 19: 647-653.

Kar S., Basu D., Das S., Ramakrishnan N.A., Mukherjee P., Nayak P. & Sen S.K. (1997) Expression of *CryIA(c)* gene of *Bacillus thuringiensis in* transgenic Chickpea plants inhibits development of podborer (*Heliothis armigera)* larvae. *Transgenic Res.* 6: 177-185.

Kar S., Johnson T.N. Nayak P. & Sen S.K. (1996) Efficient transgenic plant regeneration through *Agrobacterium* – mediated transformation of chickpea (*Cicer arietinum* L.). *Plant Cell Rep.* 16: 32-37.

Karthikeyan A.S., Sarma K.S. & Veluthambai K. (1996) *Agrobacterium tumefaciens* mediated transformation of *Vigna mungo* (L) hepper. *Plant Cell Rep.* 15: 328-331.

Kay R., Chan A., Daly M. & Mc Pherson J. (1987) Duplication of CaMV35S promoter sequences creates a strong enhancer of plant genes. *Science* 236: 1299-1302.

Khawar K. & Özcan S. (2002) *In vitro* induction of crown galls by *Agrobacterium tumefaciens* super virulent strain A281 (pTiBo542) in lentil (*Lens culinaris* Medik). *Tr. J. Bot.* 26: 165-170.

Kim J.W. & Minamikawa T. (1997) Stable delivery of a canavalin promoter-beta-lucuronidase gene fusion into French bean by particle bombardment. *Plant Cell Physiol.* 38: 70-75.

Kim J.W. & Minamikawa T., (1996I Transformation and regeneration of French bean plants by the particle bombardment process. *Plant Sci. Limerick.* 177: 131-138.

Koler F., Golz C., Eapen S., Kohn H. & Schieder O. (1987) Stable transformation moth bean *Vigna aconitifolia* via direct gene transfer. *Plant Cell Rep.* 6: 313-317.

Krishnamurthy K.V., Suhasini K., Sagare A.P., Mixner M., De Kathen A., Pickardt T. & Schieder O. (2000) *Agrobacterium* mediated transformation of chickpea (*Cicer arietinum* L.) embryo axes. *Plant Cell Rep.* 19: 235-240.

Kuchuk N.V., Shckhovsckii A.M., Komarnitskii I.K. & Gleba Y.Z. (1990) Production of genetically transformed soybean cell clones by means of electroporation of the protoplast. *Biotechnologiya.* 5: 30-31.

Lacorte C., Aragao F.J.L., Almida E.R., Monsur E. & Reech E.L. (1997) Transient expression of *GUS* and *2S albumine* gene from Brazilnut in peanut (*Arachis hypogaea* L.) seed explants using particle bombardment. *Plant Cell Rep.* 16: 619-623.

Lacorte C., Mansur E., Timmerman B. & Cordeiro A.R. (1991) Gene transfer into peanut (*Arachis hypogaea* L.) by *Agrbacterium tumefaciens. Plant Cell Rep* 10: 354-357.

Lawrence P.K. & Koundal K.R. (2000) Simple Protocol for *Agrobacterium tumefaciens*-mediated transformation of pigeonpea (*Cajanus cajan* (L) Millsp.). *J. Plant. Biol.* 27: 299-302.

Lawrence P.K & Koundal K.R. (2001) *Agrobacterium mediated* genetic transformation of pigeonpea (*Cjanus cajan* (L). Millsp) and molecular analysis of regenerated plants. *Curr. Sci.* 80: 1428-1432.

Leon P., Planckaeut F. & Walbot V. (1991) Transient gene expression in protoplast of *Phaseolus vulgaris* isolated from a cell suspension culture. *Plant Physiol.* 95: 969-972.

Lewis M.E. & Bliss F.A. (1994) Tumor formation and beta-glucuronidase expression in *Phaseolus vulgaris* inoculated with *Agrobacterium tumefaciens. J. Am. Soc. Hort. Sci.* 119: 361-366.

Li H., Wylie S.J. & Jones M.G.K. (2000) Transgenic yellow lupin (*Lupinus luteus*). *Plant Cell Rep.* 19: 634-637.

Li X.C., Lei B.J., Lu C.H., Qian H., Zhou S.J., Lu Y.B., Xie W.W. & Wang B. (1994) Molecular verification of the introduction of foreign DNA in soybean to induce earliness using the RAPD technique. *Soybean Sci.* 13: 152-156.

Li X.C., Lu C.H., Lei B.J., Qian H. & Lu Y.B. (1996) Examination of progenies of soybean, which received exogenous DNA with starch gel electrophoresis. *Soybean Sci.* 15: 166-169.

Li Z., Nelson R.L., Widholm J.M. & Bent A. (2002) Soybean transformation via pollen tube pathway. *Soybean Genet Newslett.* 29: 1-11.

Liu B.L., Yue S.X., Hu N.B., Li X.B., Zhai W.X., Li N., Zhu R.H., Mao D. & Zhou P.Z. (1989) Transfer of atrazine resistant gene from nightshade to soybean chloroplast genome and its expression in transgenic plants. Kexue-Tongbao-Chinese-Science Bulletin. 34: 1670-1672.

Liu B.L., Yue S.X., Hu N.B., Li X.B., Zhai W.X., Li N., Zhu R.M., Zhu L.M., Mao D.Z. & Zhou P. (1990) Transfer of atrazine resistant gene of black nightshade to soybean chloroplast genome and its expression in transgenic plants. Science-in-China-seriesB-Chemistry; *Life Sci Earth Sci*. 33: 444-452.

Leffel S.M., Maban S.A. & Stewart C.N. Jr. (1997) Application of green fluorescent protein in plants. *Biotechniques*. 23: 912-918.

Lei B.J., Lu C.H., Qian H., Li X.C., Lu Y.B., Zhao K., Liu G.Y. & Yang X.Y. (1995) New soybean strains of high proteins and double high content obtained from introduction of exogenous total DNA. *Soybean Sci*. 14: 203-208.

Lei B.J., Lu C.H., Qian H., Li X.C., Zhou S.J., Lu Y.B., Zhao K., Liu G.Y. Yang X.Y. Xie W.W. & Wang, B. (1996) Early-maturing lines obtained through introduction of foreign total DNA into soybean. *Acta-Agronomica-Sinica*. 22: 173-177.

Livingstone D.M. & Birch R.G. (1995) Plant regeneration and microprojectile-mediated gene transfer in embryogenic leaflets in peanut (*Arachis hypogaea* L.). *Aust J Plant Physiol* 22: 585-591.

Livingstone D.M. & Birch R.G. (1999) Efficient transformation and regeneration of diverse cultivars of Peanut (*Arachis hypogaea) L.*) by particle bombardment into embryogenic callus produced from mature seeds. *Mol. Breed.* 5: 43-51.

Locorte C., Mansur E., Timmerman A.R. & Cordeiro. (1991) Gene transfer into peanut (*Arachis hypogaea* L.) by *Agrobacterium tumefaciens. Plant Cell Rep.* 10: 354-357.

Luo G., Hepburn A. & Widhdm J. (1994) A simple procedure of expression of genes in transgenic soybean callus tissue. *Plant Cell Rep.* 13: 632- 636 .

Lurquin P.F., Cai Z., Stiff C.M. & Fuerst E.P. (1998) Half embryo cocultivation technique for estimating the susceptibility of pea (*Pisum sativum* L.) and lentil (*Lens culinaris* Medik.) cultivars to *Agrobacterium tumefaciens. Mol Biotechnol*. 9: 175-179.

Lutova L.A., Sharova N.V. & Kislin Y.N. (1996) The role of phytohormones in response to transformation of pea lines differing in symbiotic characteristics and their hybrids. *Symbiosis- Rehovot*. 21: 61-80.

Maccarone M., Dini L., Marzio L.D., Giulio A.D., Rossi A., Mossa G. & Finazzi-Agrò A., (1992). Interaction of DNA with cationic liposomes: ability of transfecting lentil protoplasts. *Biochem Bioph Res*. Co. 86: 1417-1422.

Magbanua Z.V., Wilde H.D., Roberts J.K., Chowdhury K., Abad J., Moyer J.W., Wetzstein H.Y. & Parrott W.A. (2000) Field resistance to tomato spotted wilt virus in transgenic peanut (*Arachis hypogaea* L.) expressing an antisense nucleocapsid gene sequence. *Mol. Breed.* 6: 227-236.

Mandal B., Varma A. & Malthi V.G. (1997) Systemic infection of *Vigna mungo* using the cloned DNAs of the blackgram isolate of mungbean yellow mosaic geminivirus through Agroinfection and transmission of the progeny virus by whiteflies. *J. Phytopathol.* 145: 5605-510.

Mariotti D., Fontana G.S. & Santini. (1989) Genetic transformation of grain legumes: *Phaseolus vulgaris* L. and *P. coccineus* L. *J. Genet. & Breed*. 43:77-82.

Matsumato S., Takeb I. & Machida Y. (1988) *Ecsherichia coli* lacZ gene as a biochemical and histochemical marker in plant cells. *Gene* 66: 19-29.

Matsuoka M., Kai Y. & Yoshida T. (1997) Induction of adventitious bud and gene transfer by particle bombardment in cowpea (*Vigna unguiculata*). *Bulletin of the Chugoku National Agricultural Experiment Station*. 18: 31-39.

McCabe D.E., Swain W.F., Martinell B.J. & Christou P. (1988) Stable transformation of soybean (*Glycine max*) by particle acceleration. *BioTechnol*. (New York) 6: 923-926.

McKently A.H., Moore G.A., Doostdar H. & Niedz R.P. (1995) *Agrobacterium*-mediated transformation of peanut (*Arachis hypogea* L.) embryo axis and the development of transgenic plant. *Plant Cell Rep* 14: 699-703.

McLean P., Held B., Simental J., Drong R.F. & Slightom J. (1991) Susceptibility of dry bean (*Phaseolus vulgaris*) to *Agrobacterium* infection. Transformation of cotyledonary and hypocotyl tissues. *Plant cell Tiss. Org. Cult.* 24: 131-138.

Meurer C.A., Dinkins R.D. & Collins G.B. (1998) Factors affecting soybean cotyledonary node transformation. *Plant cell Rep.* 18: 180-186.

Mohmoudian M., Yücel M. & Öktem H.A. (2002) Transformation of Lentil (*Lens culinaris* M.) cotyledonary nodes by vacuum infiltration of *Agrobacterium tumefaciens*. *Plant Mol. Biol. Rep.* 20: 251-257.

Moliner J., Himber C. & Hahne G. (2000) Use of green fluorescent protein for detection of transformed shoots and homozygous offspring. *Plant Cell Rep.* 3: 219-224.

Molvig L., Tabe L.M., Eqqum B.O., Moore A.E., Craig S., Spencer D. & Higgins T.J.V. (1997) Enhanced methionine levels and increased nutritive value of seeds of transgenic lupins (*Lupinus angustifolius* L.) expressing a sunflower seed albumin gene. *Proc. Natl. Acad. Sci.* USA. 94: 8393-8398.

Moore P.J., Moor A.J. & Collins G.B. (1994) Genotypic and developmental regulation of transient expression of a reporter gene in soybean zygotic cotyledons. *Plant Cell. Rep.* 13: 556-560.

Muthukumar B., Mariamma N., Veluthaumbi K. & Gananam A. (1996) Genetic transformation of cotyledon explant of cowpea (Vigna *unguikulata* L. walp) using *Agrobacterium tumefaciens. Plant Cell Rep.* 15: 980-985.

Nadolska-Orczyk A. & Orczyk W. (2000) Study of the factors influencing *Agrobacterium* mediated transformation of pea (*Pisum sativum* L.). *Mol. Breed.* 6: 185-194.

Nauerby B., Madsen M., Christiansen J. & Wynadaela R. (1991) A rapid and efficient regeneration system for pea (*Pisum sativum*) suitable for transformation. *Plant Cell Rep* 9: 676-679.

Öktem H. A., Mahmoudian M., Eyidogan F. & Yücel M. (1999) GUS gene delivery and expression in lentil cotyledonary node using particle bombardment. *LENS Newslett.* 26: 3-6.

Olhoft P.M., Lin K., Galbraith J., Nielsen N.C. & Somers D.A. (2001) The role of thiol compounds in increasing *Agrobacterium* mediated transformation of soybean cotyledonary node cells. *Plant Cell Rep.* 20: 731-737.

Olhoft P.M., Flagel L.E., Donaovan C.M. & Somers D.A. (2003) Efficient soybean transformation using hygromycine B selection in the cotyledonary-node method. *Planta* 216: 723-735.

Olhoft P. M. & Somers D.A. (2001) L-cystine increases *Agrobacterium* mediated T-DNA delivery into soybean cotyledonary-node cells. *Plant Cell Rep.* 20: 706-711.

Ow D.W., Wood K.V., DeLuca M., de Wet J.R., Helinski D.R. & Howell S.H. (1986) Transient and stable expression of the firefly luciferase gene in plant cell & transgenic plants. *Science.* 234: 856-859.

Owens L.D. & Smigocky A.C. (1988) Transformation of soybean cells using mixed strains of *Agrobacterium tumefaciens* and phenolic compounds. *Plant Physiol.* 88: 570-73.

Ozias-Akins P. & Weissinger A.K. (1993) Regeneration of transgenic peanut plant from stably transformed embryogenic callus. *Plant Sci* 93: 185-194.

Ozcan S. (1995) Transient expression of GUS gene delivered into immature cotyledons of pea by microprojectile bombardment. *Turkish J. Bot.* 19: 423-426.

Padgette S.R., Kolacz K.H., Delannany X., Re D.B., LaVallee B.J., Tinius C.N., Rhodes W.K., Otero Y.I., Baeny G.F., Eichholtz D.A., Peschke V.M., Nida D.L., Taylor N.B. & Kishor G.M. (1995) Development, identification and characterization of a glyphosate tolerant soybean line: *Crop Sci.* 35: 1451-1461.

Padgette S.R., Taylor N.B., Nida D.L., Bailey M.R., MacDonald J., Holden L.R. & Fuchs R.L. (1996) The composition of glyphosate-tolerant soybean seeds is equivalent to that of conventional soybeans. *J. Nutr.* 126: 702-716.

Pang S.Z., DeBoer D.L., Wan Y., Ye G., Layton J.G., Neher M.K., Armstrong C.L., Fry J.E., Hinchee M.A.W. & Fromm M. (1996) An improved green fluorescent protein gene as vital marker in plants. *Plant Physiol* 112: 893-900.

Parrott W.A., Hoffman L.M., Hildebrand D.F., Willams E.G. & Collins G.B. (1989) Recovery of primary transformants of soybean. *Plant Cell Rep.* 7: 615-617.

Parrott W.A., All J.N., Adang M.J, Bailey M.A., Boerma H.R. & Steward C.N. Jr. (1994) Recovery and evaluation of soybean plants transgenic for *Bacillus thuringiensis* var *krustaki* insectsidal gene. *In Vitro Cell Dev. Biol. Plant.* 30: 144-149.

Penza R., Akella V. & Lurquin P.F. (1992) Transient expression and histological localization gus chimeric gene after direct transfer to mature cowpea embryos. *Bio. Techniques.* 13: 576.

Penza R., Lurquin P.F. & Filippone E. (1991) Gene transfer by co-cultivation of mature embryos with *Agrobacterium tumefaciens*: application to cowpea (*Vigna unguiculata* Walp.). *J. Plant Physiol.* 138: 39-43.

Pickardt T., Saalbach I., Waddell D., Meixner M., Muntz K. & Schieder O. (1995) Seed specific expression of the 2S albumin gee from Brazilnut (*Beutholletia exelsa*) in transgenic *Vicia narbineasis. Mol. Breed.* 1: 295-301.

Pickardt T., Meixner M., Schade & Scheider O. (1991) Transformation of *Vicia narbonensis* via *Agrobacterium*-medated gene transfer. *Plant Cell Rep.* 9: 535-538.

Pigeaire A., Abernethy D., Smith P.M., Simpson K., Fletcher N., Lu C.Y., Atkins C.A. & Cornish E. (1997) Transformation of a grain legume (*Lupinus angustifolius* L.) via *Agrobacterium tumefaciens* – mediated gene transfer to shoot apices. *Mol. Breed.* 3: 341-349.

Polowick P.L., Quandt J.A. & Mahon J.D. (2000) The ability of pea transformation technology to transfer gene into peas adapted to western Canadian growing conditions. *Plant Sci.* 153: 161-170.

Potrycus I. (1991) Gene transfer to plants: Assessment of published approaches and results. *Ann. Rev. Plant Physiol. Plant Mol. Biol.* 42: 205-225.

Puonti-Kaerkas J., Sabel P. & Erikson T.E. (1989) Transformation of pea (*Pisum sativum* L.,) by *Agrobacterium tumefaciens. Plant Cell Rep.* 8: 321-324.

Puonti-Kaerlas J., Eriksson T. & Engstrom P. (1992) Inheritance of a bacterial hygromicine phosphotransferase gene in the progeny of primary transgenic pea plants. *Theor. Appl. Genet.* 84: 443-150.

Puonti-Kaerlas J., Eriksson T. & Engstrom P. (1990) Production of transgenic pea (*Pisum sativum* L.) plants by *Agrobacterium tumefaciens*- mediated gene transfer. *Theor. Appl. Genet.* 80: 246-252.

Puonti-Kaerlas J., Ottosson J.A. & Eriksson T. (1992a) Survival and growth of pea protoplast after transformation by electroporation. *Plant Cell Tiss. Org. Cult.* 30: 141-148.

Puonti-Kaerlas J. (1993) Genetic engineering in pea crop improvement. *Acta Agriculture Scandinavica* Section B. 43: 65-73.

Ramana R.V., Venu C., Jayasree T. & Sadanandan A. (1996) Direct somatic embryogenesis and transformation in *Cicer arietinum* L. *Indian J. Exp. Biol.* 34: 716-718.

Ramsay G. & Kumar A. (1990) Transformation of *Vicia faba* cotyledon and stem tissue by *Agrobacterium rhizogenes*: infectivity and cytological studies. *J. Exp. Bot.* 41: 841-847.

Ramsay G.A. (1995) A joint approach on grain legume transformation. *Grain legumes*. 10: 33.

Rathore S. & Chand L. (1997) *In vitro* transformation of pigeonpea genotypes by wild strains of *Agrobacterium tumefaciens. Intl. Chickpea and Pigeonpea Newslett.* 4: 38-39.

Riazuddin S. & Husnain T. (1993) Transformation in chickpea (*Cicer arietinum*). In: Biotechnology in Agriculture and Forestry, Vol. 23. Plant Protoplasts and Genetic Engineering IV (Bajaj, Y. P. S. (Ed.)) Spriger-Verlag. Berlin.

Rohini V.K. & Rao S.K. (2000) Transformation of peanut ((*Arachis hypogaea* L.): a non-tissue culture based approach for generating transgenic plants. *Plant Sci.* 150: 41-49.

Rohini V.K. & Rao S.K. (2001) Transformation of peanut ((*Arachis hypogaea* L.) with tobacco chitinase gene: variable response of transformants to leaf spot disease. *Plant Sci.* 160: 889-898.

Russell D.R., Wallace K., Bathe J., Martinell B. & McCabe D. (1993) Stable transformation of *Phaseolus vulgaris* via electric-discharge mediated particle acceleration. *Plant Cell Rep.* 12: 165-169.

Ryan C.A. (1990) Protease inhibitors in plants: genes for improving defense against insects and pathogens. *Ann. Rev. Phytopathol.* 28: 425-449.

Saalback L., Hofuneister J. & Bammleiin H. (1997) Plant seed as bioreactor: use of a heat stable, bacterial amylase in transgenic seeds of legumes for liquefaction of seed starch: proceedings of the 5th symposium on renewable resources. Schriftenreihe-des Bundeeeeesministeriums-fur-Ernahring-Land Wirtschaft-und-Forsten-Rcihe-A-Angewandte- Wissenschaft., Sonderheff, 182-184.

Saalbach I., Pickardt T., Machemehl F., Saalbach G., Schieder O. & Muntz K. (1994) A chimeric gene encoding the methionine rich 2S albumin of the Brazil nut (*Bertholletia excelsa* H.B.K.) is stably expressed and inherited in transgenic grain legumes. *Mol. Gen. Genet.* 224: 226-236.

Saalbach I., Pickrdt T., Waddell D.R., Hillmer S., Scheider O.N. & Muntz K. (1995a) The sulphur-rich Brazil nut 2S albumin is specifically formed in transgenic seeds of the grain legume *Vicia narbonensis. Euphytica.* 85: 181-192.

Saalbach I., Waddell D.R., Pickrdt T., Scheider O.N. & Muntz K. (1995b) Stable expression of the sulphur-rich 2S albumin gene in transgenic *Vicia narbonensis* increases the methionine content of seeds. *J. Plant Physiol.* 145: 674-681.

Saker M.M. & Kuhne T. (1998) Production of transgenic kidney bean shoots by electroporation of intact cells. *Biol. Plant.* 40: 507-514.

Santarem E.R., Trick H.N. & Essig J.S. (1998) Sonication-assisted *Agrobacterium*-mediated transformation of soybean immature cotyledons: optimization of transient expression. *Plant Cell Rep.* 17: 752-759.

Satyavathi V.V., Prasad V., Khandelwal A., Shaila M.S., Lakshmi Sita G. (2003) Expression of hemagglutinin protein of rinderpest virus in transgenic pigeon pea [*Cajanus cajan* (L.) Millsp.] plants. *Plant Cell Rep.* 21: 651-658 .

Schiemann J. & Eisenreich G. (1989) Transformation of field bean (V*icia faba* L) cells. Expression of a chimeric gene in cultured hairy root and root derived callus. *Biochemie-und-Physiologie-der- pflanzen.* 185: 135-140.

Schroeder H., Schotz A., Wardley-Richardson T., Speccer D. & Higins T. (1993) Transformation and regeneration of two cultivars of pea (*Pisum Sativum L*). *Plant Physiol.* 101: 751-757.

Schroeder H.E., Gollasch S Moore A., Table L.M., Craig S., Hardie D.C., Chrispeds M.J., Spencer D. & Higgns T.J.V. (1995) Bean alpha-amylase inhibitor confers resistance to the pea weevil (*Bruchous pisorum*) in transgenic pea (*Pissum Sativum* L). *Plant Physiol.* 107: 1233-1239.

Shade R.E., Schroder H.E., Pueya J.J., Tabe L.M., Murdock L.L., Higgins T.J.V. & Chrispeels M.J. (1994) Transgenic pea seeds expressing the a-amylase inhibitor of the common bean are resistant to bruchid beetles. *Bio-Technol.* (New York) 12: 793-796.

Sharma K.K. & Anjaiah V. (2000) An efficient method of producing transgenic plant of peanut (*Arachis hypogea L.*) through *Agrobacterium tumefaciens-mediated* genetic transformation *Plant Sci.* 159: 7-19.

Siefkes Boer H.J., Noonan M.J., Bulock D.W. & Conner A.J. (1995) Hairy root transformation system in large seed grain legumes. *Israel J. Plant Sci.* 43: 1-5.

Singsit C., Adang M.C., Lynch R.E, Anderson W.A., Wang A., Cardineau G. & Ozias-Akins P. (1997) Expression of *Bacillus thuringiensis cry1Ac* gene in transgenic peanut plants and its efficacy against Cornstalk borer. *Transgenic Res.* 6: 167-179.

Srinivasan., Mridula., Gupta N. & Chopra V.L. (1988) *Agrobacterium* mediated transformation of chickpea. *Intl. Chickpea Newslett.* 19: 2-3.

Srivatanakul M., Park S.H., Salas M.G. & Smith R.H. (2001) Transformation of parameters enhancing T-DNA expression in Kenaf (*Hibiscus canabinus*) *J. Plant Physiol.* 158: 255-260.

Stewart C.M. Jr., Adang M.J., All J.M., Boerma H.R., Cardinean G., Tucker D. & Parrot W.A. (1996) Genetic transformation, recovery and characterization of fertile soybean transgenic for a synthetic *Bacillus thuringiensis cry1Ac* gene. *Plant Physiol.* 112: 121-129.

Sunilkumar G. & Rathore K.S. (2001) Transgenic cotton: Factors influencing *Agrobacterium*-mediated transformation and regeneration. *Mol. Breed.* 8:37-52.

Sunilkumar G., Mohr L.A., Lopata F., Emani C. & Rathore K.S. (2002) Developmental and tissue specific expression of CaMV 35S promoter in cotton as revealed by GFP. *Plant. Mol. Biol.* 50: 463-474.

Takano T., Li T., Kaneda Y. & Kitamura K. (1996) Lowering the content of an allergic protein through efficient transformation in soybean seed. Report-of-the-Soy-Protein-Research-committee-Japan. 17: 14-18.

Takasaki T., Hatakeyama K., Ojima K, Watanabe M., Toriyama K. & Hinata K. (1997) Factors influencing *Agrobacterium*-mediated transformation of *Brassica rapa* L. *Breed Sci.* 47: 127-134 .

Thompson C.J., Movva N.R., Tichard R., Crameri R., Davies J.E. & Lauwereysm. (1987) Characterization of the herbicide-resistance gene *bar* from *Streptomyces hydroscopicus*. *EMBO J.* 6: 2519-2523.

Thu T.T., Mai T.T.X., Eric D., Farsi S., Tadesse Y., Angenon G. & Jacobs M. (2003) *In vitro* regeneration and transformation of Pigeonpea [*Cajanus cajan* (L.) Millsp.]. *Mol. Breed.* 11: 159-168.

Trick H.N. & Finer J.J. (1997) SAAT: sonication assisted *Agrobacterium* mediated transformation. *Transgenic res.* 6: 329-336.

Trick H.N. & Finer J.J. (1998) Sonication assisted *Agrobacterium*-mediated transformation of soybean [*Glycine max* (L.) Merrill] embryogenic suspension culture tissue. *Plant Cell Rep.* 17: 482-488.

Trieu T.A., Burleigh S.H., Kardailsky I.V., Maldonado-Mendoza Versaw W.K., Blaylock L.A., Shin H., Chiou T.J., Katagi H., Dewbre G.R., Weigel D. & Harrison M.J. (2000) Transformation of *Medicago truncatula* via infiltration of seedlings or flowering plant with *Agrobacterium*. *Plant J.* 22: 531-541.

Wang A., Fan H., Singsit C. & Ozia-Akins P. (1998) Transformation of peanut with a soybean *vsp* B promoter-uidA chimeric gene: Optimization of a transformation system and analysis of GUS expression in primary transgenic tissues and plants. *Physiol Plant.* 102: 38-48.

Warkentin T.D. & McHughan A. (1993) Regeneration from lentil (*Lens culinaris*) cotyledonary node and potential of this explant for transformation by *Agrobacterium tumefaciens*. LENS- Newsletter (ICARDA), *Lentil Expl News Service*. 20: 26-28.

Warkentin T.D. & McHughen A. (1991) Crown gall formation of the lentil (*Lens culinaris*) with virulent strain of *Agrobacterium tumefaciens. Plant Cell Rep.* 10: 489-493.

Warkentin T.D. & McHughen A. (1992) *Agrobacterium tumefaciens*-mediated beta-glucuronidase (GUS) gene expression in lentil (*Lens Culinansis* Medik.) tissues. *Plant Cell Rep.* 11: 274-278.

Weber H., Heim U., Golombek S., Borisjuk L. & Wobus U. (1998) Metabolic control of legume seed development: attaining carbohydrate metabolism in transgenic *Vicia narbonensis.* In: AEP (Ed) Opportunities for High Quality, Healthy and Added-Value crops to Meet European Demands. Proceedings 3rd European Conference on Grain Legumes, Valladolid. AEP, Paris. pp 374-375.

Xu X.L., Li X.H., Liu W.H., Li J.L. & Wang Y. (1996) Study of transferring the soybean mosaic virus coat protein (SMV-SP) gene into soybean using the Ri-plasmid. *Soybean Sci* 15: 279-288.

Xu X.L., Gao J., Liu W.H. & Li J.L. (1997) Studies on transferring B.t. k-delta endotoxin gene into soybean with Ti-plasmid primaril. *Soybean Sci.* 16: 6-11.

Yamada T., Teraishi M., Hattori K & Ishimoto M. (2001) Transformation of azuki bean by *Agrobacterium tumefaciens. Plant Cell Tiss. Org. Cult.* 64: 47-54.

Yan B., Srinivasa Reddy M.S., Collins G.B. & Dinkins R.D. (2000) *Agrobacterium tumefaciens* mediated transformation of soybean [*Glycine max* (L.) Merrill] using immature zygotic cotyledon explants. *Plant Cell Rep.* 19: 1090-1097.

Yang H., Singsit C., Wang A., Gonsalves D. & Ozias-Akins P. (1998) Transgenic peanut plant containing a nucleocapsid protein gene of tomato spotted wilt virus show divergent level of gene expression. *Plant Cell Rep.* 17: 693-699.

Yang N.S. & Christou P. (1990) Cell type specific expression of CaMV 35S GUS gene in transgenic soybean plants. *Deve. Genet.* 11: 289-293.

Young S.J., Choi E.Y., Lee H.S., Choi D.W., Oh M.H. & Kim S.G. (1995) Effect of auxin on expression of isopentyl transferase gene (ipt) in transformed bean (*Phaseolus vulgaris* L.) single cell clones induced by *Agrobacterium tumefaciens* C58. *J. Plant Physiol.* 146: 148-154.

Yue S.X., Fu J.H., Li L.C., Yuan H.L., Zhui W.X., Chen H., Su W.Q. & Wang Y. (1996) Atrazine resistant gene introduced into Heilongjiang soybean varities and its expression and heredity. *Scientia-Agricultura-Sinica.* 29: 78-83.

Zhang G.D., Zhao C.H. & Chen S.J. (1994) Genetic transformation of soybean by injecting exogenous DNA into young pods. *Soybean Sci.* 13: 268-273.

Zhang Z., Xing A., Staswick P. & Clemente T.E. (1999) The use of glufosinate as a selective agent in *Agrobacterium*-mediated transformation of soybean. *Plant Cell Tiss Org Cult* 56: 37-46.

Zhang Z.Y., Coyne D.P. & Mitra A. (1997) Factor affecting *Agrobacetrium* mediated transformation of common bean. *Journal of Am. Soc. Horti. Sci.* 122: 300-305.

Zhao L.M., Liu D.P., Sun H., Yuan Y. & Huang M. (1995) Sterile material of soybean obtained by introducing exogenous DNA. *Soybean Sci.* 14: 83-87.

Zhou J.H. & Atherly A.G. (1989) Insertion of the transposable element Ac into soybean by transformation: *Soybean Genet Newslett.* 16: 124-126.

Zhou J.H. & Atherly A.G. (1990) In situ detection of transposition of maize controlling element (Ac) in transgenic soybean tissues: *Plant Cell Rep.* 8: 542-545.

CHAPTER

23

In vitro Adventitive Embryony: A Safe Alternative to *Citrus* Budwood Importation

R.N. Khawale* and S.K. Singh

ABSTRACT

Exchange of *Citrus* germplasm is sometimes risky due to import of different unknown viruses, which may thereafter spread with the planting material. Hence, to minimize such risks an alternative method of adventitive embryony has been proposed. From the stored seed of *Citrus jambhiri* Lush and *C. reticulata* Blanco, several adventitive embryos could be regenerated from the nucellar tissues of the stored seeds, without the intervening callus stage. Embryogenesis was noted upon culture on to Murashige and Skoog medium fortified with 2.0 mg l^{-1} BAP + 0.5 mg l^{-1} kinetin + 1.0 mg l^{-1} NAA. Rooting and subsequent growth of the adventitive embryos was observed when transferred to MS medium supplemented with 2.0 mg l^{-1} IBA + 200 mg l^{-1}activated charcoal. Regenerated plantlets recorded highest 92.7% survival in glass jar filled with (peat: soilrite; 1:1). Hardened plantlets were subsequently transferred to glasshouse for further study.

Key Words: Citrus, adventitive embryony, cotyledon, budwood importation, tissue culture.

The genus *Citrus* and its relative are very important fruit crops ranking third in India after banana and mango. These crops are grown in tropical and subtropical climates and have wide distribution in the world. *Citrus* collections are generally conserved in the field gene banks in many of the *citrus* growing countries. Such collections are generally vulnerable to biotic and abiotic hazards. For germplasm exchange, budwood and seed are the most preferred tissues. However, such a practice has unknowingly led to the spread of virus complex and viroid-like diseases from one region to another within the country or inter-continental regions. The real problem is detected only when the problem takes an epidemic form. In many countries, budwood importation is restricted because of the hazards of introducing diseases and importation restricted to seed (Navarro, 1981). Furthermore, the seed is not strictly recalcitrant (King *et al.*, 1981), do not remain viable when

*Address for correspondence: Central Tissue Culture Laboratory, Division of Fruits and Horticultural Technology, Indian Agricultural Research Institute, New Delhi-110 012, India.

stored for extended period of time (Mumford *et al.*, 1982) and thus vulnerable to desiccation, which drastically reduces the germination. Although, not a large number of viruses are transmitted through seeds (Bitters & Murashige, 1967). Since *in vitro* cultured nucellar tissues, produce true-to-type plants, it has therefore been preferred as explant over the others from the seedling. Therefore, an attempt was made to develop the complete plant regeneration technique through adventitive embryony on the cotyledon from stored seeds. The protocol can be used as a safe alternative method for *Citrus* germplasm exchange and *in vitro* conservation as the method is give minimum risk of virus(es) transfer following this pathway.

MATERIALS AND METHODS

Seed extraction and storage

Fruits of the rough lemon (*Citrus jambhiri* Lush) and Nagpur mandarin (*Citrus reticulata* Blanco) collected from the orchard of Indian Agricultural Research Institute, New Delhi, India and Regional Fruit Research Station, Abohar, Punjab, respectively. Seeds were extracted from the fruits; and immediately washed in running tap water to remove the slimy layer. The agitation treatment of 0.75% sodium hypochlorite (v/v) for 15 min was given and washed in running tap water followed by air-drying under shade in laboratory condition up to 12-15 h. The seeds were then treated with carbendazim (@1.0 g /kg) and packed in polythene bag and stored in freezer at 4°C for their further use.

Culture conditions and adventitive embryos formation

Inner and outer seed coats were separated out from the stored seeds and surface sterilized by 0.75% sodium hypochlorite solution (v/v) with 0.1% of Tween 20 for 10 min followed by rinsing three times with sterile distilled water. Zygotic/nucellar embryos were then removed aseptically under the microscope and cotyledons separated out with minimum damage. Cotyledons were then cultured on Murashige and Skoog (MS) (1962) basal medium was supplemented with 30 g l^{-1} sucrose +100 mg l^{-1} m-inositol and 8% agar-agar combinations of plant growth regulators viz. (1.0-2.0 mg l^{-1}) α-benzylaminopurine (BAP), (0.5–1.0 mg l^{-1}) kinetin and (0.5–1.0 mg l^{-1}) α-naphthalene acetic acid (NAA) to induce adventitive embryos (Table 1). Before autoclaving (121°C for 15 min), the pH of the medium was adjusted to 5.7. The medium was distributed in 25 ml aliquots in 25 mm × 150 mm culture tubes. Two cotyledon segments with adhering nucellus were inoculated per tube. All the cultures were maintained in culture room under the complete darkness upto organogenesis on the explants, thereafter they were transferred to diffused light (15 μ mol m^{-2} s^{-1}).

Rooting and hardening

Various levels of NAA (0.5, 1.0 mg l^{-1}), indole-3-butyric acid (IBA) (1.0, 2.0 mg l^{-1}) and activated charcoal (AC) 200 mg l^{-1} were supplemented in MS medium for studying the *in vitro* rooting and regeneration of complete plants (Table 2). The developing adventitive embryos were excised aseptically and transferred to rooting medium. Rooted plantlets were transferred to hardening medium. Different strategies such as the sterile mixture of sand: soil (1:1) and peat: soilrite (1:1) in plastic pot with polythene cover and glass jar with polypropylene (PP) caps (Table 3). The potting

Table 1 Effect of growth regulators on adventitive embryony in Rough lemon and Nagpur mandarin

Concentration (mg l^{-1})	*Rough lemon*		*Nagpur mandarin*	
	Days to embryo induction	*No. of adventitive embryo/explant*	*Days to embryo induction*	*No. of adventitive embryo/explant*
1.0 BAP + 0.5 Kin + 0.5 NAA	50.3	7.6	–	–
2.0 BAP + 0.5 Kin + 0.5 NAA	42.6	14.7	50.8	8.4
1.0 BAP + 1.0 Kin + 0.5 NAA	42.7	13.9	49.4	18.7
2.0 BAP + 0.5 Kin + 1.0 NAA	40.6	28.7	43.2	24.3
2.0 BAP + 1.0 Kin + 1.0 NAA	46.2	19.2	44.9	20.5
C.D. at 5%	2.8	5.4	2.1	3.2

'_' No response

Table 2 Rooting and shoot elongation of adventitive embryo as affected by auxins and activated charcoal

Concentration (mg l^{-1})	*Rough lemon*		*Nagpur mandarin*	
	Days to root induction	*Rooting (%)*	*Days to root induction*	*Rooting (%)*
0.5 NAA	29.0	58.7(50.0)*	30.6	49.3(44.6)*
1.0 NAA	22.4	54.3(47.4)	24.1	44.7(41.9)
1.0 IBA	30.7	39.4(38.8)	30.0	65.4(53.9)
2.0 IBA	28.2	78.8(62.3)	29.2	67.0(54.9)
1.0 NAA + 200 AC	26.0	68.4(55.8)	26.6	60.5(51.1)
2.0 IBA + 200 AC	20.2	95.7(78.0)	21.7	70.2(56.9)
C.D. at 5%	2.34	6.25	2.75	7.65

*Transformed values

Table 3 Hardening strategies for *in vitro* grown adventitive embryo of Rough lemon and Nagpur mandarin

Treatment	*Hardening success (%)*		*Days to hardening*
	Rough lemon	*Nagpur mandarin*	
1. Plastic pot with polythene cover (Sand: Soil; 1:1)	65.1 (53.8)*	56.9 (48.9)*	34.7
2. Plastic pot with polythene cover (Peat: Soilrite; 1:1)	72.2 (58.8)	65.2 (53.8)	30.8
3. Glass jar with polypropylene cap (sand: Soil; 1:1)	76.8 (61.2)	71.2 (57.5)	29.7
4. Glass jar with polypropylene cap (peat: Soilrite; 1:1)	92.7 (74.3)	90.5 (72.0)	22.4
C.D. at 5%	3.5	3.8	3.1

*Transformed values

mixture moistened with one-fourth MS salts minus organics. Plants under *in vitro* hardening were watered regularly after two weeks of transfer at two-three days interval and with covers removed gradually. Hardened plants were then transferred to greenhouse and subsequently to the field during the planting season.

The experiments were laid out in complete block design with minimum thirty explants per treatment. The experiments were replicated thrice. The percentage data was subjected to arc sin% transformation before analysis.

Observations

Adventitive embryony

This response was noted with different frequency for the cotyledonary explant and also the growth regulators supplemented played a major role. When the cultures were maintained for a longer period on a medium comprising MS basal medium fortified with 2.0 mg l^{-1} BAP+ 0.5 mg l^{-1} kinetin + 1.0 mg l^{-1} NAA higher number of adventitive embryos were formed for both the genotypes, i.e. Rough lemon (*C. jambhiri)* and Nagpur mandarin (*Citrus reticulata* Blanco) (Table 1). Days taken for adventitive embryo induction were in the range of 40.6-50.8 for different media tested. The frequency of adventitive embryony was recorded highest (28.7 per explant) in Rough lemon then in Nagpur mandarin (24.3). It was invariably noted that adventitive embryony took place only when the cultures were kept initially in dark; whereas those maintained in complete light or in light/dark regimes did not show any encouraging response. The different stages of adventitive embryo induction and germination in Rough lemon are shown in Figure 1.

Growth of adventitive embryos and hardening

Rooting and shoot growth from adventitive embryos were significantly affected by auxin(s) and AC supplementation. Results indicated that MS basal medium fortified with 2.0 mg l^{-1} IBA and 200 mg l^{-1}AC recorded the highest rooting per cent (95.7) with good embryo growth in Rough lemon. After 45 days of rooting, plantlets were subjected to different hardening strategies for their *ex vitro* transfer (Table 3). Of the different treatments, glass jar filled with peat: soilrite (1:1) recorded the highest (92.7%) plantlet survival for Rough lemon and Nagpur mandarin (90.5%), respectively. The plastic pots with polythene cover was not found good for hardening purpose as number of plantlets showed high wilting symptoms. Similarly, the potting mixture comprising sand: soil (1:1) also did not result in proper hardening of *in vitro* raised plants and accordingly it resulted in lower number of plantlets survival, irrespective of the hardening strategies. Successfully hardened plantlets were transferred to greenhouse for further study. Hardening and plantlets development in Rough lemon is shown in Figure 2.

Discussion

Adventitive embryony using cotyledon explant has been reported by several workers in number of crops. Pooler & Scorza (1995) reported that cotyledons taken from first and second year of stored peach seeds showed difference in regeneration rate and also frequency of complete plantlet regeneration. Earlier, Kato (1986) reported cotyledon culture for the micropropagation in different tea genotypes. He observed direct adventitive embryo regeneration on the surface of cotyledon

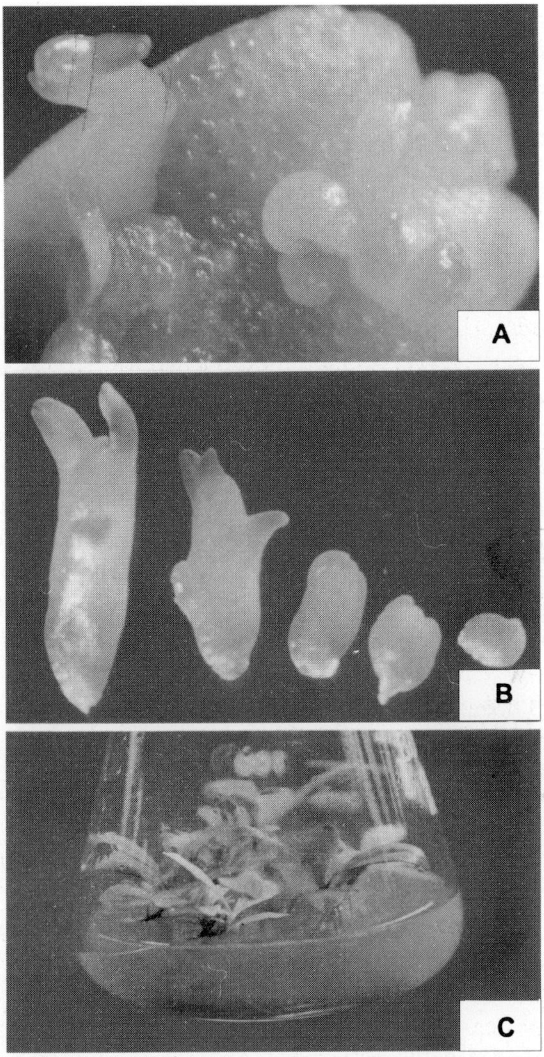

Fig. 1 *In vitro* adventitive embryony in Citrus (a) Embryo induction from nucellar tissue (b) Different stages of embryo development (c) Embryo germination-cum-rooting.

without callusing stage. Furthermore, this pathway has additional advantages as the frequency of appearance of variation could easily be avoided. Plantlets could be regenerated through adventitive embryony from nucellar tissue adhering to cotyledons in the two *Citrus* genotypes, i.e. rootstock species Rough lemon (*Citrus jambhiri* Lush) and scion variety Nagpur mandarin (*Citrus reticulata* Blanco). Cotyledons from stored seeds of both the genotypes when isolated and cultured on different growth regulator supplemented MS medium gave good response. The MS medium

Fig. 2 Rooting and hardening of *in vitro* rooted plantlets (a) peaceful development from germinated embryos (b) Plantlets under *in vitro* hardening stage in glass jar (c) Hardened plantlets ready for field transfer.

supplemented with 2.0 mg l^{-1} BAP + 0.5 mg l^{-1} kinetin + 1.0 mg l^{-1} NAA recorded highest adventitive embryo induction irrespective of the genotypes. The response was found to be effective when the cultures were kept under the dark. As the BAP concentration was reduced, the number of adventitive embryos formation declined.

When the adventitive embryos were subjected to the MS medium fortified with 2 mg l^{-1} IBA + 200 mg l^{-1} activated charcoal, root and shoot emergence were noted and the embryo thus became a complete plantlet. Highest rooting percentage was also observed on the same medium. Treatment comprising single auxin did not show proper rooting and hence the response was in general low. There was a significant improvement in rooting, when both the auxins (NAA and IBA) were used along with the activated charcoal. Efficacy of use of auxins in combination has earlier been reported by Mohanty *et al.* (1998).

The transfer of *in vitro* regenerated plantlet to greenhouse and further to the field is a critical stage in adopting such a protocol for commercial plantlet production. The results of the present study were encouraging, as good success could be obtained. Of the four strategies employed, glass jar with polypropylene cap containing sterile peat: soilrite (1:1) resulted in highest success of *ex vitro* plant survival in the two *Citrus* genotypes.

Previously, different regeneration systems were reported in *Citrus* using a variety of explants. However, regeneration of *C. jambhiri* and *C. reticulata* through nucellar tissue adherent to cotyledons holds promise as large number of plantlets can be produced, without much danger of carrying any viral disease(s). Also it is well known that since the adventitive embryos arise directly from the nucellar cells which do not carry viral load (Dhillon *et al.*, 1993). Furthermore, this pathway gets the support as large numbers of virus(es) are not found to be transmitted through seeds (Bitters and Murashige, 1967). Hence, the technique developed can be effectively used for budwood importation using seed as a material. The protocol developed in the two genotypes could be employed for undertaking attempt on their *in vitro* conservation. Different workers have reported that embryos as well as nucellar cells are ideal tissues to be successfully cryopreserved and later be regenerated into plants (Kobayashi *et al.,* 1990; Marin & Duran-Vila, 1988, 1991; Marin *et al.*, 1993).

From the results it is amply clear that the regeneration system developed in the *Citrus* species viz. *C. jambhiri* and *C. reticulata* holds promise for cultivar collection, budwood, seed transportation, conservation and even genetic manipulation using *Agrobacterium tumefaciens*.

REFERENCES

Bitters W.P. & Murashige T. (1967) A place for tissue culture in citrus research; *California Citrograph* 52: 266-72.

Dhillon G.S., Kaundal G.S. & Cheema S.S. (1993) Nucellar embryony for propagating citrus; *Indian Hort.* pp 44-45.

Kato M. (1986) Micropropagation through cotyledon culture in *Camellia japonica* L. and *C. sinensis* L; *Jap. J. Breed.* 36(1): 31-38.

King M.W., Soetisna V. & Roberts E.H. (1981) The dry storage of citrus seeds. *Ann. Bot.* 48: 865-872.

Kobayashi S., Sakai A. & Oiyama I. (1990) Cryopreservation in liquid nitrogen of cultured navel orange (*Citrus sinensis* Osb.) nucellar cell and subsequent plant regeneration; *Plant Cell Tis. Org. Cul.* 23: 15-20.

Marin M.L. & Duran-Vila N. (1988) Survival of somatic embryos and recovery of plants of sweet orange (*Citrus sinensis* (L.) Osb.) after immersion in liquid nitrogen; *Plant Cell Tissue Org. Cul.* 14: 51-57.

Marin M.L. & Duran-Vila N. (1991) Conservation of *Citrus* germplasm; *in vitro. J. Amer. Soc. Hort. Sci.* 116 740-746.

Marin M.L., Gogorcena Y., Ortiz J. & Duran-Vila. (1993) Recovery of whole plants of sweet orange from somatic embryos subjected to freezing thawing treatments; *Plant Cell Tis. & Org. Cul.* 34: 27-33.

Mohanty S., Deka P.C. & Bhattacharya S. (1998) Micropropagation of *Citrus sinensis* cultivar Musambi. *Ind. J. Agril. Sci.* 68(2): 114-116.

Mumford P.M. & Panggabean G. (1982) A comparison of the effect of dry storage on seeds of *Citrus* species; *Seed Sci. Techno.* 10: 257-266.

Murashige T. & Skoog F. (1962) A revised medium for rapid growth and bioassays with tobacco tissue culture. *Physiol. Plant.* 15: 473-497.

Navarro L. (1981) *Citrus* shoot tip grafting *in vitro* (STG) and its applications: A review; *Proc. Int. Soc. Citricul.* pp 452-456.

Pooler M.R. & Scorza R. (1995) Regeneration of peach [*Prunus persica* (L.) Batsch] rootstock cultivar from cotyledons of mature stored seed; *Hort. Science.* 30(2): 355-356.

CHAPTER

24

Micropropagation of *Gmelina arborea* Roxb. Using Mature Tree Tissues – Optimization of *in vitro* Responses

Dhiraj Naik, Varsha Vartak and Sujata Bhargava[1]

ABSTRACT

Nodal sector explants from *Gmelina arborea* Roxb. mature trees, growing in and around Pune, namely, Ambed, Parvati Pachgaon and Botanical garden were cultured on modified MS medium supplemented with 1 μM of BAP and Kinetin each. *In vitro* responses of the nodal segments for axillary shoot elongation and number of multiple shoots were compared in between the localities as well as over subcultures. Locality dependent variation with respect to axillary shoot elongation and number of multiple shoots was significant. When subculture effect was compared, significant difference only for number of multiple shoots was found. Nodal segments of the trees from Botanical garden were inoculated every month to study seasonal variation. February-March that is late winter was found to be the best season to initiate the cultures. Nodal segments inoculated during this period showed maximum elongation of axillary buds with a very little formation of callus. Addition of silver nitrate and reduction in the cytokinin concentration had a positive effect on shoot elongation as well as on number of multiples. Reduction in leaf fall was also noticed. Transferring the cultures from test tubes to 250 ml conical flasks proved to be effective in increasing number of multiple shoots per explant. The shoots, which failed to elongate upto the third subculture, were micrografted on seedling rootstocks grown *in vitro*. In the successful grafts, shoot elongation up to three nodes was attained. Induction of roots was carried out by giving a pulse treatment of IBA to the *in vitro* grown shoots.

Key Words: *Gmelina arborea*, micropropagation, mature explants, micrografting.

It is desirable to micropropagate trees using tissues from mature trees since the desired phenotype is expressed at the physiologically mature state (Bonga and Von Aderkas, 1987). However, micropropagation of trees from mature tree tissues is difficult for a number of reasons. Firstly, the apical or axillary meristems from terminal branches of trees show poor response in culture due to

*Address for correspondence: Botany Department, University of Pune, Pune – 411 007, Maharashtra, India.

[1]The authors are grateful to DBT and UGC (SAP-DRS Program), New Delhi, for financial support.

the fact that these meristems arise from a large number of divisions of the shoot apical meristem and hence lose their juvenility and to some extent their totipotency (McCown, 2000, Benson, 2000). Besides age, tree tissues are also known to exhibit season-dependent dormancy cycles, which influences their *in vitro* response (Benson, 2000). The presence of phenolics in mature tree tissues is another deterrent for *in vitro* growth (McCown, 2000). A further problem during micropropagation of trees arises because *in vitro* formed shoots often show leaf shed, shoot tip necrosis or a lack of elongation on subsequent subculture, leading to loss of micropropagated plants (Kulkarni and D'Souza, 2000).

Different strategies have been used to overcome these problems and successful micropropagation of trees from mature tissues has been reported (Munoz *et al.*, 1999). Adventitious sprouts seen from the cut end of coppiced trees are more responsive since they are formed from newly organized meristems and are therefore more juvenile (Munoz *et al.*, 1999). Rejuvenation of mature tree tissues through grafting them on seedling rootstocks is also known to occur, and such rejuvenated tissues show a better response *in vitro* (Giovannelli and Giannini, 2000; Luo and Gould, 1999). Media manipulations have been successfully carried out to overcome the problems of phenolics, leaf shed and shoot tip necrosis (Aderkas and Bonga, 2000).

In the present paper we report different methods used for optimizing *in vitro* responses of mature tree tissues of the tropical deciduous tree *Gmelina arborea* Roxb, an important timber yielding tree.

MATERIALS AND METHODS

Plant material: Mature trees from three localities in Pune, namely, Ambed, Parvati Pachgaon and Botanical garden were used as a source of plant material. Nodal segment explants were prepared from terminal branches as well as adventitious sprouts arising from the stumps of coppiced trees.

In vitro culture: Nodal segments were prepared by cutting the stem about 0.5 cm above and 1 cm below the node. Two axillary buds and about 0.5 cm petiole were included in the nodal segment. None of the nodal segments showed axillary sprouts at the time of excision. These nodal segments were surface sterilized as described in Thakar and Bhargava (1999). Nodal segments were treated for 3 h with 1% w/v bavistin (Carbendazim, BASF, India) solution to which a few drops of Tween 20 were added. The explants were rinsed with 70% alcohol, and treated with 0.1% mercuric chloride for 10 minutes. Surface sterilized nodal segments were trimmed at the cut edges and segments of size approximately 2 cm × 0.5 cm were placed in culture tubes (6 cm × 1 cm) containing 20 ml MS medium (Murashige and Skoog, 1962), supplemented with 3% sucrose, 1 μM kinetin and 1 μM BAP. The cultures were incubated at a light intensity of 19 $\mu mol\ m^{-2}\ s^{-1}$ (cool fluorescent tube lights of 40W) for a 16 h photoperiod, and at a temperature of 25 ± 2°C. After 4 weeks, nodal segments, prepared from *in vitro* grown axillary shoots (> 5 cm long) as well as the multiple shoots (also of size > 5 cm) were subcultured in 250 ml conical flasks, on a modified MS (MS') medium (half strength MS macroelements but full strength of $CaCl_2.2H_20$, full strength MS minor salts and organic supplements, 3 % w/v sucrose) supplemented with 0.1 μM BAP and 5 mg/l silver nitrate. The response of the nodal segments was studied for 4 subcultures, with an average of 20 nodal segments per subculture. About 5-6 cm long *in vitro* grown shoots were used for induction of rooting. The shoots were separated at the base from the nodal segment and the basal portions immersed in aqueous solution of 500 or 1000 µM IBA (filter sterilized) in autoclaved culture tubes for 24 h. Subsequently, the shoots were transferred to ½ MSM (half strength MS macroelements with complete omission of NH_4NO_3 and addition of full strength $CaCl_2.2H_20$, half strength MS minor salts and organic supplements, 1.5% w/v sucrose) and with 1mg/l activated charcoal. (Sharma and Ramamurthy, 2000). The medium was solidified with 0.8% w/v agar. Incubation conditions were similar for all cultures. Data from two experiments was analysed using ANOVA to study the locality- and subculture-dependent variation in the elongation and multiple shoot induction responses.

Micrografting: Shoot tip grafting was done using 15 days old (4-5 cm in height) seedlings, grown *in vitro* on ½ MSM, as rootstock. The stunted axillary shoots, produced from the nodal segments of mature trees that failed to elongate (4 weeks old and about 2-3 cm in height), were used as microscions by cutting the basal end of the shoot in a 'V' shape (wedge). Seedlings were decapitated about 1.5-2.5 cm above the cotyledonary node, keeping the cotyledons intact. A central incision was made in this decapitated seedling and the wedge-shaped microscion was inserted gently and firmly in it. Axillary shoots arising from the cotyledonary node of seedlings were removed and only the grafted scion was allowed to grow. After 4 weeks, the elongated scion was used as a source of nodal segments, which were cultured as mentioned earlier. The experiment was carried out thrice.

OBSERVATIONS

Adventitious shoots from coppiced trees of three localities did not show enhanced sprouting as compared to the explants isolated from terminal branches, but the number of tubes showing elongated axillary shoots was higher (Table 1). In addition, the use of the MS medium supplemented with silver nitrate and transfer to bigger culture vessel (250 ml conical flask) for subculture stopped leaf shed and shoot tip necrosis that was observed in cultures subcultured on the initiation medium (results not shown).

The *in vitro* responses were monitored over four subcultures. Multiple shoots were seen to arise after culturing nodal segments of the *in vitro* elongated axillary shoots. As seen in Table 2, the elongation response showed significant variation with respect to locality but not subculture, whereas the induction of multiple shoots showed significant locality- and subculture-dependent variation. The locality and subculture interactions were insignificant. Elongation and multiple shoot formation showed an increasing trend over subcultures (Fig. 1) and this response was unlike that observed in the responses of seedling nodal segments used as explants (Naik *et al.*, 2002).

The adventitious shoots from coppiced trees in the Botanical garden were used as a source of explants and were cultured every two months to observe the seasonal variation in the *in vitro* responses. Sprouting was seen to be highest during October to January, but the newly sprouted axillary shoots did not show a good elongation response. On the other hand, the sprouting response was lower in the period February to May, but the number of tubes showing elongated axillary shoots was higher (Table 3).

Table 1 Comparison of *in vitro* growth responses of *Gmelina arborea* nodal segments isolated from shoots arising from coppiced trees and Terminal branches of matrure trees

Locality	*Source of Explant*	*% Sprouting*	*% Elongation*
Ambed	A*	69	33
	B**	60	13
Parvati Pachgaon	A	34	28
	B	28	11
Botanical garden	A	49	34
	B	24	7

*A: shoots arising from coppiced trees

**B: shoots arising from terminal branches.

Table 2 Anova table for locality- and subculture-dependent variation in the responses of nodal segments isolated from mature trees. The experiment was carried out twice with at least 30 cultures per experiment

Response	*Source of variation*	*Degrees of freedom*	*F Value*	*P value*
Elongation	locality	2	4.57	0.028
	subculture	4	0.69	0.607
	interaction	8	0.52	0.825
Multiple shoot formation	locality	2	17.41	0
	subculture	4	14.66	0
	interaction	8	1.26	0.331

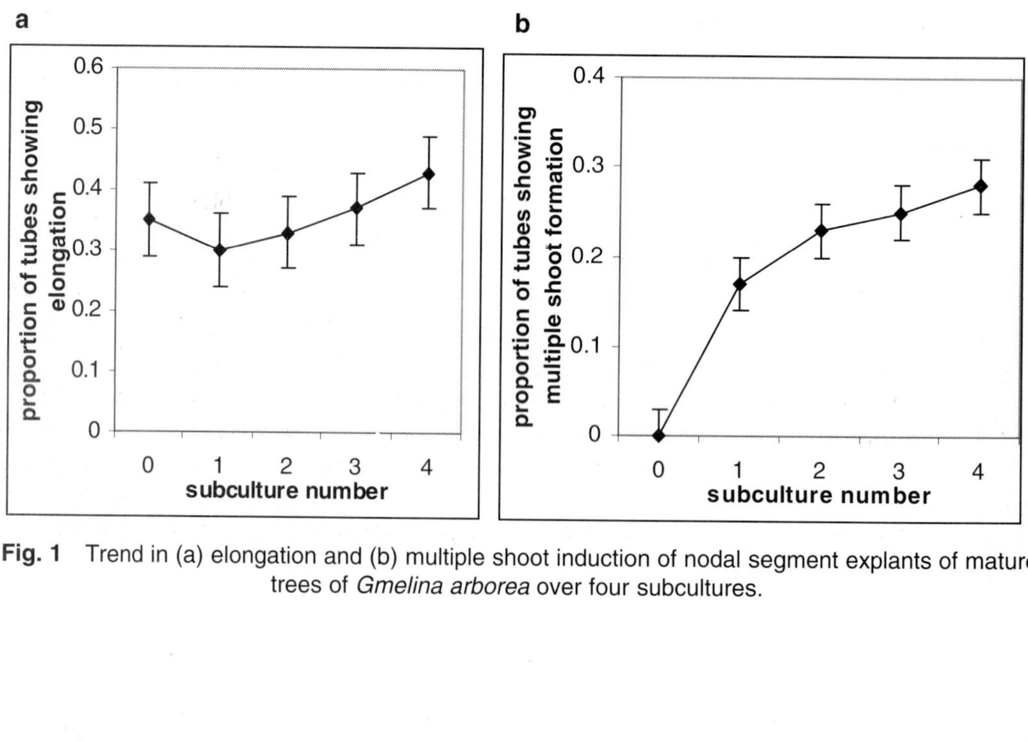

Fig. 1 Trend in (a) elongation and (b) multiple shoot induction of nodal segment explants of mature trees of *Gmelina arborea* over four subcultures.

Table 3 Season-dependent response of nodal segments isolated from mature trees. The experiment was carried out twice with at least 30 cultures per experiment.

Month of culture initiation	*% sprouting (in cultured nodal segments)*	*% elongation (in sprouted nodal segments)*
Oct-Nov	93 ± 9.9	30 ± 0.6
Dec-Jan	95 ± 0.1	28 ± 0.8
Feb-Mar	80 ± 12.1	58 ± 11.8
Apr-May	56 ± 13.4	48 ± 1.1
Jun-Jul	39 ± 10.9	30 ± 12.5

Micrografting was attempted using the unelongated axillary shoots arising from mature tree nodal segments as scions and seedlings as rootstocks. About 50%grafts showed successful unions and after four weeks, significant elongation of the axillary shoots was observed as compared to the elongation of the axillary shoots grown *in vitro* (Fig. 2). Nodal segments were isolated from the grafted scions and cultured on the same medium as for nodal segments prepared of coppice shoots. All nodal segments of the grafted scions showed sprouting (Table 4) which was an improvement over that observed in cultured nodal segments of coppice shoots (see Table 1).

Table 4 Responses of *in vitro* grafts of axillary shoots from mature tree nodal sectors on seedling root stock

Expt. number initiated	*Number of grafts grafts union*	*Successful from elongated scions*	*Sprouting of nodal sectors*
1	86	48	100%
2	52	29	100%
3	42	21	100%

DISCUSSION

Micropropagation of *Gmelina arborea* from axillary meristems of physiologically mature trees showed very poor response as compared to axillary meristems of seedlings (Thakar and Bhargava,1999). Juvenility is an important consideration in the ability of tree tissue to grow *in vitro* (Jones, 1994). The adventitious shoots arising from the coppiced tree stumps showed a better *in vitro* response because these shoots arose from newly formed meristems and were hence more juvenile than the axillary meristems of terminal branches of the tree. In addition to age, season and biotype-dependent variation was also observed. Genotype dependent variation in *in vitro* responses have been observed in tree as well as herbaceous species (McCown, 2000), indicating a genetic component in the morphogenetic potential of plants. It is known that growth and development of trees is influenced by seasons. Seasonal variation in growth and development is known to be affected by fluctuations in the endogenous hormone levels (Lavender and Silim, 1986). *Gmelina arborea* is a tropical deciduous tree and shows season-dependent leaf fall, flowering and vegetative

Fig. 2 Micrografting of stunted axillary shoots from nodal segments of mature trees.
(a) Nodal segment showing formation of stunted axillary shoots at the end of 4 weeks.
(b) Stunted axillary shoot (scion) grafted on a decapitated seedling rootstock.
(c) Elongation of the grafted axillary shoot (scion) after 4 weeks.
Arrows indicate the graft junction.

growth. This could be the reason for the differences in *in vitro* responses of nodal segments isolated in different seasons.

The lack of elongation of axillary shoots when nodal segments of the *in vitro* grown axillary shoots obtained from mature tree nodal segment explants was a major impediment in developing a micropropagation procedure for this tree species. Attempts to achieve shoot elongation, by altering the medium composition, was not successful. Hence micrografting was carried out as an alternative strategy to induce elongation. Micrografting has been reported to bring about rejuvenation of the scions and is often used to induce growth in recalcitrant tree species (Thimmappaiah *et al.*, 2002). After micrografting, the scions showed a dramatic elongation of the axillary shoots indicating that the signals received from the seedling rootstock were effective in inducing elongation, which was not possible by media manipulation. The rejuvenation and / or elongation factor(s) however remain elusive and further experiments would be needed to identify these factors.

REFERENCES

Aderkas P. & Bonga J. (2000) Influencing micropropagation and somatic embryogenesis in mature trees by manipulation of phase change, stress and culture environment, *Tree Physiology*. 20: 921-928.

Benson E. (2000) *In vitro* plant recalcitrance: An introduction, *In Vitro Cell. Dev. Biol.-Plant.* 36: 141-148.

Bonga J.M. & von Aderkas P. (1987) *Cell and Tissue Culture in Forestry*, Dordrecht, The Netherlands: Martinus Nijhoff).

Giovannelli A. & Giannini R. (2000) Reinvigoration of mature chestnut (*Castanea sativa*) by repeated micrograftings and micropropagation, *Tree Physiology*. 20: 1243-1248.

Jones O. (1994) Physiological change and rejuvenation of temperate trees from micropropagation. In: *Physiology, Growth and Development of Plants in Culture*, pp. 323-331, Lumsden P.J., Nicolas J.R. & Davis W.J. (Eds) (Dordrecht, The Netherlands : Kluwer).

Kulkarni K. & D'Souza L. (2000) Control of *in vitro* shoot tip necrosis in *Butea monosperma*, *Current Sci.* 78: 125-126.

Lavender D.P. & Silim S.N. (1986) The role of plant growth regulators in dormancy in forest trees, *Plant Growth Regulation*. 6: 171-191.

Luo J. & Gould J. (1999) *In vitro* shoot tip grafting improves recovery of cotton plants from culture, *Plant Cell Tissue Org. Cult.* 57: 211-213.

McCown B. (2000) Recalcitrance of woody and perennial plants: Dealing with genetic predeterminism, *In Vitro Cell. Dev. Biol.-Plant*. 36: 149-154.

Munoz A., Encina C., Perez E. & Alfaro F. (1999) Micropropagation of adult avocado, *Plant Cell Tissue Org. Cult.* 58: 11-17.

Murashige T. & Skoog K. (1962) A revised medium for rapid growth and bioassays with tobacco tissue culture, *Plant Physiol.* 15: 473-497.

Naik D., Vartak V. & Bhargava S. (2002) Provenance- and subculture-dependent variation during micropropagation of *Gmelina arborea* Roxb, *Plant Cell Tissue Org. Cult.* (in press).

Sharma S.K. & Ramamurthy V. (2000) Micropropagation of 4-year old elite *Eucalyptus terticornis* trees, *Plant Cell Rep.* 19: 511-518.

Thakar J. & Bhargava S. (1999) Seasonal variation in antioxidant enzymes and the sprouting response of *Gmelina arborea* Roxb. nodal sectors cultured *in vitro*, *Plant Cell Tissue Org. Cult.* 59: 181-187.

Thimmappaiah, Puthra G.T. & Shirly R.A. (2002) *In vitro* grafting of cashew (*Anacardium occidentale* L.), *Scientia Horticulturae* 92: 177-182.

CHAPTER

25

Challenges in Cashew Tissue Culture: An Overview

M. Anuradha, Shashikiran Nivas, Smitha Hegde, K. Rajendra, Vinitha Cardoza, Icy D'Silva and L. D'Souza[1]

INTRODUCTION

Cashew introduced from Brazil is now grown all over the tropics especially along the coastal regions. The tree is useful for revegetation of wastelands and for controlling erosion of degraded areas (Nayar, 1993). The pseudocarps are used for preparing jams and juice. Cashew juice has been found to be a good substratum for single cell protein culture (Layokun, 1986). The juice is also a good source of alcohol. The shell oil is used as raw material for polymer-based industries like paints, varnishes and laminates. Two important chemicals derived from the shell oil are cardol and anacardic acid which are useful pharmaceutical industry. The cashew kernels are delicious and nutritious and are highly valued in the world nut market. The export of the kernels fetches India considerable amount of foreign exchange (Balasubramaniam, 1987). The manual processing of the nuts provides work for a large number of unskilled rural laborers—95% of whom are women. The raw nuts produced in India are not enough to meet the demand of the export market for kernels and to ensure regular employment for the workers of the cashew industry.

This is because the trees grown at present are low yielding. The average yield of a ten-year-old tree is 2 kg per year. Some high yielding hybrids have been established and some elite high yielding trees have been identified in recent years (Nambiar & Iyer, 1988). If planting material prepared by vegetative multiplication of these trees is available new areas can be planted with such superior material and low yielding trees of old plantations can be gradually replaced with high yielding clones. According to the estimates of the industry the present yield of raw nuts is 734,000 tons. The projected demand for the year 2000 is 1,120,000 tons. Not only the area under cultivation has to be increased but the yield capacity of the planting material has to be improved in order to meet the target. According to the Five Year Plan, 43,000 hectares of wastelands are to be brought under

*Address for correspondence: Laboratory of Applied Biology, St. Aloysius College, Mangalore 575003
Email: labens@sancharnet.in

cashew cultivation and 10,000 hectares are to be replanted. The conventional methods of vegetative propagation like layering and grafting are not able to meet the demand for such large amounts of elite planting material.

MATERIALS AND METHODS

Mature and immature nuts of cashew collected from 5-10-year-old trees in an orchard in Mangalore were surface sterilized in a solution of 0.1% mercuric chloride and 0.1% sodium lauryl sulphate for 10 minutes, followed by three rinses with sterile distilled water. They were cut open in a laminar air flow cabinet and the entire seed from each nut was carefully dissected out. The seeds were germinated aseptically on plain agar in screw capped bottles. The cotyledonary nodes with the cotyledons intact were used for the experiments. The explants were cultured as described previously (D'Silva & D'Souza, 1992) for multiple bud initiation.

Immature nuts were used to induce nucellar somatic embryos. The seeds were cut vertically into two halves and were cultured, after discarding the zygotic embryo, with the nucellar tissue touching the surface of the medium. For callus and proembryo induction the basal MS medium was supplemented with 100 mg/l m-inositol, 20 g/l sucrose and 0.25-5 mg/l picloram, 0.25-5 mg/l 2,4-D and 0.25-5 mg/l 2,4,5-T singly and gelled with 8 g/l agar. For the development of embryos, MS medium with 0.5 mg/l picloram in combination with putrescine (0.5, 1, 2 mg/l) supplemented with 30 g/l sucrose, gelled with 2.5 g/l phytagel was used. For maturation, the embryos were cultured on MS medium with a combination of ABA (0.5, 1, 2, 5.mg/l) and sucrose (20, 30, 40, 50 g/l). For germination, the mature embryos were cultured on MS medium with 20 g/l sucrose, devoid of growth regulators.

RESULTS AND DISCUSSION

Tissue culture techniques which have given good results with a number of other plants have been tested for *in vitro* multiplication of cashew. Most of the work reported has been done with juvenile tissues (Philip, 1984; Jha, 1988; Bessa & Sardinha, 1994). In our work multiple buds have been obtained from the cotyledonary axis (D'Silva *et al.*, 1992). Murashige and Skoog medium supplemented with 30 g/l sucrose and 5 mg/l BAP was found to be optimal for the induction of the shoots. Proliferation of the buds and elongation of shoots could be got by reducing BAP (1.0 mg/l) in the medium after initiation of the buds and by dissection of the explants.

Cashew shoots form roots with intervening callus on medium having NAA. Such roots do not often have vascular connection with the shoot. When these shoots are transferred to soil they do not survive as the roots are not functional. The callus gets easily infected and the infection spreads to the rest of the plant. Good rooting without intervening callus was obtained with 0.5 mg/l IAA and 1.0 mg/l IBA (D'Silva & D'Souza, 1991).

Plants obtained by this method have been transferred to soil and are growing well. They have flowered and have borne fruit. However these results have limited use for large scale multiplication of trees grown for their yield as the genetic make up of the donor seedlings is not known.

Work with adult tissues of cashew has proved to be very difficult. Contamination is a serious problem. Material collected from the field has invariably bacterial and fungal contaminants some of which make their presence felt only when the explants are several weeks in culture. Browning is another major problem especially in case of tissues of adult plants. Anacardic acid and cardol are highly destructive to the tissues. Only a partial control can be achieved by the usual methods of controlling browning such as pretreatments with antioxidants, addition of adsorbents to the culture medium and culturing the explants in the dark (D'Silva & D'Souza, 1993). Adult tissues of cashew are very recalcitrant. Only a few buds could be induced from axillary buds of mature trees. These buds were very small and could not be isolated and rooted. Embryogenic callus was induced from the nucellus in all the auxins tested within a week in culture. Callus produced with 2,4-D and 2,4,5-T could not be maintained as it lost its embryogenic nature and turned brown within 2 weeks in culture. Proembryos developed from the callus obtained on picloram-supplemented medium after a month in culture. The proembryos transferred to 0.5 mg/l picloram with 1 mg/l putrescine gave rise to globular embryos. The globular embryos when transferred to MS medium with 0.5 mg/l ABA with 30 g/l sucrose developed into heart and cotyledonary embryos after three months in culture. For germination the somatic embryos were transferred to MS medium with 20 g/l sucrose, devoid of growth regulators. The embryos germinated in this medium within 3 days of transfer. However they could not be converted to plants (Cardoza & D'Souza, 2002). Considerable amount of work is still needed to study the responses of tissues of adult cashew trees to tissue culture techniques.

Acknowledgements

The facilities provided by the Mangalore Jesuit Educational Society and the financial support given by the DST are gratefully acknowledged.

REFERENCES

Balasubramanian PP 1987. Cashew can be still promisive in India, *The Cashew* **1** 13-19.

Bessa AMS & Sardinha RMA 1994. *In vitro* multiplication of cashew (*Anacardium occidentale* L.) by callus culture, *Abstracts VIII Int. Cong. Plant Tiss. Cell Cutl.* 23.

Cardoza V & D'Souza L 2002 Induction, development and germination of somatic embryos from nucellar tissues of cashew (*Anacardium occidentale* L.), *Scientia Hortic.* **93** 367-372.

D'Silva I & D'Souza L 1992 *In vitro* bud proliferation of *Anacardium occidentale* L., *Beitr. Biol. Pflanzen* **67** 273-279.

D'Silva I and D'Souza L 1993 Controlling contamination and browning of *in vitro* cultures of cashew, *J. Plantation Crops* **21** 22-29.

D'Souza L, Hegde S, Augustine AC, Rajendra K & D'Silva I. 1994 Micropropagation of an important tropical nut tree *Anacardium occidentale* L.(Cashew), *Abstracts VIII Inlt. Cong. Plant Tiss. Cell Cult.* 165.

Jha TB 1988 *In vitro* morphogenesis in cashewnut, *Anacardium occidentale* L., *Ind. J. Expt. Biol* **26** 505-507.

Nambiar MC & Iyer RD 1988 Tissue culture in cashew: Potential, priorities and applications, *Proc. Natl. Sem. Plant Tiss. Cult.*, pp 149-155.

Nayar KG 1993 Cashew cultivation and wasteland development, *Cashew Bulletin* **30** 2.

Philip VJ 1984 *In vitro* organogenesis and plantlet formation in cashew (*Anacardium occidentale* L.), *Ann. Bot.* **54** 149-152.

Index